MW00760155

PUBLIC WORKS CONTRACTING

Start to Finish

DANIEL G. ANDERSEN, P.E.

BNi Building News

ANAHEIM • NEW ENGLAND • WASHINGTON, DC • LOS ANGELES

BNi Building News

EDITOR-IN-CHIEF
William D. Mahoney, P.E.

TECHNICAL SERVICES
Rod A. Yabut

DESIGN
Robert O. Wright

BNI PUBLICATIONS, INC.
1-800-873-637

LOS ANGELES

10801 National Blvd., Ste.100
Los Angeles, CA 90064

NEW ENGLAND

172 Taunton Ave. Ste.2, PO
Box 14527
East Providence, RI 02914

ANAHEIM

1612 S. Clementine St.
Anaheim, CA 92802

WASHINGTON,D.C.

502 Maple Ave. West
Vienna, VA 22180

ISBN 1-55701-372-1

Copyright © 2000 by BNI Publications, Inc. All rights reserved. Printed in the United States of America. Except as permitted under the United States Copyright Act of 1976, no part of this publication may be reproduced or distributed in any form or by any means, or stored in a data base or retrieval system, without the prior written permission of the publisher.

While diligent effort is made to provide reliable, accurate and up-to-date information, neither BNI Publications Inc., nor its authors or editors, can place a guarantee on the correctness of the data or information contained in this book. BNI Publications Inc., and its authors and editors, do hereby disclaim any responsibility or liability in connection with the use of this book or of any data or other information contained therein.

TABLE OF CONTENTS

INTRODUCTION

This book is designed to be a comprehensive reference manual for the small, start-up contractor or any construction company that desires to become more organized in its operations and management activities. Material covered within this book flows from start to finish. There is a substantial variety in the way construction companies manage their own particular projects. The methods outlined in this book are straightforward, general approaches to the aspects of contracting that are encountered within the public works and commercial construction industry. As a company gains experience in contracting it will find many modifications to the general methods outlined here and it will incorporate these changes into its operations. Items that will produce changes from these general methods are each companies management philosophies, available workforce, equipment and the type of construction undertaken.

This book should also be helpful to the owner whose background is from the labor crafts. In general, those individuals that have experienced the contracting business from working on the physical construction jobsites have had to learn the management side of business in order to have a durable, long lasting, successful company.

This book should also be of special interest to all of the engineers and supervisors employed by construction companies, who are players on the jobsite or office management teams, since it covers all the major aspects of organizing and managing a public works or commercial construction company, from start to finish.

ACKNOWLEDGEMENTS

I would like to thank Jim Acret for the chapter on *Construction Law*, Arthur F. O'Leary for the chapter on *Construction Documents Used by Contracting* Agencies, and Silas Birch for the chapter on *Inspection*. The professional experience and insight by these gentlemen has helped create a well rounded book from all perspectives, for the construction contractors, engineers, and supervisors.

CHAPTER ONE

SETTING UP THE COMPANY

BUSINESS PLAN

Getting all of the company paperwork in order before contracting begins can be a frustrating and time consuming ordeal. Yet, careful planning and the selection of competent insurance, banking, bonding and accounting organizations will help to avoid many problems that can occur during construction operations. This set-up period of your company is a good time to start a relationship with a banking institution. The bank is going to want to know your business plan. Therefore, you should have a written business plan before you start talking to the bank. The business plan should be a professional-appearing document that details the following:

1.) THE KIND OF JOBS YOU WANT TO GET. For example your company might specialize in street and road construction or utility work. Just be specific in describing the type of work you envision your company undertaking.

2.) THE TYPES OF MARGINS (OVERHEAD AND PROFIT) YOU EXPECT THE COMPANY TO REALIZE. Do you expect your company to make 10% margins as they relate to direct project costs or do you expect the company to make 25%? Detail some reasons why you expect to make whatever margins you list; past experience with other companies or a detailed analysis of a particular market, for example.

3.) COSTS ASSOCIATED WITH START-UP AND CONTINUING MANAGEMENT AND OPERATIONS OF THE COMPANY. This analysis is basically the overhead of the company. This includes such things as corporation expenses, licensing, office and yard space, management labor, office equipment and supplies, telephone and electricity along with accounting and legal fees.

These are the costs that a company incurs to operate the business.

4.) LABOR AND EQUIPMENT PROCUREMENT PLANS. Detail how you plan to obtain a workforce; through the local labor unions or do you have some craft personnel who are willing to come to work for your company? With respect to equipment, have you set up accounts at some local equipment rental yards or have you discussed the purchase or lease of equipment with your local equipment dealers?

5.) RESEARCH THE LOCAL CONSTRUCTION MARKET FOR THE WORK YOUR COMPANY PLANS TO UNDERTAKE. Detail approximately how many bidders there are typically on the type of projects your company plans to undertake. Is the market for this work flooded with contractors, or on the other hand, are the contracting agencies only receiving one or two bids per letting? What is the short and long term availability for the type of work your company would construct?

6.) FINALLY, IT IS A GOOD IDEA TO DEVELOP A PROJECTED CASH FLOW (ON A TIME LINE BASIS) THAT INCORPORATES THE PROJECTED MARGINS. This is so that you and the bank (along with the bonding, insurance and accounting companies) have an idea of how much the company needs to produce in order to reach and exceed the break-even point based on your expected overhead and job profit. It is a good idea to do this on a one year and five year basis. Incorporated in your five year cash flow is your projection of the growth you expect and want your company to achieve.

FIGURE 1-0 Business Plan

XYZ GENERAL CONTRACTORS, INC.

BUSINESS PLAN

Type of Construction:

Public Works Construction

Proposed Market:

XYZ General Contractors, Inc. initially proposes to bid, acquire, and construct projects within the parks, recreation, and site improvement market. Several metropolitan area contracting agencies have passed substantial bond measures to fund this type of work over the next several years.

Typical profit margins for this type of work is approximately 15% of the direct cost. With the additional amount of parks and recreation work that is going to be available, we believe profit margins of at least 20% should be able to be realized.

Proposed Market Competition:

For the past several years construction, both public works and building construction, has been on the steady increase in the metropolitan area. The last couple years have shown relatively limited competition within the Public Works Construction market, especially during the normal construction season (May - November). During this time of the year many contracting agencies actively seek qualified contractors to bid on the projects they are letting. XYZ General Contractors, Inc. plans to fill this gap in the lack of qualified contractors that the contracting agencies are searching for.

Start-up Costs:

XYZ General Contractors, Inc. should experience a typical overhead rate for this type of company, which is in the range of 10% over time. In the initial year of operations we expect this rate to be approximately 12% - 15% of the direct costs to cover expenditures such as initial office equipment and supplies along with the initial corporate expenses and legal fees.

XYZ General Contractors, Inc. initially plans to rent any large equipment necessary for construction of our projects. At the point in time where the company has sufficient profitability and cash flow, we will begin to invest in our own fleet of heavy equipment.

XYZ GENERAL CONTRACTORS, INC.
BUSINESS PLAN
PAGE 2 OF 2

XYZ General Contractors, Inc. has talked with several key potential employees. These individuals have committed to working for us once we have been awarded a contract. With this employee commitment, we will have no problems with the projects we are awarded.

First Year Goal:

To develop a respected Public Works Construction corporation that can financially stand alone and fund new projects. XYZ General Contractors, Inc. wants to be one of the contractors that contracting agencies will rely upon when they need a qualified firm to construct a project on-time, within budget and in a partnering manner.

Long Term Goal:

To develop a Public Works Construction corporation that grows in volume at a conservative, but steady, rate while maintaining reasonable margin percentages.

To develop a respected reputation with the construction industry through ethical and service oriented business operations.

To develop an operations staff that shares the goals of the corporation. XYZ General Contractors, Inc. would like to keep personnel turnover at a minimum through a competitive compensation package along with short and long-term production incentives.

Company Projections

Contract Volume	Overhead	Profit
Year 1 $500,000	12.5% - $50,000	9.0% - $ 50,000
Year 2 $750,000	10.0% - $60,000	13.6% - $ 90,000
Year 3 $1,250,000	8.0% - $76,000	21.8% - $224,000
Year 4 $1,500,000	8.0% - $92,000	20.8% - $258,000
Year 5 $1,600,000	8.0% - $98,000	20.9% - $277,000

The above percentages are a function of direct cost. As you can see from the projections, XYZ General Contractors, Inc. plans on growing to the point were operations could be handled by a small management team. At the end of this five year time frame, we will re-evaluate the progress of the company and decide if we would prefer to stay at this level or increase the size to the management team and company volume respectively.

BANKING/LINES OF CREDIT/ACCOUNTING

With respect to a bank, look for one that not only has competitive lending and interest rates, but also understands the operations of a small construction company whether it be public works or commercial. Most banks offer very similar products with very similar rates. Therefore, the choice of a banking institution often comes down to personalities and the overall comfort level that you feel with people within the bank. You should negotiate an open line of credit with your bank in the amount of approximately 5-10% of your estimated yearly gross sales. This line of credit will become important when you need to finance the start of a project. Your line of credit should be able to incorporate at least two months of the labor payroll for your projects. This two month period is the amount of time it could take to receive your first progress payment on any particular project. (Discussed in more detail in Chapter Ten) If your company is in a monetary pinch, the payment of supplier invoices may be delayed temporarily, but the payroll to your employees cannot be delayed. For many start-up construction companies, it is impossible to negotiate this kind of initial line of credit with your bank; therefore, you may need to secure this line of credit with personal property until a point in time when this line of credit may be secured with company assets. In any case you will need to have on hand a minimum of this two-month labor expenditure, whether it be from your company bank account, personal bank account, line of credit or some other outside financing.

A start-up construction company will also have to open lines of credit with many material suppliers and rental agencies. You should send out a letter to each material supplier and rental agency that you want to do business with that describes your company, who manages it, what kind of projects you plan to construct and what kind of payment terms they can expect (Figure 1-1). After you complete their credit application (Figure 1-2), most of the material supply companies will grant a start-up contractor at least a minimal line of credit with which to get started while a history of payment can be developed.

FIGURE 1-1   Material Credit Account Letter

ABC ASPHALT PAVING COMPANY
1234 E. Delta Ave.
Gresham, Oregon 97777
Phone (503) 663-6633
Fax (503) 663-6634

February 29, 20--

EQUIPMENT RENTAL SERVICE
5678 N. Pacific Street
Gresham, Oregon 97777

RE: Credit Account

Credit Manager:

ABC Asphalt Paving Company at this time is requesting a credit account with your organization. We are a new asphalt paving contractor that plans to conduct business within the metropolitan area.

ABC Asphalt Paving Company plans to do asphalt paving on contracts with the local governmental agencies, while also serving the local home and business owners. At this time we believe we will need an account limit of approximately $2500 per month for equipment rental to service this market.

Please find attached a copy of our credit information and please contact us if you need any further information. We thank you in advance for your consideration in this matter.

Sam D. Maple
President
ABC ASPHALT PAVING COMPANY

FIGURE 1-2 Material Credit Account Application

CREDIT INFORMATION

ABC ASPHALT PAVING COMPANY
1234 E. Delta Ave.
Gresham, Oregon 97777
Phone (503) 663-6633 Fax (503) 663-6634

Business Type: Public Works Contractor

President:	Sam D. Maple	SS# 522-64-2994
Secr./Treas.:	Molly S. Maple	SS# 522-15-5270

Bank Name: Northwest National Bank
2222 Main Street
Gresham, Oregon 97777
Attn: Susie Johnson   (503) 663-8990

Checking Account   # 12-345-678
Savings Account   # 90-123-456

Federal ID#:   91-5551212

Oregon CCB#:   1346787

Washington CCB#: ABCASPHC43KW

References:			
	CDF Rock Prod.	Portland	(503) 222-3333
	ACF Geotextiles	Portland	(800) 333-5555
	GFH Rentals	Portland	(503) 444-2222
	JKL Equipment	Portland	(503) 888-1212
	STU Pipe	Portland	(503) 777-5454

Sam D. Maple
President
ABC ASPHALT PAVING COMPANY

Once you have a lending institution in place and credit accounts applied for, consideration should be given to an accounting firm. As with a bank, choose an accounting firm that is familiar with the operations of a small construction company. Prior to the start of physical work, a construction contractor should consult with and retain an accountant to make sure its financial recording keeping system is in order. If your company does not have a financial record keeping system, then the accounting firm can set you up with one. These professionals can also assist you with the best method to purchase, rent, lease, borrow money and track costs for your optimum tax advantage. It is often a good idea for the start-up construction company to have an outside firm handle the payroll. There are many such companies available who can take care of all your payroll paperwork, from tax withholdings and reporting to actually printing the checks. For the start-up contractor these payroll companies can be very economical as compared to initially hiring a person to take care of this system in-house. This can also take a big load off your mind during start up.

INITIAL EMPLOYEES

Since you cannot manage and operate an entire construction company by yourself, you will need to hire employees. This should also be a carefully planned action because you will need to give certain valued employees control over particular areas of the company in order for them to be effective. The small start-up contractor should look for an accounting/controller type individual, a project manager and several qualified field foremen. Although, with many small contractors, the company owner ends up filling one or more of these positions until the company is large enough to afford more management personnel.

There are many decisions to be made everyday to efficiently manage a construction company, so a company owner should hire people to fill as many of these initial positions as the budget will allow. With these positions filled, a contractor will have the start of a well-rounded company. Therefore, due to the importance of these positions, in-depth interviews and research into applicant's references will pay dividends to the company later because you have hired competent employees with similar views to yours on how a company should be managed. Many construction personnel are hired based solely on their resume experience. In your interviews, you should find out how these individuals view employee and subcontractor relations, prompt payment and attention to suppliers and subcontractors, craft and management overtime, relationships with the project owners, company benefit and profit sharing programs along with all of the basic past job experience information.

Make up a 25 to 50 question written test that touches on all of the above views that you, as a company owner or manager, wish your company to project. Give this test to your interviewing supervisors to see if they will manage operations similar to the image of the company. Knowing up front if a potential supervisor for your company has a similar attitude to yours on how a company should be operated can save you much frustration along with any potential damage to your companies' reputation at a later time.

Experienced and qualified individuals within the construction business will want to be rewarded for the hard work it takes to operate and manage a successful construction company. Therefore, you should have a plan in place to monetarily reward the people who help make your company a success. This can be done by incentives such as company profit sharing or stock in the corporation. If you have chosen the right people to help you build your business, these incentives will keep your employees motivated towards company profitability. This will result in a win-win situation for the employees and the company along with maintaining a quality company reputation with your suppliers, subcontractors and contracting agencies.

INITIAL CONTRACTING/PAPERWORK/DOCUMENTATION OUTLINE

Following is an outline of the paperwork and documentation that will need to be in order prior to the start of contracting:

a.) Corporation/Department of Revenue Documents

b.) State Contracting License

c.) Insurance

    Automobile

    Workers Compensation

    General Liability

    Umbrella Liability

d.) Bonding

    Bid Bonds

    Payment & Performance Bonds

e.) Labor Unions vs. Open Shop

f.) Company & Safety Policies

CORPORATION/DEPARTMENT OF REVENUE DOCUMENTS

The new contractor should consult an attorney for the best way to organize his or her business (corporation, sole proprietorship, limited liability partnership, etc.) and meet all federal, state and local requirements. The corporation is the most formal type of business structure and is its own separate legal entity that is owned by stockholders and provides many benefits to company owners. A corporation provides shareholders with only the liability of the money they have put into the company with respect to creditors, while sole proprietorships and partnerships are subject to unlimited personal liability. Corporations can also offer the advantage of allowing tax deductible benefits such as health and life insurance and travel and entertainment as well as providing an increased tax shelter for retirement plans for its shareholders. If you do want to incorporate your company, there are a few simple steps:

SELECT A CORPORATE NAME. The first thing a contractor must do is choose a name for the corporation. The name must not be registered by any other company in the state of incorporation or be similar to any other name that might confuse the public.

DECIDE ON THE STATE OF INCORPORATION. Most companies decide to incorporate in the State where they intend to do business. Some companies although decide to incorporate in Delaware because of low filing fees and tax advantages.

CHOOSE THE TYPE OF CORPORATION. The company can be a general business corporation, close corporation, a professional corporation or a not- for-profit corporation. A general corporation or a "C" corporation is a profit making venture that may have unlimited stockholders, while the liability of the stockholders is usually limited to the amount of investment in the business.

A close corporation is limited to 30 to 50 stockholders, but may vary from state to state. This type of corporation is best suited to one-person corporations or a small group of individuals who work like a partnership without the legal formalities and record keeping of a general corporation. Professional corporations are formed by individuals who provide services for which a professional license is required, such as doctors, lawyers, engineers and architects. Not-for-profit corporations are formed for charitable, religious, social or fraternal purposes. Not-for-profit corporations are made up of members who follow conditions outlined in the corporate bylaws. An "S" corporation is a special tax designation applied for by corporations already in existence. Any of the corporate structures may elect S corporation status, which eliminates federal corporation income taxes and avoids double taxation problems. Double taxation occurs when the corporation is taxed and then the stockholder is taxed again upon a distribution. With an S corporation, the corporations tax documents are an insert to the stockholders tax return. S corporations can only have 35 stockholders, must be domestic corporations with U.S. citizen stockholders and permit only one class of stock.

CHOOSE CLASSES OF STOCK. The two basic kinds of stock are common and preferred. Common stock holders have voting rights to select directors and are entitled to share in the profits of the company. Generally, preferred stockholders do not have voting rights in return for preferred rights when it comes to the distribution of dividends.

DECIDE ON THE DIRECTORS. Directors are the individuals who have the responsibility of managing the company and establishing the policies as stated in the corporate bylaws.

DETERMINE THE CORPORATE ADDRESS. The corporate address can be a business address, home address or a post office box. Many companies prefer to use a post office box to ensure privacy as well as keep their personal and business correspondence separate.

SELECT A REGISTERED AGENT. Most states require that a person or entity located in the state of incorporation be available during regular business hours to receive legal notices and other official documents. In most cases your attorney will act as the companies registered agent.

Your company will need to contact your local Internal Revenue Service office to obtain the forms that you will need to complete in order for your company to get its Federal Tax Identification Number. This process will generally take about three to four weeks.

If your company plans to hire employees, then you will need to contact the Department of Revenue to register for reporting employee withholdings and unemployment taxes for both the State and Federal levels. It would also be a good idea to contact your local labor bureau for any additional information that your company might need on employment laws for your area and business situation. These bureaus often provide free seminars or workshops for business people working within their jurisdiction to assist in compliance with their policies. As was mentioned prior in this chapter, it is often a good idea to have an outside payroll firm handle all of these weekly labor checks for your company during this initial period.

CONTRACTING LICENSE

A contracting license can be obtained through your local board. Most states require the applicant to complete a written exam to be eligible for a contractor's license along with already having a commitment for bonding and insurance. There will generally be a periodic renewal process for all of the licenses discussed that will be simply a matter of returning the appropriate documents and fees. The Contracting Board should be listed in the State Government section of your local telephone book.

Another license that will be required is a county or city business license, depending on the area and type of work you plan to construct. These business licenses can usually be obtained by application, without a written exam as opposed to the contracting license. Again, these county and city offices should be listed in your local telephone book, and your project bidding documents will also let you know if any of these additional licenses are required.

Many city, county, state or federal contracting agencies require bidders to be pre-qualified before construction contracts can be bid on or will be awarded to an individual contractor. Typically, this pre-qualification is a matter of submitting an application that lists the experience of the companies personnel, what equipment the company has available, a list of previously completed projects along with the required fees.

For the start-up contractor, this can be a problem since you have no previously completed projects. In this instance, many contracting agencies have a specified contract amount for which the bidders would need to be pre-qualified in order to submit a proposal; say $50,000 for example. For projects that are worth under $50,000, the bidders do not need to be pre-qualified; projects over $50,000, the bidders would need to be pre-qualified. This break point in the contract amount will vary depending on the contracting agency involved.

INSURANCE

Automobile, Workers Compensation, General Liability and Umbrella Liability insurance coverage can usually be obtained through one insurance agent. Set up meetings with several insurance agents that specialize and have extensive experience in the type of work of your company. After you meet with several agents choose the one that you feel most comfortable with and one that can obtain your company equitable rates.

Proof of all the above types of insurance are typically required by contracting agencies. Your chosen insurance agent will provide you with the appropriate certificates upon request, for submission to either the Contracting Agency or Prime Contractor, depending on whether your company is the General Contractor or a Sub-Contractor. Each particular project's specifications will denote how much of each type of insurance that will be required of the General Contractor. Following is a brief explanation of each of the types of insurance that you will need:

Automobile - This insurance policy is virtually the same as your personal automobile insurance. The insurance company will pay all sums that your company legally must pay as damages, to which the insurance applies, caused by an accident and resulting from the ownership, maintenance or use of a covered automobile.

Worker's Compensation - This insurance policy is to protect your employees in the event that they are injured while working on one of your projects. The employment must be necessary or incidental to your work in a state or territory listed in your particular policy.

General Liability - This insurance policy covers bodily injury or property damage that occurs on premises that you own or rent or because of operations on your construction projects.

Umbrella Liability - This insurance policy covers bodily injury, property damage or personal injury in excess of the underlying limits of the three above mentioned insurance policies.

As you can see there are several "gray" or "overlapping" areas with regards to insurance. This is a very good reason to have a single agent handle all of your insurance needs; they will have personal knowledge of your company and its policies and could save you much time and conflict in the event of an insurance claim. A typical Certificate of Insurance is shown in Figure 1-3.

FIGURE 1-3 Certificate of Insurance

ACORD™ CERTIFICATE OF LIABILITY INSURANCE		DATE (MM/DD/YY)
PRODUCER	THIS CERTIFICATE IS ISSUED AS A MATTER OF INFORMATION ONLY AND CONFERS NO RIGHTS UPON THE CERTIFICATE HOLDER. THIS CERTIFICATE DOES NOT AMEND, EXTEND OR ALTER THE COVERAGE AFFORDED BY THE POLICIES BELOW	
	COMPANIES AFFORDING COVERAGE	
INSURED	COMPANY **A**	
	COMPANY **B**	
	COMPANY **C**	
	COMPANY **D**	

COVERAGES

THIS IS TO CERTIFY THAT THE POLICIES OF INSURANCE LISTED BELOW HAVE BEEN ISSUED TO THE INSURED NAMED ABOVE FOR THE POLICY PERIOD INDICATED. NOTWITHSTANDING ANY REQUIREMENT, TERM OR CONDITION OF ANY CONTRACT OR OTHER DOCUMENT WITH RESPECT TO WHICH THIS CERTIFICATE MAY BE ISSUED OR MAY PERTAIN, THE INSURANCE AFFORDED BY THE POLICIES DESCRIBED HEREIN IS SUBJECT TO ALL THE TERMS, EXCLUSIONS AND CONDITIONS OF SUCH POLICIES. LIMITS SHOWN MAY HAVE BEEN REDUCED BY PAID CLAIMS.

CO LTR	TYPE OF INSURANCE	POLICY NUMBER	POLICY EFFECTIVE DATE (MM/DD/YY)	POLICY EXPIRATION DATE (MM/DD/YY)	LIMITS	
	GENERAL LIABILITY				GENERAL AGGREGATE	$
	___ COMMERCIAL GENERAL LIABILITY				PRODUCTS-COMP/OP AGG	$
	___ CLAIMS MADE ___ OCCUR				PERSONAL & ADV INJURY	$
	___ OWNER'S & CONTRACTOR'S PROT				EACH OCCURRENCE	$
	___ ___				FIRE DAMAGE (Any one fire)	$
					MED EXP (Any one person)	$
	AUTOMOBILE LIABILITY					
	___ ANY AUTO				COMBINED SINGLE LIMIT	$
	___ ALL OWNED AUTOS				BODILY INJURY (Per Person)	$
	___ SCHEDULED AUTOS					
	___ HIRED AUTOS				BODILY INJURY (Per Accident)	$
	___ NON-OWNED AUTOS					
	___ ___				PROPERTY DAMAGE	$
	GARAGE LIABILITY				AUTO ONLY - EA ACCIDENT	$
	___ ANY AUTO				OTHER THAN AUTO ONLY:	
					EACH ACCIDENT	$
	___ ___				AGGREGATE	$
	EXCESS LIABILITY				EACH OCCURRENCE	$
	___ UMBRELLA FORM				AGGREGATE	$
	___ OTHER THAN UMBRELLA FORM					$
	WORKER'S COMPENSATION AND EMPLOYERS' LIABILITY				WC STATUTORY LIMITS OTHER	
	THE PROPIETOR/ ___				EL EACH ACCIDENT	$
	PARTNERS/EXECUTIVE ___ INCL				EL DISEASE-POLICY LIMIT	$
	OFFICERS ARE: ___ EXCL				EL DISEASE-EA EMPLOYEE	$
	OTHER					

DESCRIPTION OF OPERATIONS/LOCATIONS/VEHICLES/SPECIAL ITEMS

CERTIFICATE HOLDER	**CANCELLATION**
	SHOULD ANY OF THE ABOVE DESCRIBED POLICIES BE CANCELLED BEFORE THE EXPIRATION DATE THEREOF, THE ISSUING COMPANY WILL ENDEAVOR TO MAIL ___ DAYS WRITTEN NOTICE TO THE CERTIFICATE HOLDER NAMED TO THE LEFT, BUT FAILURE TO MAIL SUCH NOTICE SHALL IMPOSE NO OBLIGATION OR LIABILITY OF ANY KIND UPON THE COMPANY, ITS AGENTS OR REPRESENTATIVES.
	AUTHORIZED REPRESENTATIVE
ACORD 25-S (1-95)	© ACORD CORPORATION 1988

BONDING

Many insurance agents, can also provide contractors with construction bonds. For newer or smaller contractors, bonding can be one of the major hurdles that will be encountered. In relatively recent times, during the 1980's, bonding companies were forced to complete the projects of many failing small construction companies. This caused many bonding companies to avoid small businesses unless they had a substantial amount of liquid capital and an impeccable record of successfully completed construction contracts.

Bonds that are available to small and emerging companies are usually at a higher rate (percentage of the contract amount) than that of large and established companies. This is related directly to the risk they are undertaking with the companies available liquid capital to fund projects and their completed project track record. A bonding company will typically bond a companies projects up to a limit of approximately 5 to 10 times the companies working capital.

Your local office of the Small Business Administration (SBA) can be a place to obtain assistance in bonding if you cannot obtain equitable bonding within the general market. Following is a brief explanation of the types of bonds that you will typically need:

BID BONDS - When your company is bidding a project as a Prime (General) Contractor, most agencies will require that a Bidder's Bond be submitted with your proposal. This bond is usually provided to the agency in an amount of 5% to 20% of your bid price (percentage amount will be specified for each individual project in the contract specifications). This bond insures the owner or contracting agency that if your company has the apparent low bid and decides it does not want to accept the contract, that the contracting agency can the keep the amount of your bid bond as damages associated with having to contract with apparent second low bidder.

In the event your company makes a massive judgment or calculation error in your estimate resulting in a large spread in total bid price (substantially more than that of the bid bond) between your bid and the second bidder; your company might want to consider giving the contracting agency your bid bond. This could be a method of minimizing your losses as compared to realizing large losses during construction. If a construction company does make and can prove an obvious calculation or judgment error in the estimate, typically the contracting agency, in good faith will allow the contractor to get out of the contract without any damages. Nevertheless, due to the possibility of damages being assessed against your company if you submit an inaccurate low bid, you should be very accurate and careful during project estimating.

FIGURE 1-4 Bid Bond Form

PROPOSAL BOND

Know all men by these presents, that _____

_____

a surety company duly organized under the laws of the State of _____

having is principal place of business at _____

in the State of _____, and authorized to do business in the State of

_____ is held and firmly bound unto the State of

_____,

in the full sum of ten (10) percent of the total amount of the proposal for the work hereinafter described, for the payment of which, well and truly to be made, we bind ourselves, our heirs, executors, administrators and assigns, and successors and assigns, firmly by these presents.

The condition of this bond is such that, whereas _____
(Bidder)
is herewith submitting its proposal for the following work, to wit:

_____

_____

_____

_____

said proposal, by this reference being made a part hereof;

Now therefore, if the said proposal submitted by the said bidder be accepted by _____

_____, and the contract for said work be awarded to said bidder, and if the said bidder enters into and executes the said contract and furnishes bond

as required and within the time fixed and set by the contract agreement, then this obligation shall be void; otherwise to remain in full force and effect.

Signed and sealed this _____ day of _____, _____

_____	_____
Bidder	Surety Company
_____	_____
Signature	Signature

PERFORMANCE BONDS - This bond is submitted to the contracting agency during award of the contract after your company has been declared the low bidder. This bond says that if your company does not perform or complete all aspects of the project, the bonding company will pay the contracting agency the sum of the performance bond. The redemption sum of these bonds will usually be either 50% or 100% of the contract price.

FIGURE 1-5 Performance Bond Form

PERFORMANCE BOND

Know all men by these presents: That we _____

as principal, and _____

as surety, are jointly and severally held and bound unto _____
                                                                    Owner
For the payment of which we jointly and severally bind ourselves, our heirs, executors, administrators and assigns, firmly by these presents.

THE CONDITION OF THIS BOND IS SUCH

That, whereas the said principal herein has made and entered into a certain contract, copy of which is attached hereto, which contract, together with the applicable plans, Standard Specifications, special provisions, and schedule of contract prices, is by this reference made a part hereof, whereby the said principal agrees to do in accordance with the certain terms, conditions, requirements, plans and specifications which set out in said contract and all authorized modifications of the contract which increase the amount of work and the amount of the contract. Notice to the surety of any of the immediately foregoing are waived.

Now, therefore, if the principal herein shall faithfully and truly observe and comply with the terms, conditions and provisions of the said contract, in all respects, and shall well and truly and fully do and perform all matters and things by him undertaken to be performed under said contract, upon the terms set for therein, and with the time prescribed therein, or as extended as provided in the contract, and shall indemnify and save harmless _____, and members thereof, its managers, employees, and agents, against any direct indirect damages of every kind and description that shall be suffered of claimed to be suffered in connection with or arising out of the performance of the said contract by the said Contractor or its subcontractors and shall in all respects perform said contract according to law, then this obligation is to be void, otherwise to remain to full force and effect.

Nonpayment of the bond premium will not invalidate this bond nor shall _____

_____ be obligated for the payment thereof.

Witness our hands this _____ day of _____, _____

_____   By _____
            Principal   Authorized Signature

_____   By _____
            Surety   Authorized Signature

PAYMENT BOND - Again, this bond is submitted to the contracting agency during award of the contract. This bond says that if your company does not pay its employees, material suppliers or subcontractors that the bonding company will make payment on these liens. This bond is also typically in the amount of 50% or 100% of the contract price.

FIGURE 1-6 Payment Bond Form

PAYMENT BOND

Know all men by these presents: That we _____

as principal, and _____

as surety, are jointly and severally held and bound unto _____
                                                              Owner
For the payment of which we jointly and severally bind ourselves, our heirs, executors, administrators and assigns, firmly by these presents.

THE CONDITION OF THIS BOND IS SUCH

That, whereas the said principal herein has made and entered into a certain contract, copy of which is attached hereto, which contract, together with the applicable plans, Standard Specifications, special provisions, and schedule of contract prices, is by this reference made a part hereof, whereby the said principal agrees to do in accordance with the certain terms, conditions, requirements, plans and specifications which set out in said contract and all authorized modifications of the contract which increase the amount of work and the amount of the contract. Notice to the surety of any of the immediately foregoing are waived.

Now, therefore, if the principal herein shall make payment promptly, as due to all Subcontractors and to all persons supplying to the Contractor and his subcontractors, equipment, supplies, labor or materials for the prosecution of the work, or any part thereof, provided for in the said contract, and shall pay all contribution of amounts due its workers compensation carrier and the State Unemployment Compensation Trust Fund from such Contractor or subcontractors incurred in the performance of said contract, and pay all sums of money withheld from the Contractor's employees and payable to the Revenue Department; and shall pay all other just debts, dues and demands incurred in the performance of the said contract and shall pay _____ such damages as may accrue under said contract, then its obligation is to be void, otherwise to remain in full force and effect.

Nonpayment of the bond premium will not invalidate this bond nor shall _____

_____ be obligated for the payment thereof.

Witness our hands this _____ day of _____, _____

_____   By _____
              Principal   Authorized Signature

_____   By _____
               Surety   Authorized Signature

Obviously, if your company fails and requires the bonding company to become involved and make payments, the bonding company will come after your company through legal channels for compensation for their efforts and dollars spent. Therefore, one can see why bonding companies only want to do business with construction companies that have liquid assets available and a perfect record of successfully completed construction contracts. The bonding companies will also require the owners of the company to sign a personal indemnity agreement that basically says, if the bonding company has to bail the contractor out of a project then they have first rights to your personal assets as payment when your company cannot pay any debts for which the bonding company has insured.

LABOR UNIONS VS. OPEN SHOP

There are advantages and disadvantages to operating your construction company using labor organizations or operating non-union (open shop). Either way you choose to go, the cost per manhour will be virtually the same on a publicly funded project due to the Davis-Bacon Act. The Davis-Bacon Act of 1931 set a uniform minimum "prevailing wage" for all construction crafts in all parts of the United States. These wages, set by the U.S. Department of Labor, ensure similar pay for similar work as it relates to other workers in a particular area. The Davis-Bacon Act is related to Public Work Projects or in other words publicly financed contracts.

Your particular project specifications will give the applicable wage rates that each bidder will be required to meet if it is publicly funded; if it is not, then you can pay your employees whatever the market will bear as long as it is above the federally set minimum wage. As a contractor or manager, some care should be given to this open market wage decision. Many contractors have lost quality employees because they attempted to pay a substantially different hourly wage between their private and public projects.

One difference between union and non-union on prevailing wage rate projects with respect to employees payment, is that you can pay the "fringe benefit" to the employee if you are non-union or to union trust funds if you are signatory to the labor organization. A non-union contractor can set up insurance and pension funds for the employees that meet all of the governmental requirements to avoid paying payroll taxes on this fringe benefit amount. The Associated General Contractors of America (AGC) is a good source for non-union contractors to obtain these fringe benefit programs at a reasonable rate.

For many privately funded projects, each bidder is free to use whatever wage rates they choose as discussed above. If you are a union contractor, there should be some pre-bid negotiations with your local chapter relating to wage rates to be used as they can often times offer the contractor some help in these instances.

There are labor unions for every craft that is used on a construction project. (Laborers, Carpenters, Masons, Equipment Operators, Ironworkers, Teamsters, Electricians, etc.) Union employees can only do work that is within their jurisdiction. For example, Laborers cannot operate heavy equipment or do electrical wiring. In a non-union situation, an employee can do work associated with the Laborers for a portion of a shift while doing work associated with the Operating Engineers for the remainder of the shift. During that shift the employee would get paid the appropriate wage rate relating to the type of work he was doing during that particular portion of the shift.

An advantage to operating your company with Labor Organizations is that if you are going to construct a project out of your local area, you could call the appropriate union hall and they will supply your company with the necessary manpower to meet your requirements. If you are operating non-union, you would have to search out your own employees or move employees from your local area.

The labor organizations, along with the Associated General Contractors of America, have apprenticeship programs that help the employees gain the knowledge and skills of the construction industry before they are assigned to projects. If these apprentices are sent out to a project, they get paid a reduced wage rate corresponding to how far they are along in their particular program. Once the apprentice if fully trained, or completed their program, they will get paid full journeyman wage rates. Many of the public contracting agencies have set contract goals or requirements for the hiring and training of apprentices. Therefore, having your company belong to one of these organizations can be very beneficial.

Many residential contractors operate non-union because of the large amount of privately financed projects that they construct with no set labor wage rate requirements. A great deal of Commercial contractors will operate with the use of labor organizations if they construct mostly publicly funded projects, while most will operate open shop if they construct mostly privately financed projects. On the other hand many of the Public Works contractors operate with labor organizations because of the typically higher skill of labor that can be provided. And also due to the fact that most public works contracts that are publicly financed, requiring prevailing wage rates, these contractors are already having to pay rates equal to that of the unions. Wage rates are obviously not the only reason a contractor will choose to operate with or without a labor organization, other factors come into play such as apprenticeship training, workers compensation, employee pensions and available workforce which will all be discussed later in this book.

COMPANY & SAFETY POLICIES

Every construction company should develop and distribute to their employees a Company Employment Policy Manual. Items that should be included in this policy manual are the following:

1.) Sheet for employee history information,

2.) Explanation of employee time reporting and the company pay period and reimbursement of expenses,

3.) Explanation of how the company will handle legal holidays, vacation and sick leave,

4.) The company's policy and disciplinary action relating to absence and tardiness,

5.) Equal employment opportunity policy,

6.) The company's policy on sexual harassment,

7.) The company's compliance with the programs relating to Workman's Compensation and Unemployment Insurance,

8.) The company's policy regards out of town subsistence or covered moving expenses,

9.) And the company's Substance Abuse Testing program if you choose to have one.

10.) The company's safety policy,

11.) It also is convenient to include an IRS form - W-4, Employee's Withholding Certificate, so that everything related to the hiring of an employee is in one pamphlet.

Construction companies should realize the need to develop and administer a safety program. The contractor has a legal and ethical obligation to keep the jobsites safe for its employees and for the public. Contractors are also realizing that keeping the jobsites safe benefit them in terms of the premiums paid for insurance coverage. The lower you can keep your insurance premiums, the less you have to add to your proposal estimates for overhead as it relates to these expenditures; therefore, making your company more competitive in the marketplace.

If your company is large enough, a full time Safety Director based in your home office is a good investment. In most small construction companies the execution of the safety program falls upon the jobsite supervisors while still being the final responsibility of the owner or manager. Whether your company is large or small, open and direct communication between the field operations and the home office is critical. Communication between jobsite employees and the supervisor is also very important because the workers often see possible safety hazards that are not seen by a supervisor who is not physically doing the work.

Following is a basic list of items that should be covered in your safety policy:

1.) Project inspection by the jobsite supervisor,

2.) Daily visual inspections by each crew's foreman,

3.)   Weekly safety meetings to be held at the jobsite to discuss potential hazards concerning the work and progress and future work. Include the owner representatives and your subcontractors in these meetings,

4.)   Physical qualifications or limitations of employees,

5.)   Safety training,

6.)   Medical/First Aid supplies, Material Safety Data Sheets and their location on the jobsite. Also telephone locations with the associated numbers and routes to the nearest medical facility in the event of an emergency,

7.)   Jobsite housekeeping and storage of materials,

8.)   Required employee safety equipment (example: hard hats, safety glasses, earplugs, etc.),

9.)   General rules relating to the operation of equipment, vehicles and machinery,

10.)   General rules relating to working at heights, in trenched excavations, near power lines, etc.

Many construction companies now implement Safety Incentive Programs. These programs basically consist of safety awards for periods of accident free work. Most contractors will provide T-shirts, coffee cups, hats, gift certificates and other items like these as monthly or bi-monthly safety awards. These programs assist in involving the employee in the overall safety of the project since they are rewarded for the crew's safe working environment. The cost to the contractor is minimal as compared to the increase in insurance premiums if there was to be an accident at the jobsite, not to mention the direct safety of your employees.

If your company plans on being a member of a labor organization, they will also greatly assist you in developing and maintaining a quality safety program that can be continually updated as the marketplace changes. These labor organizations can also provide training in first aid, CPR, and safe excavation, along with many others. Anyone on your projects who is involved in a supervisory role should take these classes.

CHAPTER TWO

TYPES OF CONTRACTS
AND ASSOCIATED RISKS

In construction contracting there are many ways in which the contracts can be drafted and issued. The three basic types of contracts that we will discuss are:

Negotiated Contracts,

Cost Plus Contracts,

Fixed Price Contracts.

NEGOTIATED CONTRACTS

Negotiated contracts occur when the contracting agency negotiates directly with one or more construction companies. These contracts are much more common in the private construction industry. Private owners will choose a construction company based on their reputation and the overall quality of similar projects that they have completed in the past.

Typically the owner will provide the construction company with a set of plans and specifications for the project. In this case the owner would have already spent a great deal of time and effort with an engineering/ architectural firm to get to the point of a complete set of plans and specifications.

The owner or contracting agency may want the contractor to design and construct the project per some specified guidelines - design/build contract. This type of construction process can also occur when the owner is on a fast track to complete a project. In the case of design/build project, the owner provides the contractor with a general idea of what they want to accomplish and the contractor hires an engineering/architectural firm to work with them through the entire process from design, permitting, construction and close-out. If a contractor works on a great deal of design/build projects, they may have engineers and architects on staff to complete the design end of the job in-house. A design/build contract may fall into any of the three categories discussed in this chapter.

In either of the above cases, the contractor will prepare an estimate on the cost to construct the project. The owner and contractor will then have a meeting to discuss the estimate and mutually agree to a contract price to complete the project per the supplied (by the owner or by the contractors team) plans and specifications. Where more than one contractor is involved in this process, the owner typically asks for them to submit proposals for review.

In the case of owner-supplied plans, the proposals can be per the plans and specifications supplied by the owner, or they can incorporate designs by the contractor that may be more economical to the contracting agency. The owner will review these proposals and select a contractor based on the company's reputation and history, economics of the proposal and any differing design aspects.

For the types of contracts discussed in this book, negotiated contracts would have mid to low range risks associated with them. This is because the contractor has the opportunity to discuss and negotiate as many costs as he can before construction starts, but after that point he must adhere to the contract amount mutually agreed upon with the owner.

The contract amount on virtually any type of contract can be changed if the work or site conditions turn out to be different than that of the contract plans and specifications via a change order to the original contract. For this reason your company should work hard to develop a clearly defined scope of work for a negotiated project so that when differing conditions do occur, everyone involved knows the extra work is beyond the scope of the original contract.

Also, when working within the private sector, it may be more difficult to get change orders from the owner unless your have a clearly defined scope of work within your legal contract. This is due to the fact the many private owners have already leveraged themselves financially and will fight the contractor vigorously to have the contract stay within or below the original budget.

Depending on the style of work involved in the project, the labor and equipment risk, how clearly defined the scope of work is and many other factors, the typical profit margin would range from 5% to 15% of the project cost.

COST PLUS CONTRACTS

There are times when the owner and contractor cannot come to an agreement on the cost of construction (the contractor believes the project will cost more than the owner believes the contract will cost). At these times the contract may be handled on a cost plus basis. A cost plus contract is basically written as the "direct cost" of the project plus a negotiated profit or margin.

This type of contract may also be helpful if the project needs to be constructed immediately, without time to fully complete the contract plans and specifications or execute the bidding process. The contracting agency can choose a reputable contractor or one that they have worked successfully with in the past and write this contractor a cost plus contract in order to expedite construction.

A General Contractor writing Subcontracts can also utilize cost plus contracting. Often times the work of a subcontractor overlaps or depends on work that the General Contractor is to construct. A cost plus contract can be written in this instance where the Subcontractor could not accurately submit a Fixed Price proposal to the General Contractor due to an operation that relies on the production of the General Contractor in a related operation.

The "direct cost" of the project typically includes the following items:

1.) Wages paid to employees,

2.) Fringe benefits paid to the employees or labor organizations,

3.) Taxes and Insurances that are percentage of your labor costs (Federal Unemployment, State Unemployment, F.I.C.A., General Liability, Workers Compensation and any local taxes),

4.) Equipment costs (Negotiated hourly, weekly or monthly rates for each piece of equipment on the project owned by the contractor or invoice cost of outside rented equipment) and any associated taxes and insurance,

5.) Permanent material costs (Concrete, asphalt, steel, lumber, etc.) and any associated taxes,

6.) Miscellaneous tools and materials (Concrete forming lumber, snap-ties, nails, electrical tools, etc.) and any associated taxes and insurance,

7.) Utility costs (Water, telephone, electricity and toilets),

8.) Subcontractor costs with any associated taxes.

Cost Plus contracts have the lowest amount of risk associated with them; therefore, they generally also produce the smallest profits or margins. As long as the contractor accurately negotiates the direct costs and margins, he can reasonably assure a small but safe percentage profit.

It is very important to have a clear understanding with the owner on what "direct costs" exactly include. Typical profits or margins for these types of contracts range from 2% to 15%; therefore, a contractor cannot pay direct costs out his negotiated profit. When negotiating the profit margin on a cost plus contract, the construction company should evaluate what the owner is designating as direct costs and adjust the profit margin to get the proper return on investment.

Often, a contractor does not accurately track the direct costs or fails to negotiate project overhead items into the direct costs and actually completes the job at a financial loss. Many construction companies operate with an overhead (indirect company operation costs) rate of approximately 10%. The contractor, at a minimum, has to cover overhead and depending on how many indirect costs are attributed to the direct cost of the contract, you can determine the margin rate you need to charge the owner in order make your desired profit on the project.

FIXED PRICE CONTRACTS

Fixed Price contracts are the most extensively used in public works construction contracting. As in many negotiated contracts, the agency has plans and specifications prepared by an engineering/ architectural firm, or by their own agency, prior to sending the project out to bid. Contractors then prepare estimates and proposals on the cost to complete the work. These proposals are submitted to the owner at a specified date, time and place while also conforming to the agency's format and with all of the required documentation. The qualified construction company with the lowest bid will be awarded the contract if he has met all of the necessary requirements. No negotiations take place between the owner and the contractor after the proposals have been opened. The contract is written in the amount of the proposal to the lowest responsive bidder.

In private industry the bidding process can include just a few contractors chosen by the owner, again from past experience, recommendation, or from the contractors reputation within the industry. For publicly financed projects, any qualified construction company may submit a proposal. These types of contracts are not typically of the design/build nature due to the differences they produce in the proposals submitted to the owner; although design/build projects may end up as a fixed price contract after negotiations are complete.

Fixed Price contracts have the most risk for the contractor associated with them of the three types of contracts discussed here. If the contractor's estimate is accurate and the project is managed correctly, then they can also be the type of contract with the most profit associated with them. On the other hand, if the contractor left some costs out of his estimate, if the project is managed poorly or if some aspect of the project is built incorrectly and has to be re-built, the contractor would then have to absorb these additional costs with no reimbursement from the owner. Typical profit margins on this type of contract would range from 5% to 25%; again depending on the labor and equipment risks, time of year construction is to occur, along with many other factors discussed throughout this book.

The more risk associated with a contract, the higher the profit. Margins or profits attached to projects vary greatly with the style of work being constructed. Commercial/ Residential general contractors typically have low labor and equipment risk (and consequently low margins) associated with their projects since subcontractors accomplish a majority of the work. Subcontractors and materials incorporated into a project are fixed costs. Therefore when these areas are the major portions of a project, a low risk is being taken by the general contractor as to what the final costs of the project will be. On the other hand, a public works contractor typically has labor and equipment as the major cost on a project. These areas of cost are dependent upon the estimated production rates and are not fixed. So more risk is being taken as to what the job will eventually cost and a greater profit margin should be added to the base cost.

Projects, whether Commercial/Residential or Public Works, will have varying combinations of cost associated with the categories of labor, equipment, materials and subcontractors. With this in mind, the contractor should analyze how much relates to each of these project categories when choosing the appropriate margin to add on to the direct costs of the project.

CHAPTER THREE

FINDING PROJECTS TO BID

A construction contractor cannot obtain work if they do not know that a particular project, one that they would like to construct, is available to bid. The contractor can find out about projects that are advertised for bid in many different places. The following are the areas that will be discussed in this chapter:

Plan Centers,

Local Business Journals,

National/Regional Construction Magazines,

Contracting Agency Information Packages.

PLAN CENTERS

Every major city within the United States will have at least one construction plan center. For a periodic membership fee, contractors can join their local plan center and have access to a large quantity and variety of construction projects plans and specifications that are available to bid on by local and regional contracting agencies. Most plan centers are organized like a library. A contractor can "check-out" a particular project's plans and specifications for review at the plan center. The plan centers usually also have a mailing that is sent out to its members detailing what projects they have available for review and their corresponding proposal due date. For contractors who bid a lot of projects, the plan center can be an economical and centralized alternative for review of plans and specifications. When a contractor orders a bid package directly from the owning agency, there is typically a fee ranging from $10.00 to $300.00 for the plans and specifications. Rarely is this fee refundable upon return of the package, or is the proposal package free of any costs. One can easily see that a contractor can save a fair amount of plan ordering fees by using a plan center if the contractor analyzes a great number of plans and specifications per month.

General Contractors who need subcontract proposals on certain portions of projects that they are bidding can call the plan centers to find out what other specialty or general contractors have reviewed the projects plans and specifications from a list kept by the plan center. They can then call or mail out solicitations to those contractors listed by the plan center to see if they will be providing subcontract proposals on the areas of work in which the subcontractors specialize.

LOCAL BUSINESS JOURNALS

Most major cities within the United States also have published daily business journals for which a contractor may subscribe to. These daily business journals will generally have news articles describing important events within the construction industry and the general business community along with a section that lists newly advertised construction projects for bid. The advertisements for these construction projects will normally detail the contracting agency, a general description of the project, bid due date, any special bidding requirements, and pre-qualification information, along with how to obtain a copy of the plans and specifications.

There are also publications available that are more regional than a particular cities business journal. These publications will advertise construction projects that are available to bid within a larger region or even multiple states. The news articles are more regional along with the construction projects they list for bid, while they still detail the same information on the projects out for bid.

Both of the above mentioned types of publications would most likely include a classified advertisement section. Within the construction industry this is a good place to buy and sell equipment, search for employees, advertise for subcontract proposals on projects you are bidding as a general contractor along with buying and selling construction materials.

NATIONAL/REGIONAL CONSTRUCTION MAGAZINES

There are many National and Regional Construction Magazines to which a contractor can subscribe. Some of these magazines have a section dedicated to the advertisement of upcoming projects to bid. One problem with using these types of magazines for your sole source of obtaining advertisements is that they are generally published monthly, although some are weekly. Most projects are advertised approximately one month before the bids due date; therefore, a contractor might not have sufficient time to prepare an estimate if he relies solely on ordering his plans and specifications from the advertisements in these publications. Often these type of publications will only detail a few of the construction bid opportunities that are available, and usually those projects are the very large jobs. It is a good idea to subscribe to several of these construction magazines and many are free. This will keep you up to date on what is happening within the regional and national industry, outside of your own company.

CONTRACTING AGENCY INFORMATION PACKAGES

When a contractor does a lot of work with particular contracting agencies, they can sometimes receive direct mailing information on projects that are coming out for bid from within that agency. Most government agencies (example: Departments of Transportation, USDA Forest Service, and Environmental Protection Agency) will provide contractors with information on upcoming projects. Contractors can then simply review the mailings sent out by these agencies, pick the projects that appear to be of the style they want to construct and call/fax/ e-mail the agency to request a copy of the plans and specifications.

Most of these governmental agencies charge a small fee for their proposal packages while others send them out free of charge.

With the heavy use of computer systems within business today, many agencies have their advertisements and bid packages available on-line. A contractor, through a modem, can dial up the agencies contract line and view what projects the agency is advertising for bid. If there is a project that appears inviting, the contractor can download the specifications and bid package into his personal computer for printing a hard copy of this information right in his office. These government agency information lines may also contain past bid tabulations, contract award information along with much more information on how to do business with that particular agency.

CHAPTER FOUR

DECIDING WHAT PROJECTS TO BID

Every business day there are contracting agencies advertising many construction projects to be bid on and to be awarded to construction contractors. Obviously, one contractor cannot bid on every project advertised; therefore, we will discuss several ways to decide which jobs are worth your estimating time and efforts:

Advertisement Description/Proposal Bid Items,

Prime (General) Contracting,

Subcontracting,

Resource Availability/Location - Equipment & Manpower,

Pre-Qualification.

ADVERTISEMENT DESCRIPTION/PROPOSAL BID ITEMS

When contracting agencies advertise for bids on construction projects they generally will include a brief description of the work to be completed. For example they might state that the project is "Asphaltic Concrete Overlays", "Reinforced Concrete Sewer Pipe Installation", "Structural Steel Highway Overcrossing" or "Traffic Signal Installation" as that projects particular description of work. Most construction companies are somewhat specialized or tend to look at specific types of projects in which they can reasonably compete against other contractors. The brief description given by the agencies advertisements will rule out many of the projects a contractor sees.

If the general description you see in an advertisement appeals to the type of work your company wants, then you might order the plans and specifications from the owner or check them out at your local plan center with which you have a membership. When the plans and specifications arrive, you will want to review the project bid items. Most proposals will have a number of bid items that simply break the overall proposal into quantifiable portions of work; although, many projects are bid on a lump sum basis and do not provide this quantifiable breakdown. For example a list of bid items for a city street rehabilitation project might look like Figure 4-1.

FIGURE 4-1   Bid Item List

RECONSTRUCTION OF 14TH AVENUE

Bid Item	Quantity	Units	Unit Price	Total Price
1) Mobilization	1	L.S.	$_____	$_____
2) Traffic Control	1	L.S.	$_____	$_____
3) Clearing & Grubbing	2	Acres	$_____	$_____
4) Demolition/Remove	1	L.S.	$_____	$_____
5) Roadway Excavation	975	C.Y.	$_____	$_____
6) Structure Excavation	90	C.Y.	$_____	$_____
7) Aggregate Base	500	C.Y.	$_____	$_____
8) Asphaltic Concrete (B)	350	Tons	$_____	$_____
9) 12" Concrete Pipe	100	L.F.	$_____	$_____
10) 24" Concrete Pipe	80	L.F.	$_____	$_____
11) Type I Inlet	1	EA.	$_____	$_____
12) 48" Dia. Manhole	12	V.F.	$_____	$_____
13) Concrete Curb	1,000	L.F.	$_____	$_____
14) Concrete Sidewalk	500	S.Y.	$_____	$_____
15) Temporary Signs	128	S.F.	$_____	$_____
16) Permanent Signs	160	S.F.	$_____	$_____
17) Chain Link Fence	1,000	L.F.	$_____	$_____

Total Proposal $_____

If your construction company owns or has access to an excavator, bulldozer and dump trucks, this might be a project that you would consider bidding as a prime contractor. Your company could do the clearing & grubbing, demolition/removal, roadway excavation, structural excavation, aggregate base, concrete pipe, inlet and manhole while subcontracting the traffic control, asphalt concrete, concrete curb/sidewalk and chain link fencing. Specialty contractors typically do these subcontracted items, although a general contractor may complete these items within his own company if he has the appropriate equipment and manpower.

Other items to look for in a projects advertisement are:

1.) The bonding requirements,

2.) Contract time/liquidated damages,

3.) Contractor licensing requirements,

4.) Labor or subcontractor utilization requirements,

5.) Labor wage rates and

6.) If the project has a mandatory or non-mandatory pre-bid meeting.

Make sure that your company can meet all of these contractual obligations before you spend too much pre-bid time with the project, as the contractor's time is much to valuable to waste on projects that cannot be bid because they fail to meet one of these requirements.

From the contract bid items provided, construction advertisement and a brief look at the contract plans, you should be able to decide whether the project is one in which your company can compete. Once a contractor has been through this process several times, he will become very proficient at analyzing projects from their advertisement information.

PRIME (GENERAL) CONTRACTING

Once you have reached the decision that you want to bid on a particular project, you have to decide if you want to be the Prime (or General) Contractor. The General Contractor is the company who is constructing the majority of the work and the one in which the contracting agency will sign a contract. One factor that will play a part in your decision to be a prime contractor is the overall cost of the project (most contracting agencies will provide an approximate range of what they think the project will cost). If the estimated cost of the project is too large, your company might not be able to obtain bonds for the job. Also, on most projects the contractor will have to initially fund the construction of the project until a progress payment can be made by the contracting agency. Most contracting agencies will make payments on a monthly basis. The contractor wants to make sure he can absorb at least two months payroll along with some other critical start-up costs, until he gets the first progress payment from the owner. The request for payment will be submitted after the first months work, but will generally not be paid by the contracting agency for approximately 30 days.

A substantial amount of contract specifications also require that the general contractor construct a certain percentage of the project with the use of its own forces. So if you want to be the prime contractor make sure you can cover that required percentage (generally 30% - 50%) with your own work force.

From the previous example of a city street rehabilitation bid item list, if your company specializes in baserock and asphalt paving, minor concrete construction or fencing; then you might consider providing the general contractors with a subcontract proposal to complete those items.

SUBCONTRACTING

A subcontractor is a company that will give the general contractor a bid or proposal on a certain portion of the project. The contracting agency does not sign contracts with the subcontractors, although the subcontractors are directly tied to the requirements of the contract between the owner and the prime contractor. If the particular project you are contemplating is too large to bond or initially fund, you might want to consider bidding a portion of work to other construction companies who will bid as a general contractor. Some general contractors will also require payment and performance bonds from their subcontractors, but only on that particular subcontract amount.

When submitting a subcontract proposal to the general contractors at bid time, a subcontractor should clearly define the scope of work that he is bidding on. Also detailed in the subcontract proposal should be the following:

1.) The amount of time it is estimated that it will take your company to complete its portion of work,

2.) If any services (traffic control for example) will need to be provided by the prime contractor for the subcontractors work,

3.) If your company has received any pre-bid project addenda's from the contracting agency,

4.) If the cost of bonds is included in the subcontract proposal or any other particular inclusions or exclusions.

It is a good idea to fax or call in your subcontract proposal the day before the general contractor is required to submit his bid. This early submission helps the general contractors have time to analyze all of the subcontract and material quotes that are coming in to his office. Although, it is a good idea not to submit your quote too early, as numbers seem to get passed around to the competition at times. The practice of bid shopping is unethical; nevertheless, this does occur within many construction markets. On large projects the general contractors may receive hundreds of quotes to analyze prior to closing and submitting their bid package. If the subcontractor has many of the project bid items included in his proposal, a scope letter, sent to the prime contractors approximately one week prior to the bid date, simply listing the items he intends to quote along with any special requirements will let the general contractor know that a company will be covering those particular bid items. The subcontractor would then send his quote to the prime contractor the day before the bid as mentioned above. An example of a subcontract proposal that relates to our previous street rehabilitation project is shown in Figure 4-2.

FIGURE 4-2   Subcontract Proposal

ABC ASPHALT PAVING COMPANY
1234 E. Delta Ave.
Gresham, Oregon 97777
Phone (503) 663-6633 Fax (503) 663-6634

April 23, 20--

GENERAL CONTRACTOR, INC.
5678 W. Water St.
Vancouver, WA 98888

RE: City Street Rehabilitation – Vancouver, WA

Bids Due: April 24, 20-- @ 10:00 AM

Estimator:

With respect to the above mentioned project, ABC Asphalt Paving Company would like to offer the following baserock and asphalt paving quote. We are very experienced in this type of work and would like to offer your company references upon request.

Bid Item	Units	Unit Price	Total
Price			
1.) Mobilization	1 L.S.	$ 1,000.00	$
1,000.00			
6.) Aggregate Base	500 C.Y.	$ 15.00	$
7,500.00			
7.) Asphalt Concrete	350 Tons	$ 40.00	
	$14,000.00		

TOTAL PROPOSAL $22,500.00

Notes:
1.) One mobilization is included above, additional mobilizations will be at $750.00/each.
2.) Subgrade to be within 0.2 ft. of grade before ABC Asphalt Paving Company is to place baserock.
3.) ABC Asphalt Paving Company estimates it will take 3 working days to complete the above work.
4.) Bond costs are not included in the above prices, but ABC Asphalt Paving Company is bondable at an additional 1.5% if required.
5.) ABC Asphalt Paving Company has received 2 addenda's on the project.

Exclusions:
1.) Traffic control.
2.) Surveying and testing.

If you have any questions regarding this proposal or our company, please contract us at our home office and we thank you in advance for considering ABC Asphalt Paving Company for this project.

Joe Maple, Estimator

RESOURCE AVAILABILITY/LOCATION – EQUIPMENT & MANPOWER

Every construction company has two main resources available to it: Equipment and Manpower. The availability and location of these two resources can play a major role in whether a contractor decides to bid on a particular project. A contractor should not underrate these two resources, as they are the main ingredients that keep your company operating, so you should take the appropriate care of both of them. You need to ask yourself this question prior to the bid: does the company have or can we get the supervisors, work force and equipment necessary to construct this project? If the company does not have immediate access to these resources you might consider advertising for qualified supervisors, joining a labor organization or renting the additional equipment needed to construct the project.

Also consider the location of the project. Are the costs associated with moving your labor force and equipment to the project site going to make your bid non-competitive in the marketplace? Often a project will come out for bid that is near a project that your company is currently working on or one that you are currently bidding. Construction companies will at times split or separate portions of their mobilization and indirect project expenses into the individual projects bids or proposals as a method of gaining an advantage on the competition. In other words, using the same mobilized equipment, field office, construction yard, project supervision, etc. to construct multiple projects at the same time and in the same vicinity reduces the amount of these direct and indirect costs that are attributable to the individual projects.

With time, the answers to the questions discussed in this chapter come relatively quickly and easily to a person when they gain experience within a particular construction company and the construction industry in general.

PRE-QUALIFICATION

Most governmental contracting agencies will require a contractor who plans to bid on their projects as a prime contractor to be pre-qualified with their agency. This pre-qualification means that the contractor allowed to construct a particular style of work up to a specified contract size. For example a contractor might be pre-qualified with an agency for Sewer Construction up to $500,000, or Concrete Construction up to $250,000, or Excavation and Clearing up to $400,000.

Your company should take these contracting agency pre-qualifications very seriously. If your company is not pre-qualified for a particular style of construction and for the appropriate amount, your bid will be rendered non-responsive by that contracting agency if you chose to submit one. There are also specific dates that the contractor must be pre-qualified prior to, in order for his bid to be considered by the contracting agency. For example, the agency might require the contractor to be pre-qualified 10 business days prior to the bid date.

Many of the county or city agencies will utilize the pre-qualification form of their particular State. Therefore, the contractor who plans to do work for governmental agencies should make sure he is pre-qualified with the States in which he would like to do business in and keep copies of these forms for which to submit to the local county or city agencies. Figure 4-3 details a typical pre-qualification form.

These pre-qualification applications should also be very accurately completed. Many government agencies are very strict on the dollar amount and the style of construction for which they will pre-qualify construction contractors. It is often very hard for a young construction company to achieve the pre-qualification amount they need to bid and compete in the marketplace. Many government agencies want the contractor to have completed some projects of similar size and style prior to obtaining that pre-qualification status with their particular agency. The question becomes, how does a contractor complete a particular style or size of a project if no agency will pre-qualify him to construct one? Some government agencies need to realize that there is another entity that has just as much or more interest in your companies ability to construct a certain size or style of project; your bonding company. Government agencies require bid, performance and payment bonds be submitted prior to construction of a project; therefore, your bonding company has to make a choice prior to your company bidding any particular project if you have the capability of constructing the project. Your bonding company has more to loose if your company fails on the project than does the government agency.

FIGURE 4-3 Pre-Qualification Form

CONTRACTOR'S PRE-QUALIFICATION

A. Bidder: _____

   Address: _____

   Phone: _____ Fax: _____

B. Date Submitted: _____

C. Bidder is a/an: Individual: _____
                    Partnership: _____
                    Corporation: _____
                    Joint Venture: _____

D. If a Joint Venture, name of other Joint Venture participants:

   _____

   _____

E. Project: _____

F. Authorized Signature: _____

G. Experience: Enter the dollar amount / years of experience in each class of work

	Max. Dollar Amount	Years Experience
Earthwork	$_____	_____
Asphalt Paving	$_____	_____
Concrete	$_____	_____
Masonry	$_____	_____
Steel	$_____	_____
Landscaping	$_____	_____
Site Utilities	$_____	_____

Electrical $_____ _____

Building Construction $_____ _____

Alteration/Remodeling $_____ _____

Painting $_____ _____

Plumbing $_____ _____

HVAC $_____ _____

Roofing/Insulation $_____ _____

Sheet Metal $_____ _____

Flooring $_____ _____

Irrigation $_____ _____

H. Key Personnel　　　　　　　Title　　　Year of Experience

_____　_____　_____

_____　_____　_____

_____　_____　_____

I. Major Equipment Owned　　　　　Estimated Book Value

_____　_____

_____　_____

_____　_____

_____　_____

_____　_____

J. Reference of Similar Projects　Contract Amount　Date Contact　Phone #

_____　_____　_____　_____

_____　_____　_____　_____

_____　_____　_____　_____

CHAPTER FIVE

ESTIMATING THE PROJECT

Once a contractor has decided that he is going to bid on a particular project, it is time to put together an estimate on what he believes the job is going to cost to construct and for him to decide how much profit the company wants to make on the project. Listed below is a step-by-step procedure for compiling an estimate and keeping it organized so that changes can be accounted for at any time during the process:

Getting a Feel for the Project,

Estimate Breakdown/Quantity Take-offs,

Hand vs. Computer Program Estimating,

Conceptual vs. Past History Estimating,

Initial Estimate,

Estimate Scheduling,

Pre-bid Jobsite Visit,

Reviewing, Adjusting and Closing the Estimate.

GETTING A FEEL FOR THE PROJECT

Once the project specifications arrive at your office from the contracting agency, read through them thoroughly. Mark out, or highlight important sections of the specifications such as bond amounts, insurance requirements, contract time and liquidated damages, applicable labor wage rates, working shift restrictions, traffic lane closure regulations, etc.

Typically the specifications special provisions (specifications that are special to that project) will be broken up into sections that will cover each type of construction operation or bid item within the project. In the case of a commercial construction project the contract specification will generally be broken down into the particular AIA codes. Read through all of these sections for any special materials to be used or certain methods with which the work is to be constructed.

When you have a basic handle on the specifications and requirements of the project, start looking through the contract plans. The plans will include drawings for the layout of the project, applicable cross-sections and profiles, recommended staging sequence along with all the details necessary for construction. It is a very difficult job for the contracting agency to provide an absolutely accurate and completely detailed set of contract plans and specifications. If you have any questions about a particular plan or specification, call the contracting agencies representative listed in proposal package. Many construction specifications will require that questions be transmitted in written form. It is better to ask many questions, if necessary, than to have any confusion or conflict with the contracting agency during construction. If you point out to the contracting agency an area of bidding confusion or lack of plan information, they will generally develop and send to all of the projects plan holders an addendum to the contract plans and specifications. This addendum will then be incorporated into the bidding documents and the contract by the contracting agency and successful bidder. On some larger or highly detailed projects, many addenda may be issued by the contracting agency. When multiple addenda like this occur, one of them will often extend the bid date so that all of the bidders can incorporate into their estimates the changes brought about by the addenda. It is imperative that you make sure before your bid proposal is submitted, that your company has received all of the projects addenda. Generally, if your company does not document receipt of all of the addenda in you bid package, your proposal will be rendered non-responsive and will not be accepted. In most cases, the bid proposal will have a section for the contractor to list the number of addenda they have received and evaluated in their estimate and proposal.

When reading through the plans try to get a feel for how the actual or physical work for each item will be constructed. Decipher what needs to follow first in order for other items to follow and any special methods of construction that will need to be incorporated.

Many plans and specifications will refer to the "standard drawings and specifications". These are drawings and specifications that are used on virtually all of the projects constructed by a particular contracting agency.

These books are published and sold through each individual agency and are necessary information if you plan to construct projects with that agency. Many contracting agencies, such as Counties or Cities, may use another larger agencies standard plans and specifications, their respective state's for example.

Therefore, when you are examining the projects special provisions, you should make a check to see what standard plans and specifications the contracting agencies incorporates into their projects and make the necessary arrangements to have access to them.

ESTIMATE BREAKDOWN/QUANTITY TAKE-OFFS

Once you have a reasonable feel for the project to be built, its time to start the estimate breakdown and do the quantity take-offs. First, breakdown each bid item within the projects proposal into the operations that will be necessary to construct that item. If the project is bid on a lump sum basis (without unit price and quantity breakdowns per bid item), then make up bid items for your own use and organization. The bid item breakdown will assist you in keeping the work in quantifiable operations of work that can be easily tracked and modified later in the estimate and during construction. In other words, estimate a lump sum project the same as you would for a bid item project, the only difference being you will make up your own bid items that will total to a lump sum final bid. Following are of examples of how a contractor might breakdown a particular bid item:

No. 1 Bid Item - 24" Concrete Pipe

1.) Sawcut existing pavement,

2.) Remove existing pavement,

3.) Trench excavation,

4.) Bed and lay pipe,

5.) Backfill,

6.) Patch existing pavement.

No. 1 Bid Item - 24" Concrete Pipe (Alternate Breakdown)

1.) Demo/Patch existing pavement,

2.) Excavation/Bed & Lay Pipe/Backfill.

No. 2 Bid Item - Concrete Retaining Wall

1.) Fabricate forms,

2.) Form footings and walls,

3.) Install reinforcing steel,

4.) Place concrete,

5.) Wet finish and cure concrete,

6.) Strip forms,

7.) Dry finish.

No. 2 Bid Item - Concrete Retaining Wall (Alternate Breakdown)

1.) Fabricate forms,

2.) Form/Strip footings and walls,

3.) Install reinforcing steel,

4.) Place/Wet finish/Cure concrete,

5.) Dry finish.

Each individual contractor will breakdown their bid items differently; its up to you to find an acceptable method that suits your company and one that you can use for the same items on every project you bid. Breakdown the bid items into operations that are easily distinguishable or that happen at different times or by different crews. Its important to develop a uniform method for your company so that any estimator on your staff can pick up an estimate and be able to follow the work of whoever had developed the estimate previously. Often times within a company the individual who estimates the project is not the person who constructs or manages the project in the field. Also, if an inexperienced person estimates the project it will need to be reviewed and adjusted by an experienced company manager; therefore, one can see the need to have everybody within your company using a uniform estimating method.

Now that the bid items are broken down, you need to develop quantity take-offs for each operation you have just broken-down within the bid items. This is done by measuring and reading dimensions off of the plans, reading the project specification for requirements, and calculating this information into the quantities you are trying to develop. Some agencies will provide in the plans, specifications or proposal sheets, some of the quantities you are trying to calculate. It is a good idea to go ahead and do your own quantity calculations to see how they compare with the contracting agencies quantities. You will want to take-off these quantities as they relate to your bid item breakdown and in measurements that will be easily taken in the field once the project is under construction. The operations and quantities that you obtain during the estimate will be used throughout the project until it is complete. The quantity take-offs are the very basis of every estimate, so you should take the necessary time and effort to be as accurate as possible since everything that follows these take-offs, within the estimate, will be based on those numbers that you calculate.

From the above examples of a bid item breakdown, following is what the quantities and units might look like from your take-off:

No. 1 Bid Item — 24" Concrete Pipe

    1.) Sawcut existing pavement   — 3,000 Inch-Feet

    2.) Remove existing pavement   — 278 Square Yards

    3.) Trench excavation   — 463 Cubic Yards

    4.) Bed and lay pipe   — 500 Lineal Feet

    5.) Backfill   — 324 Cubic Yards

    6.) Patch existing pavement   — 93 Tons

No. 2 Bid Item — Concrete Retaining Wall

1.) Fabricate forms	— 2,500 Square Feet
2.) Form footings and walls	— 4,500 Square Feet
3.) Install reinforcing steel	— 7,500 Pounds
4.) Place concrete	— 121 Cubic Yards
5.) Wet finish and cure	— 1,250 Square Feet
6.) Strip forms	— 4,500 Square Feet
7.) Dry finish	— 4,000 Square Feet

HAND VS. COMPUTER PROGRAM ESTIMATING

Once you have broken down the project into quantitative operations, it's time to start the actual cost estimate. There are two ways in which the estimate can be formulated: by hand, or by the use of a computer program. Before personal computers, contractors were forced to compile their estimates by hand. When estimating very small construction contracts or projects with very few operations, hand estimating can still be the most cost effective means of obtaining project cost estimates. Once the projects that you are estimating get many bid items and operations, it is easier and more accurate to use a computer program. There are a variety of construction estimating programs on the market, or a program can be developed in your own office that exactly fits the needs of your particular company and style of work that you construct. There is construction software available on the market that will incorporate your estimates into your companies job costing, accounting, scheduling and total project management.

Whether your company uses hand or computer software estimating, the estimate should be broken into some particular cost categories such as the following:

EQUIPMENT - This would include both the costs of company owned equipment and outside rented equipment. If your company rents equipment, breakdown the daily, weekly or monthly rental cost from your rental agencies rate sheet into an hourly cost. Be sure to include the hourly costs of insurance and FOG (fuel, oil, grease) in this outside rental rate. If your company owns its equipment the rate should be broken down into an hourly cost from past history information such as equipment payments, insurance, licensing, tires or tracks, FOG, maintenance and repairs, along with any associated taxes.

S.T. & M. - This is the cost for small tools and miscellaneous. Equipment that is too small to charge a rental rate under the equipment category falls into this category. (*Skil* saws and blades, drills, hand tools, etc.) Also materials that do not become a permanent part of the project are covered under this category. (Concrete form lumber, nails, snap-ties, etc.) This rate should be per unit of whatever operation you are estimating, although the rate can be a percentage of labor costs. Since labor costs vary from job to job, most companies will use the rate per unit method. Example: S.T. &M. for forming concrete footings and walls might be at $2.00/S.F., or S.T. & M. for trench excavation might be $0.25/C.Y.

LABOR - This hourly cost for labor would include payment to the employee depending on craft, fringe benefits, Federal and State unemployment, F.I.C.A., local city or county labor taxes, workers compensation and general liability. All of these costs are functions of labor expenditures and should be compiled into a percentage of the hourly rate per craft.

An example of this would be as follows:

Labor Accruals

Federal Unemployment	@ 0.0080
State Unemployment	@ 0.0540
F.I.C.A.	@ 0.0452
City Transportation Tax	@ 0.0051
Workers Compensation	@ 0.1650
General Liability	@ 0.0045
Total Labor Accrual	0.2818

EXAMPLE WITH A BENEFIT PACKAGE

Carpenter Rate		$17.50/Hr.
Labor Accrual	@ 28.18%	$ 4.93/Hr.
Fringe Benefit		$ 4.75/Hr.
Total Carpenter Rate		$27.18/Hr.

EXAMPLE WITHOUT A BENEFIT PACKAGE

Carpenter Rate		$17.50/Hr.
Fringe Benefit Paid To The Employee		$ 4.75/Hr.
Total Paid To Employee		$22.25/Hr.
Labor Accrual	@ 28.18%	$ 6.27/Hr.
Total Carpenter Rate		$28.52/Hr.

MATERIALS - This category is the direct cost of any material that is installed permanently into the project such as concrete, framing lumber, aggregates, reinforcing steel, drainage pipe, precast manholes, etc., along with any associated taxes. Remember to include the applicable material waste or compaction percentages in the material quantities from your neatline (without waste or any extra) take-offs. For example, a contractor typically has to purchase approximately 5% more concrete than the neatline take-off quantity. Also, if your operation is embanking soil or placing baserock, you will need to account for an additional 10% - 15% that will be lost in compaction of the material, as compared to the neatline take-off quantity. Subcontractors - The cost of any subcontractors that you plan to use on a particular project should be separated into an individual category.

Each operation of each bid item should have a cost breakdown into one of the above mentioned categories. This breakdown will make it much easier to adjust your estimate at a later date, along with keeping track of the costs in a quantifiable form.

CONCEPTUAL VS. PAST HISTORY ESTIMATING

To develop the actual costs for the estimate you will need to obtain the production rates for each operation of the estimate that will be constructed. In other words, how long will it take to finish each operation with a particular crew and with the equipment your company has available. There are two ways in which to come up with these rates: conceptually and from your companies past history.

Conceptual production rates occur when you are estimating some work that you or your company has never constructed before. It literally means constructing the operation in your mind and deciphering how long the operation will take to build, or at what kind of production rate the operation can be built. As you can imagine this kind of estimating has the most risk associated with it and this risk should be taken into consideration when you are assigning a profit or risk margin to the estimate.

The other way to obtain these production rates with which work can be constructed is from your companies past history. From these completed project reports you can see at what rate your company constructed similar work to that which is being estimated. This, of course, is the less risky of the two methods to develop construction production rates. As your company completes more varied projects, you can use the developed rates from these jobs to bid more accurately and confidently in the future.

This author believes that estimating projects with your company's documented past labor production rates is the safest and most conservative approach. Many contractors estimate simply using unit prices from their construction experience. This approach can work to some degree, but it is also dangerous due to the changing material and labor costs through time. Also, unit prices for the same operation can differ greatly depending on the conditions of the individual project sites. Labor production rate estimating can easily adjust for all changes in cost and productions in a conservative and accurate manner.

INITIAL ESTIMATE

Once you have gotten a feel for the project, and completed the estimate breakdown and quantity take-offs, it's time to put together the initial estimate using your hand or computer software estimating method, along with your conceptual or past history production rates. Also at this time you should "plug" in your best educated guess for the prices of your permanent materials and subcontractor quotes which you will receive later in the estimating process. These "plugs" will be adjusted later when material suppliers and subcontractors submit the actual prices and quotes to your company prior, to the bid date. Following is a sample of estimating a few of operations using our previous Concrete Retaining Wall example:

FABRICATE FORMS - 2,500 S.F. @ 25 S.F./Manhour = 100 Mhrs.

With a crew of   1 Carpenter Foreman
                 2 Carpenters
                 1 Laborer

this yields 25 Crewhours.

Equipment -	Pick-Up	25 Hrs. @	$ 5.00/Hr.	= $	125.00
	Generator	25 Hrs. @	$ 3.00/Hr.	= $	75.00
S.T. & M. -	2,500 S.F. @ $2.00/S.F.			= $	5,000.00
Labor -	Carp. 4M	25 Hrs. @	$30.00/Hr.	= $	750.00
	Carpenter	50 Hrs. @	$25.00/Hr.	= $	1,250.00
	Laborer	25 Hrs. @	$20.00/Hr.	= $	500.00

Subtotal of the above	= $	7,700.00
Materials (Permanent)	= $	0.00
Subcontractors	= $	0.00

Total Cost to Fabricate Forms   **$ 7,700.00**

FORM FOOTINGS/WALLS - 4,500 S.F. @ 15 S.F./Mhr. = 300 Mhrs.

With a crew of   1 Carpenter Foreman
                 2 Carpenters
                 1 Laborer
                 1 Operator

this yields 60 crew hours.

Equipment -	Pick-up	60 Hrs. @	$ 5.00/Hr.	= $	300.00
	Generator	60 Hrs. @	$ 3.00/Hr.	= $	180.00
	Crane	60 Hrs. @	$25.00/Hr.	= $	1,500.00
S.T. & M. -	4,500 S.F. @ $0.50/S.F			= $	2,250.00

Labor -	Carp. 4M	60 Hrs. @	$30.00/Hr.	= $	1,800.00
	Carpenter	120 Hrs. @	$25.00/Hr.	= $	3,000.00
	Laborer	60 Hrs. @	$20.00/Hr.	= $	1,200.00
	Operator	60 Hrs. @	$27.00/Hr.	= $	1,620.00

Subtotal of the above		= $	11,850.00
Materials (Permanent)		= $	0.00
Subcontractors		= $	0.00

Total Cost to Form Footings/Walls   **$ 11,850.00**

INSTALL REINFORCED STEEL - 7,500 Lbs. @ 1,500 Lbs./Mhr. = 5 Mhrs.

With a crew of   1 Operator (Handling)
                   1 Laborer (Handling)

this yields 2.5 crew hours

Equipment -	Crane	2.5 Hrs. @	$25.00/Hr.	= $	62.50
S.T. & M. -	7,500 Lbs. @ $0.01/Lb.			= $	75.00
Labor -	Operator	2.5 Hrs. @	$27.00/Hr.	= $	67.50
	Laborer	2.5 Hrs. @	$20.00/Hr.	= $	50.00

Subtotal of the above		= $	255.00
Materials (Permanent)		= $	1,875.00
Subcontractors (Installation)		= $	1,500.00

Total Cost to Install Reinforcing Steel   **$ 3,630.00**

PLACE CONCRETE - 121 C.Y. @ 3.5 C.Y./Mhr. = 34.57 Mhrs.

With a crew of   1 Laborer Foreman
                   2 Laborer

this yields 11.5 crew hours

Equipment -	Pick-up	11.5 Hrs. @	$ 5.00/Hr.	= $	57.50
	Generator	11.5 Hrs. @	$ 3.00/Hr.	= $	34.50
S.T. & M. -	121 C.Y. @ $2.00/C.Y.			= $	242.00
Labor -	Lab. 4M	11.5 Hrs. @	$22.50/Hr.	= $	258.75
	Laborer	23.0 Hrs. @	$20.00/Hr.	= $	460.00

Subtotal of the above			= $	1,052.75
Materials -	Concrete 127 C.Y. @ (With Waste In Quantity)	$58.00/CY.	= $	7,366.00
Subcontractors			= $	0.00

Total Cost to Place Concrete   **$ 8,418.75**

WET FINISH AND CURE - 1,250 S.F. @ 50 S.F./Mhr. = 25 Mhrs.

With a crew of   1 Mason Foreman
1 Mason

this yields 12.5 crew hours

Equipment -	Pick-up	12.5 Hrs. @	$ 5.00/Hr.	= $	62.50
S.T. & M. -	1,250 S.F. @ $0.10/S.F.			= $	125.00
Labor -	Mason 4M	12.5 Hrs. @	$30.00/Hr.	= $	375.00
	Mason	12.5 Hrs. @	$27.00/Hr.	= $	337.50

Subtotal of the above   = $ 900.00
Materials (Permanent)   = $ 0.00
Subcontractors   = $ 0.00

Total Cost to Wet Finish and Cure $ **900.00**

STRIP FORMS - 4,500 S.F. @ 32.5 S.F./Mhr. = 138.46 Mhrs.

With a crew of   1 Laborer Foreman
2 Laborers
1 Operator

this yields 34.5 crew hours

Equipment -	Pick-up	34.5 Hrs. @	$ 5.00/Hr.	= $	172.50
	Generat.	34.5 Hrs. @	$ 3.00/Hr.	= $	103.50
S.T. & M. -	4,500 S.F. @ $0.15/S.F.			= $	675.00
Labor -	Lab. 4M	34.5 Hrs. @	$22.50/Hr.	= $	776.25
	Laborer	69.0 Hrs. @	$20.00/Hr.	= $	1,380.00
	Operator	34.5 Hrs. @	$27.00/Hr.	= $	931.50

Subtotal of the above   = $ 4,038.75
Materials (Permanent)   = $ 0.00
Subcontractors   = $ 0.00

Total Cost to Strip Forms $ **4,038.75**

DRY FINISH - 4,000 S.F. @ 125 S.F./Mhr. = 32 Mhrs.

With a crew of   1 Mason Foreman
1 Mason

this yields 16 crew hours

Equipment -	Pick-up	16 Hrs. @	$ 5.00/Hr.	= $	80.00
	Generator	16 Hrs. @	$ 3.00/Hr.	= $	48.00
S.T. & M. -	4,000 S.F. @ $0.15/S.F.			= $	600.00
Labor -	Mason 4M	16 Hrs. @	$30.00/Hr.	= $	480.00
	Mason	16 Hrs. @	$27.00/Hr.	= $	432.00
Subtotal of the above				= $	1,640.00
Materials (Permanent)				= $	0.00
Subcontractors				= $	0.00
Total Cost to Dry Finish				$	**1,640.00**

You will notice that we made a subtotal of the equipment, S.T. & M. and labor categories. These are the categories of your estimate that have "risk" associated with them. Permanent materials and subcontractors should be fixed costs after you receive quotes from the material suppliers and subcontractors; while the other categories are your best estimate concerning how long it will take to complete the operation and are subject to change during construction. It is useful to know the total dollar value of your risk categories, or subtotal, as you can assign the profit or risk margin as a percentage of this total.

As you can see from the fabricate forms, form footings/walls, place concrete, wet finish and cure along with the dry finish operation, we are using our own company workforce to complete the entire operation. Yet, in the install reinforcing steel operation, we are using our workforce to unload the steel, while purchasing the rebar for a subcontractor to install in the footings/wall.

At this time it is a good idea to check the bid item unit prices of your initial estimate. From our above example the total concrete retaining wall cost is $38,177.50, or approximately $315 per C.Y. Most contractors who have been involved in the construction industry have developed a rough idea of the approximate unit price of the commonly constructed bid items. Doing a quick double check of the unit prices at this time to see if any of them are way out of the ballpark, can save you much frustration later, along with the possibility of a grossly bad estimate.

From the above example you can obtain a total cost for each operation and similarly for each bid item and finally for the total project. This total for the project is what we will call the total "direct cost".

If a contractor is hand estimating, a bid item summary and a proposal summary sheet might look like those detailed in Figures 5-1 and 5-2 respectively.

FIGURE 5-1　Item Summary Sheet

XYZ CONSTRUCTION, INC.
ITEM SUMMARY SHEET
DATE: June 11, 1997

BID ITEM:　Concrete Retaining Wall

DESCRIPTION	QUANTITY	UNITS	MHRS	Unit Price	EQUIPMENT	Unit Price	S.T. & M.	Unit Price	LABOR	Unit Price	SUBTOTAL	Unit Price	MATERIALS	Unit Price	SUBCONTR	Unit Price	TOTAL
Fabricate Forms	2500	S.F.	100.0	0.080	$200.00	2.000	$5,000.00	1.000	$2,500.00	3.080	$7,700.00	0.000	$0.00	0.000	$0.00	3.080	$7,700.00
Form Ftgs/Walls	4500	S.F.	300.0	0.440	$1,980.00	0.500	$2,250.00	1.690	$7,620.00	2.630	$11,850.00	0.000	$0.00	0.000	$0.00	2.630	$11,850.00
Install Rebar	7500	LBS.	5.0	0.008	$62.50	0.010	$75.00	0.016	$117.50	0.034	$255.00	0.250	$1,875.00	0.200	$1,500.00	0.484	$3,630.00
Place Concrete	121	C.Y.	34.6	0.760	$92.00	2.000	$242.00	5.940	$718.75	8.700	$1,052.75	60.876	$7,366.00	0.000	$0.00	69.576	$8,418.75
Wet Finish/Cure	1250	S.F.	25.0	0.050	$62.50	0.100	$125.00	0.570	$712.50	0.720	$900.00	0.000	$0.00	0.000	$0.00	0.720	$900.00
Strip Forms	4500	S.F.	138.5	0.061	$276.00	0.150	$675.00	0.688	$3,087.75	0.898	$4,038.75	0.000	$0.00	0.000	$0.00	0.898	$4,038.75
Dry Finish	4000	S.F.	32.0	0.032	$128.00	0.150	$600.00	0.228	$912.00	0.410	$1,640.00	0.000	$0.00	0.000	$0.00	0.410	$1,840.00
BID ITEM TOTAL	121	C.Y.	635.1	23.149	$2,801.00	74.107	$8,967.00	129.492	$15,668.50	226.748	$27,436.50	76.372	$9,241.00	12.397	$1,500.00	315.517	$38,177.50

FIGURE 5-2   Project Summary Sheet

XYZ CONSTRUCTION, INC.
PROJECT SUMMARY SHEET
DATE: June 11, 1997

PROJECT: Lauralville Expressway   OWNER: City of Lauralville   BID DATE: June 20, 1997

DESCRIPTION	QUANTITY	UNITS	MHRS	Unit Price	EQUIPMENT	Unit Price	S.T.&M.	Unit Price	LABOR	Unit Price	SUBTOTAL	Unit Price	MATERIALS	Unit Price	SUBCONTR	Unit Price	TOTAL
Mobilization	1	L.S.	15.0	500.000	$500.00	975.000	$975.00	375.000	$375.00	1850.00	$1,850.00	0.000	$0.00	0.000	$0.00	1850.00	$1,850.00
Flagging	150	M.H.	150.0	0.000	$0.00	0.500	$75.00	22.500	$3,375.00	23.00	$3,450.00	0.000	$0.00	0.000	$0.00	23.00	$3,450.00
Erosion Control	1	L.S.	10.0	175.000	$175.00	100.000	$100.00	255.000	$255.00	530.00	$530.00	250.000	$250.00	0.000	$0.00	780.00	$780.00
Excavation	250	C.Y.	32.0	3.904	$976.00	2.000	$500.00	3.264	$816.00	9.17	$2,292.00	0.000	$0.00	0.000	$0.00	9.17	$2,292.00
Concrete Wall	121	C.Y.	635.1	23.149	$2,801.00	74.107	$8,967.00	129.492	$15,668.50	226.75	$27,436.50	76.372	$9,241.00	12.397	$1,500.00	315.52	$38,177.50
Backfill	135	C.Y.	20.0	3.541	$478.00	0.500	$67.50	3.778	$510.00	7.82	$1,055.50	7.500	$1,012.50	0.000	$0.00	15.32	$2,068.00
Fencing	435	L.F.	0.0	0.000	$0.00	0.000	$0.00	0.000	$0.00	0.00	$0.00	0.000	$0.00	15.000	$6,525.00	15.00	$6,525.00
PROJECT TOTAL			862.1		$4,830.00		$10,684.50		$20,999.50		$36,614.00		$10,503.50		$8,025.00		$55,142.50

Now that you have the total direct cost, you will then need to calculate a total for the "indirect costs" which include:

1.) Mobilization of employees, equipment and materials to and from the project site,

2.) Temporary job utilities such as telephone, electricity, portable toilets, etc.

3.) Construction yard space, project office and storage container rent.

4.) Supervision, engineering and office management.

The total of the direct and indirect costs is what we will call the "total base cost".

ESTIMATE SCHEDULING

Now that you have the initial estimate complete and have arrived at a total base cost, you have enough information to schedule the project to be constructed. The schedule can be broken down into many different forms depending on the size and complexity of the project. For larger projects the activity side of the schedule might be the bid items. For more complex projects you might have to break down the activity side of the schedule into the operations of the bid items, depending on their importance. The other side, or scale, of the project schedule should be time or duration. For larger jobs the time scale might be in months, while for smaller jobs the time scale might be in days.

For small construction projects, like those that are built by small, emerging contractors, a simple hand completed bar schedule should be sufficient. For large, complex projects a form of scheduling called the "critical path method" (CPM) is most often used. There are many computer programs on the market that utilize the critical path method of scheduling. These programs make a much simpler operation of complex or large item scheduling as compared to trying to create a critical path schedule by hand. Computer programs are available that can assist the contractor in accurately scheduling the project, providing easy updates and even help with the accounting and project management. The basic idea behind critical path scheduling is that each activity has a specified duration and has other activities that are interrelated with it. In other words, a particular activity cannot be started until another particular activity has been completed. An example might be that the concrete forms of a retaining wall cannot be stripped until the concrete has been placed in the forms and cures. Figure 5-3 is what a simple bar schedule would look like for our previous example of a concrete retaining wall.

FIGURE 5-3 Retaining Wall Schedule

XYZ CONSTRUCTION, INC.
Pre-Bid Schedule
Concrete Retaining Wall

ACTIVITY	WK 1	WK 2	WK 3	WK 4	WK 5	WK 6
Mobilization	XX					
Fabricate Forms	X	XX				
Form Footing/Wall	XX	XX	XXX			
Place Reinf. Steel		XX	X			
Place Concrete		X	X			
Wet Finish/Cure		XX	X	X		
Strip Forms			X	XX		
Dry Finish				XX		
Demobilization				XX		

If your initial schedule reveals that the project cannot be constructed within the allowed time frame, then you need to consider crew overtime. This overtime can be realized in additional working days each week or additional hours each working day. The contractor may also need to consider overtime on particular operations if it is to your economic advantage to start a following operation early or to even complete the entire project early. Although, the estimator needs to keep in mind that field operations typically do not maintain the same production rate on overtime hours. The field crew is more physically and mentally exhausted and therefore cannot typically keep up the same rate of production. On publicly funded projects any hours worked over 8 each day and hours worked over 40 each week are to paid at overtime rates. So for the above reasons the duration of the schedule needs to be analyzed carefully so that the estimate can be adjusted accordingly.

PRE-BID JOBSITE VISIT

The best time to visit the jobsite is this time, when the initial estimate and project schedule have been completed. Many contractors visit the jobsite as soon as they have obtained the project plans and specifications. The contractor needs to complete the initial estimate and schedule prior to jobsite visits; this gives you the opportunity to learn substantially more about what is involved in the project and how it is to be constructed. If you visit the jobsite before this initial estimating work there will typically be many unanswered questions once you begin the estimate because you did not know the details of construction.

For project locations that are a short distance away from your office, several pre-bid jobsite visits may be in order. In this instance, the contractor might look at the site before, during and after the initial estimate if time allows in his particular schedule. If the project location is a long distance away from your office, make the site visit after the initial estimate, and make it count, so that you may obtain all the information you will need to accurately adjust your estimate.

During the jobsite visit, the contractor should try to identify all of the existing conditions that could effect the construction operations within the estimate. Also look for what the conditions will look like at the time of actual construction. If you are visiting the site in the summer time, but the job is to be constructed in the winter, identify any site problems that might be encountered during the actual construction season.

For an example of what a contractor might look for on a jobsite visit, we will use our previous example of a 24" Concrete Pipe.

1. Since most sewer pipes are located under a street; how many traffic lanes are available? This would relate to how much traffic control your company will have to perform. Can you construct the work with a simple shoulder closure, or will you have to use flaggers and one-way traffic through the work zone? Also, you will have to answer the questions "where are we going to put traffic while we are not working and

how are we going to protect the public from the trench when it is not completely backfilled?"

2. What is the condition of the existing pavement? What type of pavement exists and how deep is the structural section of the road? This will relate to the production rate at which you estimated the sawcutting and pavement removal.

3. Investigate the area to try and find out what kind of soil there is. This would relate to the production rate at which you estimated trench excavation. Can the excavated trench material be used elsewhere on the project, or will you have to haul the material off the jobsite? If the trench is beyond 4 feet deep (or your current local trenching code) you will also need to consider the type of shoring or trench layback needed to ensure the safety of your crew from trench sidewall cave in. Many project plans and specifications will include boring logs, completed by the owner prior to the bid, that details the existing soil conditions and the water table. Many times, these boring logs are not included in the bidding documents but are available at the contracting agency. If you have any questions regarding the availability of boring logs, call the contracting agency representative.

4. Is there space available to store any materials at the jobsite prior to their use on the project? If there is no space available for safe, temporary storage, then the contractor will have to schedule materials in as they are needed, which at times can require considerable coordination with the material suppliers.

5. Are utilities available near the jobsite? For construction of the project you might need an electrical power or telephone drop, water or sanitary facilities.

6. Look for any other existing conditions that could effect the production rates you have estimated for the project.

Following is what a contractor might look for on a jobsite visit using our previous example of a concrete retaining wall:

1. What is the access like to the concrete retaining wall site? Will you have to construct an access to the wall for your equipment and manpower? Would it be more economical to construct access for the concrete trucks, or utilize a pump truck to get concrete to the wall?

2. Is there water close to the site for water curing of the concrete? If water is not available at the site you might have to utilize a curing compound; but is this allowable by the contract specifications? If curing compounds are not allowable, then the contractor will have to bring water onto the jobsite and use some sort of water retention medium on the concrete surface for wet curing.

3. Is the concrete retaining wall in an area where you would have to be concerned with vandalism of the fresh concrete or the cured surfaces? If vandalism is a concern, then

the contractor might want to consider an after work hours guard service or construct a protection device. Is there any existing structure or embankment that would limit the immediate space around the retaining wall? If the area directly adjacent to the wall is limited, will the forming system that you planned on utilizing still work in this situation?

REVIEWING, ADJUSTING AND CLOSING THE ESTIMATE

Since you have looked at the jobsite and identified the existing conditions that will affect the production rates with which you have estimated the project, it is time to adjust the initial estimate. This simply means considering the conditions that you have identified and modifying your estimated production rates, crew and equipment. As you can probably see the use of a personal computer will facilitate this procedure, while also assisting with the accuracy of the adjustment, as compared to hand estimating. With your new adjusted estimate, you should now check to see if any of the changes effect your initial estimated construction schedule; if so, adjust the schedule and indirect costs.

Now that you have your adjusted base cost its time to set up a method to "close out" your estimate. "Close out" means the final adjusting of equipment, labor, materials and subcontract quotes on bid day. We suggest using quote comparison sheets to compare quotes submitted to your company for each type of material and subcontractor used within the estimate. Using our previous concrete retaining wall example, you would have a material sheet for concrete and reinforcing while having a subcontract sheet for the installation of reinforcing steel. These quote comparison sheets should include a section for the quantity and price that you have previously plugged into your estimate. Another section should be an area for you to insert the prices that are submitted to your company by the material suppliers and subcontractors. Once you have all the prices and quotes submitted, you can choose the most cost effective price and make an adjustment to your base price which is equal to the difference between your plugged price and the quoted price. This difference in price can be either an add or a deduct to your base cost.

Figure 5-4 is a sample of a quote comparison sheet for a retaining wall utilizing our previous example of the purchase of concrete and reinforcing steel along the installation of the rebar by a subcontractor.

FIGURE 5-4   Quote Comparison Sheet

XYZ CONSTRUCTION, INC.
QUOTE COMPARISON SHEET

Description	Quantity	Units	Plugged Unit Price	Plugged Total Price	Unit Price	Total Price	Unit Price	Total Price	Unit Price	Total Price
Project: Bid Date: Comparing:					Company: Contact: Phone:		Company: Contact: Phone:		Company: Contact: Phone:	
	Plugged Total				Total		Total		Total	
					DBE: Bonded: Tax: Comments:		DBE: Bonded: Tax: Comments:		DBE: Bonded: Tax: Comments:	

If your material and subcontract quotes are called in to your office, instead of being faxed, we suggest that you accurately document all of the information regarding the quote. This documentation will save you much confusion and the possibility of mistakes when you are comparing many quotes on a particular item. Document if the companies quoting have seen all of the pre-bid addenda's, included tax, and included bond if they are a subcontractor. If they did not include the cost of bonding, at what rate are they bondable, along with any other inclusions or exclusions to their quote. Request that the material supplier or subcontractor, via facsimile or mail, follow up the telephone quote. Again, we suggest using a standard form for these telephone quotes as detailed in Figure 5-5.

FIGURE 5-5 Telephone Quote Form

TELEPHONE QUOTE SHEET

Date: _____ Time: _____ AM ___ PM ___

Project: _____

Company: _____

Address: _____

_____

_____

Contact: _____

Telephone: _____ Fax: _____

Bid Item	Quantity	Description	Unit Price	Total Price

Tax Included? _____ If yes, at what rate? _____

Bondable? _____ If yes, at what rate? _____

Quote taken by: _____

Each material or subcontract quote comparison sheet will have an adjustment to the initial estimate unless you, luckily, happened to plug in the exact amount of the later submitted price. You will track these adjustments to your base cost using a "close out sheet".

The first row of the close out sheet is the base cost, per category, of your estimate. As the prices and quotes are submitted to your company, adjust the estimated base cost per category to obtain a new final base cost. From Figure 5-6 one can see that we added 0.5 weeks of the project Superintendent and his vehicle to the base cost, along with adjusting the three quote comparison sheets from the prior example.

FIGURE 5-6 Close out sheet total should match totals in below

XYZ CONSTRUCTION, INC.
ESTIMATE CLOSE-OUT SHEET

Project:

ADJUST. NUMBER	DESCRIPTION	EQUIPMENT	S.T. & M.	LABOR	SUBTOTAL	MATERIALS	SUBCONTR.	TOTAL

From the final base cost, a decision has to be made as to the amount of profit, or risk margin, to be added. There are many factors that should go into this decision on the part of the contractor. Below we discuss several of these considerations, many which should already have been addressed when you decided to bid on the job:

1. Competition/Number of Bidders – Review the contracting agencies plan holder list and mandatory pre-bid sign in sheets. How many other contractors are also bidding on the project and does history show them to be competitive bidders? If so, then you might have to choose a somewhat slimmer margin.

2. Manpower Availability - Does your company have the trained craft manpower immediately available to construct the project?

3. Supervisor Availability - Does your company have on its current staff, supervisors that are capable of managing the project?

4. Equipment Availability – Does your company have, or have access to, the equipment necessary for construction of the project?

5. Cash Flow Implications - Will the project have to be funded heavily up front by your company? Is it a large, labor intensive project that will drain your cash reserves and available credit before receiving the first progress payment?

6. Project Difficulty - Is the project difficult to construct in terms of construction techniques or contract time available?

7. Projected Weather Conditions - What will the weather conditions be like during the actual construction time? Is there a risk of either cold or hot weather protection to the work in progress?

8. Distance from the Home Office - Is the project sufficiently far enough away from the home office that the manager will not be able to make frequent trips to the jobsite?

9. Designing Engineer/Architects Reputation - What is the reputation of the person or firm who designed the project? Do they typically have problems or conflicts with contractors in terms of change orders, schedule, inspections, etc.?

10. Owner/Contracting Agencies Reputation - Similar to the above; what is the reputation of the owner or contracting agency? Has your company worked with them before?

One suggestion we might offer is to use a margin that is a percentage of either your Subtotal or Labor base cost. The Subtotal can be viewed as the total of the categories in your estimate that contain risk, while labor is the category in your estimate that contains the most risk. As we discussed earlier in this chapter, the materials and subcontract quotes should relatively fix the costs for these categories, while the equipment, S.T. & M. and labor are only your best estimate, and are subject to change during construction. Typical values for the percentage of profit or risk margin as compared to Subtotal might be in the 30% to 45% range depending on the type of work. Typical values for the percentage of profit or risk margin as compared to Labor might be in the 45% to 75% range depending on how labor intensive the project is. If the contractor were to choose a 50% of labor risk margin, this simply means that he could miss his estimated labor productions by half and still theoretically break even on the job. Breaking even on projects is no way to operate your company; most contractors are not in the business for the practice. Again, these are typical percentages risk that would create a profit or risk margin of approximately 10% to 25% as compared to your total base cost. Once a company gains experience in contracting you will find a relatively consistent margin percentage with which to bid for your particular style of construction and what cost category your margins are best based on.

The following choice of a margin is from our previous example of the concrete retaining wall as if it was the entire project:

Labor Cost $15,668.50   45%   Labor = $7,050   = 18.5% of Total

Subtotal Cost $27,436.50   35%   Subtotal = $9,602   = 25.2% of Total

For this project we will choose a margin of $8,000 which is approximately 21% of the total base cost.

Therefore,		
	Total Base Cost	$ 38,178
	Margin	$ 8,000
	Bond @ 1.5%	$ 693
	Total Bid	$ 46,871

Now that you have a total bid price, you will have to spread the profit or risk margin along with the cost of bonds into the base price of each of your bid items. Also, depending on how the particular contracting agency pays for mobilization, you might have to include your supervision costs in the spread of your margin into the individual bid items. Most contracting agencies will only initially pay up to 10% of the total contract price for mobilization, while any amount over 10% would be paid at the end of the project. So if your total of mobilization and indirect costs exceed this 10% of your bid, then you are better off to spread the excess amount into other bid items with your margin and bond costs, so that you would get paid for these indirect costs earlier.

There are other ways to spread margins into individual bid items that maximize future payments from the contracting agency due to possible overruns and underruns in the proposal bid item quantities. If you believe that a particular bid item will overrun in quantity, then you might want to risk putting more margin spread on this particular item. If the item does overrun in quantity, the contractor would maximize profit by gaining a larger percentage profit on the extra quantity. On the other hand, if you believe that a particular bid item will underrun in quantity, then you might want to risk putting a smaller, or no, margin spread on this particular item. In this case, if the item does underrun in quantity the contractor would again maximize profit, losing less (or none) of your margin due to part of that item not being constructed. Another way a contractor might want to unbalance his bid is to spread more of your margin into items that will be constructed first. By doing this the contractor will get paid more of his margin initially, thereby limiting the amount he might have to fund the construction of the project with his cash reserves. The contractor should be aware that unbalancing of bid items is a serious gamble in which your company may win or lose. This can happen when the bid items are deleted or when the quantity is substantially changed and the contractor has unbalanced his unit prices. This author believes a contractor should bid his unit prices close to the actual calculated unit prices with simply small adjustments made to even numbers for the final bid unit price. Grossly unbalancing a bid is not only a large gamble, but also not a good way to start off a partnering relationship with the contracting agency.

All of the above-mentioned estimating paperwork should be kept in an organized manner for adjustments to your proposal, of which there will be many. Also, the contractor should keep the entire organized estimating package for a period of several months after the bid opening, even if you are not the apparent low bidder. In many cases the apparent low bidder may have some sort of problem in obtaining the actual contract with particular agency. In those cases the contract is often awarded to another contractor at a later date.

CHAPTER SIX

CONSTRUCTION LAW

The subject of "construction law" is extensive, and it can be made to seem complicated. But it can also be simplified. Part of a lawyer's everyday job is to make complicated things seem simple, but also to make simple things complicated. You'll understand what I mean if you go to the courthouse and listen to lawyers arguing about the law. One lawyer will be telling the judge that the issue is a simple one, easily decided. The opposing lawyer will likely be talking about the complexities of the case, and arguing the exceptions to a simple rule. There are those whose business it is to make the law seem complicated, but it is our business here to simplify a basic understanding of construction law.

The topic of construction law is extensive because the construction trade itself is extensive. The construction process has many diverse activities that invite the application of different fields of law. We will break down broad fields of law: contracts, torts, and real estate, for example, into parts that have a direct impact on the business of construction so as to make those parts understandable to the average contractor. Let us begin with a brief examination of what law is and how it works.

WHAT LAW IS

Law is to a democratic society and a free enterprise economy what the sea is to a fish: it is the very element from which democracy and business draw nourishment and life. This is no exaggeration. The collapse of communism is followed all around the world by recognition that democracy and enterprise depend upon the rule of law. Former communist nations struggling to enjoy the economic benefits of the industrial revolution, are finding that they are severely hampered by a lack of laws, judges, and, yes, even lawyers! Neither democracy nor business can exist without law and law enforcement. Without them, elections and contracts are meaningless.

LAW

Law is rules of conduct. A law either requires, or prohibits, certain behavior. One law requires that a person obtain a license from the state before engaging in the contracting business, as another law prohibits false advertising by contractors. The very right to engage in the contracting business requires an understanding of law, because a licensed contractor must know what is required, and prohibited, by the contractors licensing law of your particular state or contracting jurisdiction.

STATUTES, ORDINANCES, AND REGULATIONS

Governmental power is divided between federal, state, and local government. Local government is divided between counties and cities. All governments exercise some of their power through commissions, districts, authorities, departments, and bureaus.

The United States congress and the legislatures of the 50 states pass laws known as "statutes". Cities and counties pass laws known as "ordinances". The various districts, boards, bureaus, authorities and agencies also have the power to pass laws, in this case known as "regulations". In case of conflict between federal and state, or local law, the U.S. Constitution provides that federal law is supreme. Likewise, state law controls law in the event of conflict with local law.

Examples of federal laws are OSHA (industrial safety), Davis Bacon (prevailing wage), and the Miller Act (requiring payment bonds on federal contracts). State and local laws that affect the construction industry include competitive bidding laws, contractors licensing laws, planning and zoning laws, building codes.

INTERPRETATION OF STATUTES, ORDINANCES, AND REGULATIONS

Even the most carefully drafted laws are likely to be unclear in specific situations. It is the role of the courts to interpret the laws and to apply them to specific situations. If there is a question about whether a person has violated the contractors license law, or a parking regulation, or any other law, the question will be decided by a court.

The legislative power is separate from the judicial power and therefore the courts do not make laws, but only interpret them. The function of a court in interpreting a statute, ordinance, or regulation is to determine the intention of the agency that passed the law, whether the agency be the Congress of the United States, a state legislature, a city council, or a government agency.

COMMON LAW

Statutes, ordinances, and regulations do not, and cannot, resolve all human disputes. Areas that are not covered by statutes, ordinances, or regulations are covered by the common law. Common law covers such topics as contracts, real estate, and torts: it is nothing more than the records of the decisions of judges in prior cases.

Common law had its beginnings in tradition. Legal scholars and reporters kept notes of court proceedings and reported how the judges decided cases. Other judges would consult these reports for guidance, and over the centuries there developed a rule called stare decisis. "Stare decisis" means "stand by things decided": so later judges follow the decisions of earlier judges.

Common law has proved adaptable to changing times. We now have such specialties as space law and electronics law. And courts are not immune to criticism, so occasionally they overrule earlier decisions.

When people present a court with a question of contract law, the judge resolves the question by reviewing the records of decisions of courts in similar cases. The law to be applied by the judge in such a case is determined by examining the decisions that judges made in previous, similar cases. A pilot learns the stall characteristics of an airplane by looking it up in a book and the book is based on the experiences of other pilots in the same or similar airplanes. Questions of common law are resolved in much the same manner.

APPEAL

Generally, the judicial system is divided into two branches: trial courts and appellate courts. A party who loses a lawsuit in the trial court always has the right to appeal the decision to a higher court. A Court of Appeal functions like the eraser on a pencil: it is there to correct mistakes. Between 20% and 30% of all trial court decisions are appealed to a higher court. Since a party is entitled to a trial that is free from error, a Court of Appeal will reverse the decision of a trial court if the decision is erroneous, and will order the trial court to correct the error or, if that is impossible, to retry the case.

APPELLATE DECISIONS

The decisions of the Courts of Appeal are made in writing, and the opinions are printed in law books. The law books are carefully indexed. Nowadays the opinions are not only in books, but in legal databases. Lawyers learn to look up rules of law in the opinions of the Courts of Appeal the way a student might look up a subject for a school paper in an encyclopedia.

CHANGING THE LAW

Both statutory law and the common law can always be changed by statute, and they often are. Judges also gradually change the common law so as to keep up-to-date with advances in technology, and changes in social and political thinking.

The common law is not abstract and theoretical. Every decision of every court is based on an actual dispute of some kind. The parties to a dispute present their cases and judges decide them. Thus, the development of common law depends on the self interest of the people in the lawsuit (litigants). Each party is entitled to be represented by a lawyer, and each lawyer's job is to find the cases supporting the interests of the client, and present those cases to the judge in a convincing way.

ADVOCACY

It sometimes seems as if lawyers and judges want to make the law seem complicated. To understand why this happens, you have to accept the fact that the American legal system is based on advocacy. Justice can best be achieved if parties are free to advocate positions that advance their interests. Thus, runs the theory, each party will place before the judge all of the evidence and arguments that support his position, which will enable the judge to reach a fair result.

In the process of litigation, an advocate attempts to make strong points while confusing the strong points of the opponent. Lawyers don't like to admit ignorance, and ignorance of the law may be hidden behind a confusing cloud of jargon.

While we all understand the manner in which lawyers operate, construction contractors should handle their affairs in a slightly different manner. As discussed in other areas of this book, the construction contractor should present his claims with the contracting agency in a very straight forward manner, dealing strictly with the contract specifications. Leave the complication and litigation to the lawyers in the rare event that you cannot come to a previous resolution with the contracting agency.

In this book, not every confusing exception to general rules will be discussed. The object here is to give a basic understanding of construction law that will help the reader to walk safely through the construction industry every day.

HOW LAW IS MADE

Laws are made by the people (constitutions), by Congress and state legislatures (statutes), by cities and counties (ordinances), and, to a disturbing extent, by federal, state, and local agencies (regulations).

Bismarck said those who like treaties and sausages should not see how they are made. The same could be said of laws. It is sufficient for our purposes to understand that statutes are drafted and amended by Congress and the state legislatures with the assistance and advice, and under the influence of, lobbyists. Legislators are also politicians. Politicians are sometimes tempted to write ambiguous legislation so that they can interpret it as any particular audience of voters might want it to be interpreted.

LITIGATION

There are five phases of litigation: **PLEADINGS, DISCOVERY, MOTIONS, TRIAL, AND APPEAL.**

1. PLEADINGS are formal legal statements of the positions of the parties: the claims of the plaintiff and the defenses of the defendant.

2. DISCOVERY is a process that requires parties to answer questions, either orally (depositions) or in writing (interrogatories). The purpose of discovery is to permit parties to discover the truth, and assemble evidence that may be introduced at trial. Unfortunately, discovery may also be misused to intimidate and exhaust an opponent.

3. MOTIONS, which are usually argued in open court, request that a judge make orders to make litigation easy or to decide issues in a case without trial.

4. TRIAL, is a formal examination before a competent judge, or judges, of the matter at hand in order to render a decision on the matter.

5. A COURT OF APPEAL reviews portions of the record of the trial court (including pleadings and a transcript of the proceedings at the trial) brought to its attention by counsel, reads the briefs and listens to the arguments presented by counsel, and agrees with or reverses the decision of the trial court.

The loser of a case on appeal may petition the state Supreme Court for a hearing. A state Supreme Court will agree to review only a small percentage of the cases presented to it. The criteria for review are whether the case presents an important point of law, or an unsettled point of law that needs to be clarified, or a case in which the Court of Appeal made a wrong decision that needs to be reversed.

The federal court system has three branches: trial courts (district courts), circuit Courts of Appeal, and the United States Supreme Court. Most cases are handled by state courts. Federal courts have jurisdiction only to consider questions of federal law (statutes passed by the United States Congress), disputes between states, and cases that arise under the United States Constitution.

HOW ADR FITS IN

Ambrose Bierce, in The Devil's Dictionary, defined *litigant: n. a person about to give up his skin for the hope of retaining his bones.* Bierce was alluding to the spiritual and monetary expense of litigation. The five phases of litigation are pleadings, discovery, motions, trial, and appeal. ADR (alternative dispute resolution) may remove three, four, or even all five phases.

The most popular form of ADR is arbitration, which is provided for in a majority of construction contracts and subcontracts. Arbitration completely removes phases 2 and 5 (discovery and appeal) and effectively removes phases 1 and 3 (pleadings and motions).

The next most popular form of ADR, mediation, removes all five phases of litigation.

ARBITRATION

Arbitration is a system under which the parties to a dispute appoint an arbitrator whose decision is as enforceable as the decision of a judge and not subject to appeal. Unless otherwise agreed, there is no discovery. *Motions* are usually minimal, or non-existent in arbitration. *Pleadings* in arbitration are a simple statement of a demand for relief by the claimant, and an answering statement (which is optional) by the responding party.

MEDIATION

Mediation differs from arbitration in that a mediator has no power to make an enforceable decision. The function of a mediator is to help the parties resolve their own dispute. Experience shows that when parties have a sincere desire to resolve a dispute, mediation is successful in more than 90% of the cases.

CASE IN POINT

The Arizona Supreme Court held that an arbitration agreement that was "grossly inequitable" was not enforceable by a property owner against a construction contractor. An addendum to a construction contract provided that the owner had the option of either selecting arbitration or litigation in court as the means for resolving any dispute, and the addendum also gave the owner the right to change his mind at any time up to final judgment. The court held that an arbitration agreement has to be enforceable by both sides in order for it to be enforceable by either side. The court said that the agreement was so grossly inequitable that it ran counter to the philosophy of encouraging arbitration and that, under the circumstances, the arbitration agreement was unenforceable and the parties would be required to resolve their dispute by litigation in court.

Stevens/ Leinweber/ Sullens, Inc v Holm Development & Management, Inc, 165 Ariz 25 (1990).

CHAPTER SEVEN

CONTRACT LABOR AND SUBCONTRACTOR UTILIZATION

REQUIREMENTS

Contracts that are funded through public sources can have varying kinds of labor and subcontractor utilization, project hiring, and employee protection requirements associated with them. Since the government is paying the bill for construction of the project, they incorporate requirements or goals into their contracts to protect the rights of the employees and subcontractors that work on the project. This book will cover these requirements and goals in the following format:

Disadvantaged Business Enterprises,

Equal Employment Opportunity / Affirmative Action,

Special Utilization Requirements.

DISADVANTAGED BUSINESS ENTERPRISES

Due to alleged past discrimination within business in the United States, many government agencies have set up various programs to assist Disadvantaged Business Enterprises (DBE's) in obtaining construction contracts. The goal of these programs is to help get the disadvantaged business' established to the point where they can compete in the industry without government assistance. There are basically three ways in which these governmental agencies assist DBE's in being awarded construction contracts that will meet the contract percentages or dollar volumes that have been adopted by that particular agency:

    1.) Set-aside contracts,

    2.) Subcontracting goals or requirements, and

    3.) Joint Venturing.

Set-aside contracts are those in which the government agency will only accept bid proposals from those types of contractors that they are trying to target. Some of the types of set-aside contracts are: emerging small business, small business, and disadvantaged business enterprise. Following are general definitions for these categories of contractors.

Small Business - for construction contracting, a small business is one whose average annual receipts of the company and its affiliates for its preceding three fiscal years are not greater than $17 million.

Emerging Small Business - for construction contracting, an emerging small business is one whose average annual receipts of the company and its affiliates for its preceding three fiscal years in not greater than $8.5 million or half that of a small business.

Disadvantaged Business Enterprise - a DBE contractor can be any of the following depending on the interpretation of the particular agency: Emerging Small Business, Minority Business Enterprise (MBE), or a Woman Business Enterprise (WBE).

Minority Business Enterprise - an MBE contractor is a small, disadvantaged business which:

1. is at least 51% unconditionally owned by one or more socially and economically disadvantaged individuals; or

2. in the case of a publicly owned business, at least 51% of the voting stock is unconditionally owned by one or more socially and economically disadvantaged individual; and

3. whose management and daily business operations are controlled by one or more such individuals. The above mentioned ownership must fall under one of the following categories: Asian-Pacific American, Black American, Hispanic American, Native American or any other concern certified for participation in the Minority Small Business and Capital Ownership Development Program under Section 8(a) of the Small Business Act.

Woman Business Enterprise - for construction contracting, a WBE is a small business that is at least 51% owned by a woman or women who are U.S. citizens and who also control and operate the business.

In the proposal information section of these contracts that are to be let, there will be an area detailing if the project is a set-aside and for what category of contractor. If your company falls under the stated set-aside category, then you may submit a bid to the contracting agency assuming your company meets any additional requirements of the contract. If your company does not fall under the set-aside category, then your bid will not be accepted by the contracting agency.

The other way for the contracting agency to achieve their quota of DBE contracts is through subcontracting goals or requirements on the projects in which they put out for competitive bid. The general contractors who are submitting bids to the contracting agency must document these subcontracting goals or requirements. This documentation may be required to be submitted with the bid or delivered to the contracting agency within a set duration after the bid opening, by the top one, two or three bidding contractors. This documentation that must be submitted to the contracting agency by the general contractors can be very extensive and must often be started up to three weeks prior to the bid date. Some contracting agencies require the general contractors to send direct mailings to the DBE's requesting a subcontract proposal for the work that the DBE specializes in. The general contractor must often follow up these direct mailings with telephone conversations that document exactly what transpired in the conversation.

FIGURE 7-1   Typical Subcontract Request Letter

ABC ASPHALT PAVING COMPANY
1234 E. Delta Ave.
Gresham, Oregon 97777
Phone (503) 663-6633 Fax (503) 663-6634

November 3, 20--

TO: PROSPECTIVE SUBCONTRACTORS

FROM: ABC ASPHALT PAVING COMPANY

RE: N.E. 12TH STREET RECONSTRUCTION
                 CITY OF PORTLAND, OREGON

BID DATE: NOVEMBER 10, 20-- @ 6:00 PM

SUBCONTRACT BIDS DUE: NOVEMBER 10, 20-- BY 2:00 PM

ABC Asphalt Paving Company plans to bid the above referenced project as a Prime Contractor. We would appreciate a sub-quote from your company in the areas of your expertise.

If you need further information on the project, please give us a call and we can discuss access to the contract plans and specifications along with any other project questions that you might have.

We would appreciate a response from your company as to your bidding status by checking one of the boxes below and faxing this sheet back to our office; we look forward to the opportunity to work with your organization.

Company Name: _____

- ☐ We will be quoting this project as a subcontractor
- ☐ We will be quoting this project as a supplier
- ☐ We will not be quoting this project

Thanks,

Sam D. Maple, President
ABC ASPHALT PAVING COMPANY

FIGURE 7-2   Typical Telephone Follow-up Sheet

XYZ CONSTRUCTION, INC.
Subcontract Solicitation
Telephone Follow-up Sheet

PROJECT:

CONTRACTING AGENCY:

SUBCONTRACTOR	DATE	TIME	CONTACT MADE	CONTACT PERSON	QUOTING ?	QUOTED ?	MBE WBE ESB	QUESTIONS	ANSWERS

Prime contractors are also often required to offer the DBE a bonding waiver, special financing options or many other contracting advantages. Bid packages often require the prime contractor to submit a subcontracting plan which details exactly which subcontractors your company plans to utilize, their subcontract amount, along with general information about the subcontractor.

FIGURE 7-3   Typical Subcontracting Plan

SUBCONTRACTING PLAN

Bidder Name: _____

Project Name: _____

Total Bid Amount: _____

We have elected to subcontract work in the following areas to the following subcontractors.

SUBCONTRACTOR	DIVISION OF WORK	DOLLAR AMOUNT	% OF BID	MBE ?	WBE ?	ESB ?

On occasion, contractors may have to submit to the contracting agency a narrative that describes why your company did not use the DBE subcontract quotes that were received. As one can tell, the documentation proving your "good faith effort" can be substantial; therefore, several weeks of working on this requirement is not out of the question if you have to document the dealings with up to 50 or more other companies.

The difference between goals and requirements, in the case of subcontracting, lies in the extent that the contracting agency will enforce penalties for not achieving the percentage or dollar volume set for DBE subcontracting on the project. In many cases, if the General Contractor does not meet the subcontracting "goals", the contracting agency might waive the amount they are short or negotiate with them to hire other subcontractors to meet the set goals. On the other hand, if the subcontracting "requirements" of a project are not met, the contracting agency might award the project to the next bidder who did meet the DBE requirements or throw out all bids if none of the bidders met the requirement. The distinction between goals and requirements along with their enforcements is a very gray area and, as one can probably guess, there have been numerous lawsuits filed that relate to the legalities of the DBE set-aside and subcontracting programs. To this day the fine points of these program are being worked out.

At this point in time, we suggest that if the rules are set by the contracting agency, the contractor should follow them to the best of his ability. If your company's philosophy does not agree with the rules set by a particular agency, then challenge them through legal channels prior to a bid if you choose. If you wait until you are the apparent low bid on a project in which you do not meet the DBE requirements, you are risking losing a project that you might otherwise have had the opportunity to successfully construct. One of the most important ways to handle any philosophical problems you might have with the way a governmental agency deals with these contracting requirements is to vote for those lawmakers at your local, state and national level who have similar ideas to yours on how these contractual requirements should be drafted or abolished.

With both subcontracting goals and requirements, a contractor may be awarded a construction contract in which he did not meet the set DBE percentage or dollar volume if he did what is called a "good faith effort". A good faith effort is one in which the contractor advertised for DBE subcontract participation, sent out direct mailings to DBE's, called DBE's to request bids and offered negotiable bonding and payment methods for the project, along with a follow-up on all of the above. Each individual contracting agency has a definition for what they consider to be a good faith effort. It is up to each bidding contractor to document and prove to the contracting agency that they did everything within their power to meet the DBE subcontracting goals, but failed to make the goals due to a lack of DBE quotes or that the DBE quotes were substantially above those of non-DBE's. Proving a good faith effort is very tough to accomplish with most contracting agencies; they want the contractors to meet the goals or requirements of the contract. Because of this reason, we recommend that the contractors avoid a time consuming conflict with the contracting agency and simply make the goals or requirements, if at all possible.

In subcontracting portions of a project to a DBE, most contracting agencies require that the DBE provide a "commercial useful function" to the overall project. Commercial useful function means that the subcontractor completes actual and physical work on the job in which the company has experience and that is necessary for completion of the overall project. For a DBE supplier on the project (of which often times your company can apply a percentage of the purchase total towards the DBE subcontracting goal or requirement), they must regularly produce the product and sell it in substantial quantities to the general public and/or industry at established market prices. These regulations had to be established so that general contractors would not simply pass the billings of a non-DBE subcontractor or supplier who worked on a project, through a DBE subcontractor or supplier for the purpose of meeting their contract goals. These contract goals or requirements for DBE subcontracting were established to assist actual disadvantaged businesses in obtaining contracts for work in which they specialize or for products that they produce, not to establish DBE brokers of labor and materials.

Another way for a general contractor to meet the DBE subcontracting goals of a project is to joint venture the job with a qualified DBE. A joint venture means that both companies combine a certain amount of craft employees, supervisors, equipment, bonding power and monetary backing to actually establish a new company. Both parties in a joint venture must share in the risk of a project, whether the job ends up financially with a profit or loss. Of course the general contractor is the major partner and the DBE is the minor partner, typically in a major to minor range of 90% - 10% or 80% - 20% respectively. The rules for joint venturing a project with a DBE are typically defined in the jobs specifications; if not, it would be a good idea to contact the contracting agencies representative if your company is planning to bid a project in this manner. The application for recognition of the joint venture generally needs to be submitted to the contracting agency a specific number of days before the bid opening.

EQUAL EMPLOYMENT OPPORTUNITY/AFFIMATIVE ACTION

Equal Employment Opportunity Programs (EEO) and Affirmative Action Clauses are typically found in all publicly funded contracts. These programs are set up so that contractor or subcontractor will provide equal employment opportunities to all applicants, without discrimination, and take affirmative action to assure these equal employment opportunities. On many projects, especially publicly funded projects, the contracting agency will give the contractor Equal Employment Opportunity/Affirmative Action and Davis-Bacon pay rate posters for the contractor to display on the jobsite; typically in the projects office trailer. It is generally a contract requirement that these posters be displayed where the crew can read them and so they may contact the appropriate contracting agency representative if they have a problem with any of these employment regulations.

On all Federal-aid construction contracts and all related subcontracts of $10,000 or more, the contractor must accept as his operating policy the following statement:

"It is the policy of this Company to assure that applicants are employed, and that employees are treated during employment, without regard to their race, religion, sex, color, national origin, age or disability. Such action shall include: employment, upgrading, demotion, or transfer; recruitment advertising; layoff or termination; rates of pay or other forms of compensation; and selection for training, including apprenticeship, pre-apprenticeship, and/or on- the-job training."

Key elements of any equal employment opportunity and affirmative action program, regardless of what agency you are contracting with, should include the following:

The contractor will not discriminate against any employee or applicant for employment because of physical or mental handicap in regard to any position for which the employee or applicant for employment is qualified. The contractor agrees to take affirmative action to employ, advance in employment, and otherwise treat qualified handicapped individuals without discrimination based upon their physical or mental handicap in all employment practices.

Employment practices covered by these provisions include: employment, upgrading, demotion or transfer, recruitment, advertising, layoff or termination, rates of pay or other forms of compensation, and selection for training, including apprenticeships.

The contractor agrees to post in conspicuous places, available to employees and applicants for employment, notices that state the contractors' obligation to take affirmative action with respect to equal employment opportunities.

The contractor should include in his subcontracts and purchase orders, the provisions of EEO and affirmative action that will be binding upon each subcontractor or vendor.

The contractor will designate an EEO Manager who will have the responsibility for and must be capable of effectively administering and promoting an active contractor program of EEO and who must be assigned adequate authority and responsibility to do so.

FIGURE 7-4 Typical EEO Certification Form

EQUAL EMPLOYMENT OPPORTUNITY
CERTIFICATION FORM

Company Name: _____Project: _____

Federal ID #: _____CCB #: _____Contract #: _____

The prime contractor shall submit a report for this project. Each subcontractor shall submit a separate report for this project.
Complete all categories for each employee working on the project during the reporting period.

Reporting period from: _____ to: _____

EMPLOYEE'S NAME	ZIP CODE	SSN	TRADE	LEVEL (J OR A)	RACE	SEX (M OF F)	HOURS WORKED

Level - Journeyman or Apprentice

Race - AA - African American
H - Hispanic
A - Asian American
NA - Native American
C - Caucasian

Submit forms to the Contracting Agency

Signature:_____ Print Name: _____ Title: _____

When the contract specifications require that a minimum percentage of the contractors work force are to be apprentices, minorities or women; the contracting agency will often require the contractor to submit a monthly utilization report. This report documents the contractors jobsite workforce in terms of trade, level, race, sex and hours worked so that the contracting agency can determine if the contractor is within the minimum contract percentages.

FIGURE 7-5   Monthly Utilization Report

MONTHLY SUBCONTRACTOR PAYMENT AND UTILIZATION REPORT

Company Name: _____ Project: _____

Federal ID #: _____CCB #: _____Contract #: _____

Contract Amount: _____

SUB CONTRACTOR	MBE WBE ESB	TRADE	ORIGINAL CONTRACT AMOUNT	CHANGES TO CONTRACT	CURRENT CONTRACT AMOUNT	PROGRESS PAYMENT MADE	TOTAL PAYMENTS MADE	RETAINAGE

It is hereby certified that the above referenced subcontractors have been utilized by our company in the amounts represented above and that the information provided herein is complete and accurate.

Contractors Authorized Representative: _____ Date: _____

SPECIAL UTILIZATION REQUIREMENTS

On many publicly funded projects, there are also goals for the hiring of minorities and women within your companies own workforce. These goals are typically set in the range of 4% -7%, for each category (minority or women) of your total project workforce. Your company may have to submit reports to the contracting agency that documents if your company is meeting these goals on a periodic basis.

In rare cases, working on an Indian Reservation for example, the contract documents may require your company to hire a specified percentage of Native Americans. Generally, when these situations occur, there will be a specified agency that will assist the contractor in obtaining employees to meet the requirements who are as qualified as possible.

Through the different labor and subcontractor utilization requirements of a particular contract, the construction contractor must be careful to take into account, in his bid proposal, any lost production or additional risks that he might encounter. Lost production or additional risks may be due to the quality and experience of the workforce he may be forced to utilize. Therefore, sufficient time must be spent studying the contract documents to make yourself fully aware of any specified requirements.

CHAPTER EIGHT

STANDARD CONSTRUCTION DOCUMENTS

This chapter has been written from the perspective of the contracting agency, or their designer, for the purpose of allowing the construction contractor to see the process of contract preparation. Insight into these methods and processes can assist the construction contractor in understanding, following and speaking the language of the construction contract.

Various construction industry organizations have been diligent in providing standard form documents which are consistent with the demands of changing requirements of the construction industry. For more than a century, the American Institute of Architects (AIA), through its national study committees, has responded to the needs of the industry by issuing and updating a complete range of construction forms and contracts. These documents have provided a measure of uniformity to a decentralized industry. The contents and structure of the AIA documents, and the general principles on which they are based, are widely known and accepted by the construction industry. Members of the recognized national contracting, subcontracting, and engineering societies consult with Designers and the legal counsel of the AIA document committee in developing and revising the standard documents.

Designers and other contract administrators across the nation do not always conform to a precisely defined approach to construction administration, but the widespread use of standardized form documents has unquestionably resulted in a fairly high degree of uniformity. Yet many construction contracts do not incorporate standard AIA documents. This is especially prevalent within the heavy highway construction market, while virtually all commercial building contracts do incorporate standard AIA documents.

DESIGNERS AS CONTRACT ADMINISTRATORS

Although much has been written to assist contractors in appropriate procedures needed for conducting a construction business, very little has been made available to aid construction professionals in understanding their professional functions and responsibilities during the construction period.

The chapter will also aid those who represent the owner's interests in a construction contract. This includes facility managers and construction managers. It is a necessity in their job to know what the Designer should be doing during the administration phase. Contractors who use this chapter for reference will better understand their relationship with the owner and Designer. Construction lawyers will obtain an overview of the duties of Designers, contractors, and owners when the standard AIA documents form the contract.

Large design firms generally have several projects in the construction phase at any one time and, depending on their organizational structure, may have one or more full-time construction administrators on staff. These Designers are in the construction phase of the firm's contracts on a more or less continuous basis and are generally well qualified and experienced. Some larger firms, however, if organized in separate teams for each project, may not have full-time construction administrators. This function will be fulfilled as needed by the project designer or administrator, much like principals of smaller firms.

The principals and key personnel of small and medium-sized design firms must be capable of rendering the construction phase services as and when called upon, even though this might represent only a small and infrequent part of their total involvement in the firm's obligations. Ninety percent of their time may be used up by various other pressing professional activities such as seeking new assignments; conferring with consultants, clients, and employees; designing; drafting construction documents; writing specifications; and office administration, along with other equally necessary, diverse, and demanding pursuits.

Administration of the contract provides the designer the opportunity to observe the construction as it progresses and the means to suggest adjustments when necessary to preserve the design integrity of the project. The designer will also be able to quickly recognize when contractors and suppliers have misunderstood the intent of the contract drawings and specifications. This will be beneficial to all involved, as it will often reduce the economic waste incurred in the correction of errors. It has been a tradition during the past century for designers to administer the construction contracts under which their projects were built. However, in the last two or three decades some liability insurance carriers and legal advisors to the design profession have reasoned that there would be considerably less exposure to professional liability claims if designers had less direct involvement in the construction process. Accordingly, many designers and firms declined to render services beyond completion of the construction documents. Although this approach eliminated the possibility of some types of malpractice occurrences during construction, it also resulted in a greater possibility of construction errors caused by contractors' faulty interpretation or misunderstanding of the construction documents. If the designer was not available during construction, errors or omissions in the documents were not discovered in time to be rectified in an economical manner.

In recent years, informed opinion has reverted once again to favor the designer's traditional role of monitoring the construction contract. This has several advantages which outweigh the disadvantages. With the designer more intimately involved in the conversion from drawings and specifications to physical reality, there is a greater chance of preventing contractor misconceptions and misinterpretations in a timely manner. It also affords the designer an opportunity to correct errors and anomalies in the documents before the construction progress makes them impossible, impractical, or too costly to rectify.

DESIGNER'S VS. CONTRACTOR'S SPHERES OF RESPONSIBILITY

A significant twofold outcome of this evolution is that (1) designers now recognize that the physical construction is the contractor's sole responsibility and designers no longer interfere with the contractor's determination and control of means, methods, techniques, sequences, and procedure of construction; and (2) the designer's role during the construction period is limited to counseling, administering, monitoring, observing, and reporting. It is of extreme importance that these two principles not be violated by designers in their preparation of construction documents or in construction administration.

Another way of distinguishing the differences in contractors' and designers' duties is that the contractor is responsible for carrying out the construction contract, while the designer's role is to monitor it. The monitoring process includes interpreting the contract, providing technical assistance to the contractor, observing the work to determine its compliance with the contract, reporting all significant and relevant information to the owner, certifying payments to the contractor, resolving disputes, facilitating communication, and consulting with the owner. An interesting aspect of this evolution is that the designer's construction period role used to be termed *supervision* of construction, whereas now it would be referred to as *observation* of construction. A subtle but significant difference.

CONSTRUCTION CONTRACT BENEFITS BOTH PARTIES

Owners and contractors mutually share the benefits and burdens of their construction contracts. The failure of either to realize their expected legitimate objectives is often caused by shortcomings in carrying out the physical work of the contract but is just as often caused by inadequate or inefficient administration of the contract.

The contract, is the medium of expression by which each party delineates the goals it seeks to obtain. The contractor expects to secure desirable construction employment for its forces, which can be performed under favorable physical and economic conditions, with the reasonable expectation of recovering its costs and a fair profit. The owner aspires to receiving its required construction project built according to the contract drawings and specifications, available for use on or before the agreed delivery date, and to pay no more than the stipulated price.

CONTRACT ADMINISTRATION IN THE OWNER'S BEHALF

The owner's legitimate interests in a construction contract must be assured through knowledgeable preparation of the contract in the first place. Many owners are quite competent and able to care for their own interests. They are also fully capable of negotiating on the level of the most canny of general contractors. Others will undoubtedly need the skilled assistance of their legal advisors, designers, and other construction industry experts.

After contract provisions have been completely discussed, agreed upon, and reduced to mutually acceptable written terms, and the contract has been duly signed by both parties, then only upon the diligent follow-through of capable contract administration can the owner realistically expect to fully realize the benefits offered by the contract.

Designers who administer construction contracts in behalf of owners customarily apply the principles and procedures reflected in the various forms of standard agreements of the American Institute of Architects. The owner will usually engage the designer's services for this purpose by using the appropriate owner-designer agreement and by making certain that it contains harmonious conditions requiring the contractor's cooperation and participation. The premise of this chapter is that the designer's services will have been contracted for by the owner using the standard AIA form:

- Standard Form of Agreement Between Owner and Architect with Standard Form of Architect's Services, AIA Document B141-1997 or a similar type of agreement. It is also based on the construction contract being formed substantially from one of the following two standard forms of construction agreement:

  ⇒ Standard Form of Agreement Between Owner and Contractor where the basis of payment is a Stipulated Sum, AIA Document A101-1997

  ⇒ Standard Form of Agreement Between Owner and Contractor where the basis of payment is the Cost of the Work Plus a Fee with a negotiated Guaranteed Maximum Price, AIA Document A111-1997

The construction contract must also include the AIA General Conditions:

- General Conditions of the Contract for Construction, AIA Document A201-1997

CONTRACTUAL RELATIONSHIPS

The AIA standard documents are based on the contractual relationships in the conventional contracting systems prevalent in the United States and some other countries. These relationships are illustrated in Figure 8-1. When analyzing a given situation, it is helpful to review this diagram to determine and visualize the true contractual relationships and communication links between the entities.

Although communication should flow freely along contract lines, the single exception is that the designer, as administrator of the construction contract, controls communication between the owner and contractor. All parties involved should respect the proper communication procedures. This will facilitate systematic administration while avoiding confusion and unexpected liability consequences.

FIGURE 8-1 Contractual Relationships

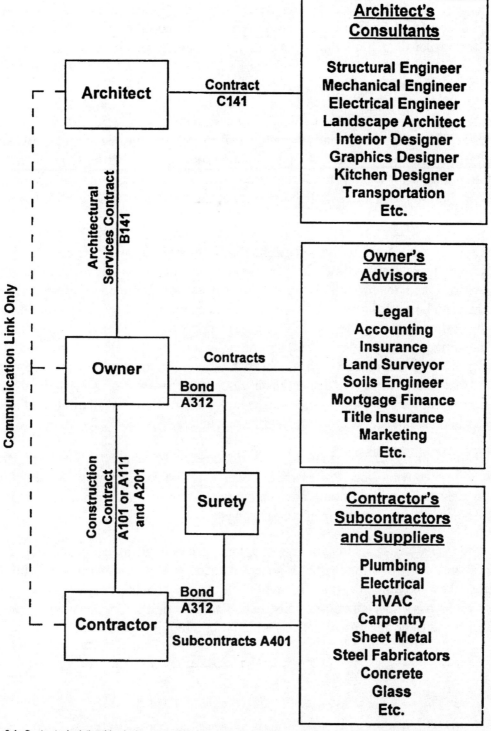

Fig. 8-1 Contractual relationships in the conventional construction contract using AIA form documents. The boxes represent the various entities involved in a typical building project. The solid lines connecting the boxes represent agreements between the entities, generally in the form of written contracts. (Note that the numbers of the appropriate AIA documents are shown.) The dashed line represents a communications link only. The solid line connecting Owner and Contractor is the *construction contract*, comprised of the contract documents. This is the contract which will be administered in the owner's behalf by the architect.

GENERAL PRINCIPLES

All recommendations and procedures will be based on the use of the standard AIA documents with the objective of providing effective and efficient service to the client in full compliance with the offerings of the design services contract. Consideration is also given to the responsible limitation of professional liability where possible and to the elimination of unnecessary risk. Unavoidable risk should be controlled and minimized. This is in the interest of both the designer and the owner.

The AIA standard documents have evolved on a continual basis over decades of practical daily usage. They are periodically republished to reflect the developing trends in construction technology, techniques, and administration as well as legal and insurance implications and professional liability consequences. The latest issue of the core construction industry documents, all coordinated and consistent with each other is the 1997 edition.

It is always advantageous to use standard form documents where possible, as the construction industry is accustomed to their use and most participants are acquainted with their provisions and the principles upon which they are founded. Customization to address the peculiarities of a specific project can be accomplished by means of deleted or added paragraphs amending the standard forms. In this way, only the deviations need be studied, rather than an entirely new and unfamiliar document. It is also recommended that the actual printed AIA forms be used, rather than retyping them, so that users can remain confident that there have been no basic document revisions. Although standard documents may be incorporated into contracts and specifications by citation, it is always preferable to include the actual documents for immediate reference by the user.

In this chapter, many of the standard AIA documents have been quoted in part or paraphrased for convenience. Professionals planning to take important administrative actions on their projects are not only advised but urged to refer to the actual documents in their entirety for the exact wording in proper context.

Although the opinions expressed in this chapter are based on many years of practical experience and on the orthodox and generally accepted interpretations of the AIA documents, they may not be directly applicable to all construction projects. Advice furnished by lawyers specializing in construction will be helpful in considering unique, complex, or unusual contractual situations which might require special analysis.

THE CONTRACT DOCUMENTS

Contract Documents Defined

There would be fewer disputes and disappointments in the construction industry if all parties to construction contracts clearly understood their obligations and were fully aware of their rights and privileges. These are to be found primarily in the contract documents.

The term *contract documents* is uniformly used in all AIA standard form agreements to designate the group of documents that comprise the construction contract between the owner and contractor. The term contract documents and the word contract may be used interchangeably.

Owners and contractors, when making claims against one another, will attempt to enforce the provisions of conversations, inferences of actions, and any scrap of paper developed in the course of their relationship. Similarly, designers frequently search the documents seeking authority to compel the owner or contractor to perform some duty. However, only those documents and promises that have been reduced to writing and incorporated into the contract can be relied upon as being legally enforceable.

Which are the Contract Documents

The AIA General Conditions (Document A201) contains a contractual definition of the contract documents (Subparagraph 1.1.1):

"The Contract Documents consist of the Agreement between Owner and Contractor (hereinafter the Agreement), Conditions of the Contract (General, supplementary and other conditions), Drawings, Specifications, Addenda issued prior to execution of the Contract, other documents listed in the Agreement and Modifications issued after execution of the Contract. A Modification is (1) a written amendment to the Contract signed by both parties, (2) a Change Order, (3) a Construction Change Directive or (4) a written order for a minor change in the Work issued by the Architect."

...And Which Are Not

The same subparagraph (A201, 1.1.1) identifies some documents which should never be considered as contract documents:

"Unless specifically enumerated in the Agreement, the Contract Documents do not include other documents such as bidding requirements (advertisement or invitation to bid, Instructions to Bidders, sample forms, the Contractor's bid or portions of Addenda relating to bidding requirements)."

Furthermore, Paragraph 3.12.4 clarifies the status of shop drawings and other submittals:

"Shop Drawings, Product Data, Samples and similar submittals are not Contract Documents."

More Non-Contract Documents

In addition, by inference, the contract documents do not include numerous other documents used in the construction process. For example, none of the following, unless specifically incorporated into the contract documents, are contract documents:

Correspondence, letters of transmittal, unsigned change orders, insurance policies, insurance certificates, performance bonds, labor and material payment bonds, title reports, topographic and boundary surveys, soil tests, material tests, laboratory reports, engineering calculations, environmental impact reports, materials certifications, inspection reports, subcontracts, purchase orders, payment requests, schedules of values, payment certificates, field observation reports, wording on shop drawing stamps, requests for information, requests for price quotations, cost breakdowns, price quotations, conference reports, certificates of compliance, manufacturers' literature, industry standards, guarantees, warranties, certificates of substantial completion, notices of completion, operating instructions, photographs, certificates of occupancy, mechanics' liens, lien releases, building permits, building codes, or zoning regulations.

The design services agreement is not a part of the owner-contractor agreement and is therefore not a contract document.

Any written agreements between the owner and contractor which pre-date the construction agreement are not contract documents unless specifically incorporated in the construction agreement. (A201, 1.1.2)

If the owner, contractor, or designer wish to consider any of the normally non-contract documents to be included in the contract, they simply must so-provide in the supplementary conditions or in the agreement before it is signed, or afterwards as an amendment to the contract.

Two More Contract Documents

In order to rise to the stature of a contract document, a writing must be agreed to by the owner and contractor and it thus becomes a part of the contract or an amendment to it.

There are two notable exceptions to this general concept: construction change directives, which are signed only by the owner and designer, not the contractor (A201, 7.3), and an designer's written order for minor changes, which is signed neither by the owner or contractor, but solely by the designer. (A201, 7.4)

Both of these are contract documents by virtue of the definitions of Modification (3) and (4) in A201, 1.1.1, quoted above.

The Project Manual

The project manual, according to Subparagraph 1.1.7, is the volume usually assembled for the work which may include the bidding requirements, sample forms, conditions of the contract, and specifications. The project manual is not a contract document, although some of its inclusions will be and some will not.

Oral Agreements Are Contracts But Not Documents

Generally, the only obligations which can be enforced by the owner or contractor against the other are those which appear as a requirement in one of the contract documents. Likewise, no rights are acquired by either unless they are conferred by some provision of a contract document.

Some actions or oral statements, however, may serve to amend the contract, even though they are not contract documents. For example, an owner might ask a contractor to move a brick pier and agree to pay an additional sum. After the contractor has removed and reconstructed the pier, the owner cannot then repudiate the oral agreement even though there is no related contract document.

Enumeration of the Documents in a Contract

Confusion in a specific contract can be lessened or possibly avoided by careful enumeration of the documents which are to be considered part of that agreement. The AIA standard form owner-contractor agreements have blank spaces designated for the purpose of listing all of the contract documents:

- Article 8 in AIA Document A101-1997, Standard Form of Agreement Between Owner and Contractor where the basis of payment is a Stipulated Sum

- Article 15 in AIA Document A111-1997, Standard Form of Agreement Between Owner and Contractor where the basis for payment is the Cost of the Work Plus a Fee with a negotiated Guaranteed Maximum Price

- Article 5 in AIA Document A107-1997, Abbreviated Standard Form of Agreement Between Owner and Contractor for Construction Projects of Limited Scope where the basis of payment is a Stipulated Sum

- Article 1 in AIA Document A105, Standard Form of Agreement Between Owner and Contractor for a Small Project where the Basis of Payment is a Stipulated Sum, Small Projects Edition, 1993

Document Review

Properly managed architectural and engineering practices employ rigorous review procedures and the documentation is routinely checked at various stages of progress. All of the construction drawings, specifications, and all other contract and bidding documents should be included in a comprehensive review process. Extensive check lists are available and many have been produced by individual offices, reflecting past problems with the hope of eliminating them in the future.

Document review should always be under the close supervision of competent qualified senior personnel. The documents should be checked for compliance with the owner's program and instructions, applicable building codes, industry standards, and customary construction procedures.

Checking should also evaluate whether or not the documents, when executed in the field, will properly carry out the design intent. The checker should make sure that the documents reflect the proper use of materials and that the structural and other systems are practical, economical, and constructible. One of the most important aspects of document review is checking for physical coordination among the various engineering and design disciplines. It is also necessary to confirm that all engineering recommendations have been carried out on the drawings and in the specifications.

The final review before release of the documents for contract bidding should be comprehensive and thorough. Regardless of the scope and quality of document review, however, it is inevitable that some level of error or inconsistency will remain undetected.

Errors of Inconsistency

Of the significant errors, some will be in the form of inconsistencies between the various elements of the contract documents. For example, the specifications might not agree with the drawings or the agreement in some respect. However, some inconsistencies and conflicts occur within a single document. For example, one section of the specifications does not agree with another, or the floor plan is not in accordance with a section.

Those errors that are discovered by or brought to the attention of the designer during the bidding period, can be easily dealt with by prompt consideration of the technical problem and issuance of an appropriate addendum to the affected document. Most of the remaining errors will more than likely be discovered during the construction period. Some will remain undetected to surface years later or possibly never.

Those differences that are caused by conflicting requirements in two or more of the contract documents could sometimes be easily resolved if the documents could be ranked in a hierarchical order of authority.

This would be simple and effective for some types of inconsistency. For example, if the specifications stated that a certain group of light bulbs are to be 300 watts and the lighting fixture schedule on the electrical drawings required 500 watts, then the higher ranked document would govern. However, this settles only the contractual obligation of the contractor, leaving other unresolved problems. For example, if the drawings were designated as governing, then the 500 watt bulbs would have to be supplied by the contractor. However, if the 500 watt designation is incorrect, the contractor must be further directed to supply 300 watt bulbs and issue a credit for the difference in cost between 500 watt and 300 watt bulbs. Yet, if the error is not discovered until after the 500 watt bulbs have already been installed, the contractor must be reimbursed for the extra costs of relamping, restocking (if allowed by the supplier), supervision, overhead, and profit. The credit for changing to smaller bulbs will be consumed or exceeded by the associated costs.

Some will point out that this problem could have been prevented entirely if the same information had not been given in two places. The bulb size should be shown in the lighting fixture schedule on the drawings and eliminated from the specifications, or vice versa. This is undoubtedly true for this example and generally represents good advice. However, other situations often arise where the inconsistency is more difficult to perceive. For example, where an industry standard, code, or regulation is cited in the specifications and the size, gauge, thickness, or some other quality shown on the drawings is inconsistent with the cited standard, code, or regulation.

Precedence of the Contract Documents

Considering the extraordinary volume of data contained in the typical set of construction documents, literally hundreds of thousands of bits, it is surprising that we do not see more errors, inconsistencies, and anomalies. Fortunately, most errors encountered are trivial and cause no more than a momentary pause and possibly a minor degree of professional embarrassment to the designer with no economic consequence to owner or contractor.

Over the past several years, the concept of establishing a hierarchy of documents has never been embraced by the professionals serving on the AIA's document committees. On the contrary, the AIA recommends that a precedence of documents not be established, but rather that all documents are complementary. What is required by one document is as binding as if required by all. Subparagraph 1.2.1 of the AIA General Conditions provides:

"The intent of the Contract Documents is to include all items necessary for the proper execution and completion of the Work by the Contractor. The Contract Documents are complementary, and what is required by one shall be as binding as if required by all; performance by the Contractor shall be required only to the extent consistent with the Contract Documents and reasonably inferable from them as being necessary to produce the indicated results."

This does not directly resolve the problem of inconsistent or conflicting requirements but instead requires the designer to make a determination taking into account all relevant factors obtained from anywhere in the documents or reasonably inferred from them. The designer's decision must determine what the contractor is obligated to do and what adjustment, if any, is to be made in the contract price or time. This is the acid test of an designer's ability to be fair to both owner and contractor.

Establishing Priority of Documents

The AIA, recognizing that occasionally some situations require a precedence of documents to be established, recommends a priority ranking of the documents followed by a requirement that in case of inconsistencies in or between drawings and specifications, the higher cost condition should govern.

Guide for Supplementary Conditions, Fourth Edition, AIA Document A511, 1987, where an order of precedence is required, suggests that the following be added to A 201, 1.2.1:

"In the event of conflicts or discrepancies among the Contract Documents, interpretations will be based on the following priorities:

1. The Agreement

2. Addenda, with those of later date having preference over those of earlier date

3. The Supplementary Conditions

4. The General Conditions of the Contract for Construction

5. Drawings and Specifications

In the case of an inconsistency between Drawings and Specifications or within either Document not clarified by addendum, the better quality or greater quantity of Work shall be provided in accordance with the Architect's interpretation."

The recommended order of precedence considers the drawings and specifications to be of equal authority, and they are last after all other documents. The final paragraph, which relates only to inconsistencies between the drawings and specifications, settles only the matter of controlling the contractor's obligation for price and time but does not settle the issue of which requirement is correct or proper to be carried out. When the error is not discovered until after the questioned work is completed, the contractor must be paid all extra costs of changing from the more stringent or costly condition to the correct one. Apparently, the paragraph does not apply to inconsistencies between or among these two documents and the others listed. It also fails to answer the question of inconsistencies within the other documents.

Most designers who feel that it is preferable to have a precedence of documents provision in their contract documents tend to favor drawings over specifications, while construction attorneys seem partial to the agreement over all other documents. Apparently, each is more trusting of the documents with which they are most familiar or have the most control over. Some designers specify that large scale drawings will take precedence over small scale drawings and that figured dimensions take priority over scaled dimensions.

Obviously incorrect typographic or clerical errors must be interpreted reasonably. No contractor can be considered credible when claiming to be misled by a 40 inch thick concrete floor slab requirement which is obviously intended to be 4 inches thick, or by a number 40 steel reinforcing bar which is meant to be a number 4 bar. Neither is the owner to be taken seriously when demanding that the contractor furnish the 40 inch thick slab or the number 40 steel bars or requests a monetary credit in lieu thereof.

Designer's Interpretations

Designers who are required to render interpretations and make decisions resolving inconsistencies or conflicts between or among the documents must stay meticulously within the contractual guidelines of A201, 1.2.1. Regardless of what the designer intended to include in the documents, the decision and all inferences must be based on what can actually be found in the contract documents.

The designer-decision maker, being also the author responsible for the inconsistent or erroneous documents, must at all costs guard against self-serving rulings. If a fair ruling must favor the contractor and go against the owner, it is usually an indication that the documents are indeed imperfect. If this creates some problem between the designer and client, then this becomes a matter the two of them must resolve. It is of no concern to the contractor and should not enter into the designer's decision regarding the owner-contractor relationship. The contractor should not be held responsible for any errors in the documents.

In determining what is reasonably inferable from inconsistent, erroneous, or incomplete documents, the design professional must exercise impartial judgment based on a credible and plausible analysis justified by the actual state of the documents. This decision will be final and binding upon the parties if not appealed to arbitration within 30 days (A201, 4.4.6). If the designer's determination is not sustainable from evidence that can be found in the contract documents, fair, knowledgeable, and reasonable arbitrators will find no difficulty in overturning or amending the designer's decision in their award.

Identifying the Contract Documents in Controversies Involving the Correct Drawings

A problem that must be sorted out and conclusively proved in many construction industry disputes is exactly which documents comprise the contract. On the surface, the issue seems fairly straightforward, as most modern construction contracts include a listing of the contract documents.

In general, the contract documents include the agreement, the general, supplementary, and other conditions, the specifications, the drawings, the addenda if any were issued prior to the contract signing, and modifications if any were issued after the contract was signed. Gathering these documents together in one place should not be too difficult. The problem usually is in organizing, identifying, and interpreting the miscellaneous conglomeration of drawings that have accumulated after the contract was signed. There are also the questions of which sheets have been superseded and identifying which are missing.

Controversial Construction Drawings. When all of the construction drawings are collected and viewed together, there is often confusion as to the various editions or versions of some, many, or all of the sheets. The owner, contractor, subcontractors, and suppliers may have differing views on which drawings comprise the official set or the contract set.

The set that has been stamped approved by the building department may differ from some or all of the other editions and may or may not be the set on which the contract was predicated.

In addition to the bound sets, there will be numerous loose sheets. The final contract set usually will include some of each.

The problem can be solved easily if the various editions are dated, but care must be taken when comparing dates so that reproduction dates are not confused with edition dates. One cannot always assume that the contract drawings will all be dated at or about the contract agreement date or that they will all be the same date. Some offices mark the relevant prints with a rubber stamp, "Contract Set", at the time they are issued.

This all is neatly resolved when the AIA General Conditions are a part of the construction contract. In the General Conditions, it is provided that the owner and contractor shall both sign the contract documents but that if one or both fail to sign them, the designer will identify the contract documents if requested. (A201, 1.5.1)

Causes of Confusion. It would seem simple enough to keep track of the contract drawings even if the owner and contractor had failed to sign them. However, identification problems also arise when changes in the drawings are made after the contract has been signed. Changes are made in the interest of reducing construction costs, to reflect unforeseen construction conditions, and accommodating changes in the owner's mind during construction. Changes are then made to changes. Partial changes are then made to partial changes. Some changes are later canceled. Some are wholly or partially reinstated or only partially canceled. Some of the changes, cancellations, or reinstatements are made hurriedly over night or over the weekend to accommodate construction progress in the field and sometimes with or without the formality of accompanying correspondence. Some changes are made over the telephone and followed later with a drawing. This tangled web gets more snarled when projects are under construction for protracted periods of time. Lack of adequate record keeping, changes in personnel, and passage of time render the problem exceedingly difficult to unravel.

Change Procedures

This all sounds like the architectural and engineering professions are not very disciplined. However, there are accepted procedures in the design community for designating changes in the documents and tracking the state of the contract. These procedures are generally understood by the construction industry. It is when the established procedures are circumvented or omitted that the confusion proliferates.

Dates on Drawings. The general date of a drawing, usually found in the title block, is the date when the drawing was started or finished. Some offices will conform all of the drawings in the original set to a single uniform date. The title block date is usually never changed after the drawing becomes a contract document.

New drawings originated after the contract is signed will usually be dated when started or finished or may possibly be conformed to a change order date. Printing dates are usually not on the originals, but are rubber stamped on the surface of the prints only. This is not the drawing date, but merely the date on which the reproduction was made. Some offices deviate from this procedure by recording printing dates on the original instead of on the surface of the print. Drawings of different printing dates may be identical if no interim changes have been made on the originals. Other dates may have been stamped on the surface of prints, such as the dates when the prints were received in various offices. These received dates may or may not establish the drawings as a part of the contract.

Drafting and Reproduction. To fully understand the dating and identification phenomenon, it is necessary to understand the relationship between drawings and their reproductions. All drawings are not made in the same way and they are not reproduced in the same way.

Original drawings are made using pencil or ink on tracing paper or sheet plastic. Whether the drawings are produced by hand drafting or by computer driven plotters, the principle is the same. The type of paper used is usually translucent so that reproductions can be made directly from the originals.

The drafted originals are always handled carefully in the designer's office. To lose or damage an original drawing would be a catastrophe, as all of its production time would be lost. Computer-produced original tracings, if lost or damaged, could be replaced quickly and inexpensively by merely replotting from the original computer program.

Reproductions. Years ago, the prints made from the originals were called blueprints as they had white lines on a blue background. It was a cumbersome wet process and the prints had to be dried. Although this process is no longer used, some still use the old term blueprint to represent any roll of printed construction drawings. The common procedure today is a dry ammonia vapor process that produces the familiar print with a white background and purple lines. The same process can also yield other colors of print called bluelines, blacklines, or brownlines. The background color can be other than white if different colors of paper are used.

If the original drawings are on opaque paper, such as with pasted up originals, the prints can be produced by xerography. Most architectural and engineering offices have dry electrostatic copiers that can produce same size prints up to 11 inches by 17 inches.

Larger xerographic prints can be made at the local reprographic shop (that used to be called the blueprinter).

Opaque originals cannot be reproduced by the ammonia vapor process. To do so would require a reproducible tracing, made by xerographically copying the opaque original onto translucent paper.

If required, the reproducible tracing can be altered by eradicating unwanted information and by drafting in new information.

Half size prints are often used for convenience in carrying, for desk top use, and for economy. A half size print of a 30 inch by 42 inch drawing is only 15 by 21, thereby saving 75 percent of the paper volume. All scales on the drawings are halved. Readability is sacrificed, so the full size drawings should be used for most technical purposes, such as estimating and the actual construction. Full size and half size prints are identical, except for size, when made from the same originals.

Record of Drawing Revisions

A common practice in architects' and engineers' offices is to identify and record all changes made in the content of drawings. Deletions, additions, and alterations made before the incorporation of the drawings into a signed construction contract are not often recorded, as it would rarely serve any useful purpose. However, after the bidding drawings are released, all changes made in the original drawings should be faithfully recorded.

The prevalent system in use consists of a Revision Block on the face of each original drawing. This is usually in the form of a rectangular table where all changes are listed by date and content. As each change is made on the drawing, it is added to the list. Each change is assigned an identifying symbol, commonly a number in a small triangle, called a delta. Typically each change recorded in the revision block would include the delta number, date, and a brief description of the change.

The change location on the drawing is also marked with the delta symbol. In order to identify the changed portion of the drawing it is usually circumscribed by a billowing line resembling a cloud and generally called a cloud. To completely understand the nature and extent of any change, one must usually compare the changed area on the revised drawing with the same area on the previous edition of the same sheet. When a new revision is to be added to a previously revised sheet, the last added cloud is removed, and the new cloud is added to identify the new revision. The cloud is usually drawn on the back of translucent originals to facilitate its removal if necessary. Naturally, all the old deltas must remain. One must review all the old editions of a particular sheet to fully grasp its change history.

When an entire new sheet is added to describe a construction change, it should be properly dated and identified in its revision block.

Since the advent of CADD (Computer Aided Design and Drafting) some offices are finding it more convenient to issue supplementary drawings to describe changes in the contract drawings. This is done by issuing computer generated copies of portions (8 1/2" x 11" or 11" x 17") of the original drawings, altered to illustrate the changes. The new drawing title or description and the date in the new title block, along with its related documentation should establish whether or not the drawing is a contract document.

Issuing Changed Drawings

As each new edition of a particular sheet is issued, the recipients should insert it in their complete set of contract drawings, ensuring that all editions of each sheet are available for comparison. Complete, up-to-date sets of the contract drawings should be maintained by the contractor's office, site office, the owner, and the designer. Major subcontractors and suppliers will also maintain complete sets.

When revised drawings are issued, they should always be accompanied by a letter of transmittal, explaining the significance of the new drawing. In some cases, the drawing is being issued provisionally, pending the contractor's quotation and the owner's acceptance of proposed cost and time changes. Other times the changed drawings are being issued for carrying out work in the field.

Alone, a changed drawing does not constitute a change in the contract. The change must have been authorized by an agreed change order or something equivalent. The drawing merely explains and illustrates the proposed or agreed change. Naturally, the change order and the changed drawing must be consistent. Neither will become a contract document until agreed upon by the contractor and owner.

There should be no problem in determining the status of the contract drawings at any moment in time, if the accepted procedures are followed.

CHAPTER NINE

SUBMITTING THE PROPOSAL

The project estimate has been completed and your company has arrived at a final bid price. Now it is time to submit your proposal to the contracting agency. The three types of proposals of which you might submit are:

Negotiated Projects,

Cost-Plus Projects,

Fixed Price Projects.

NEGOTIATED PROJECTS

A negotiated project is one in which a contracting agency negotiates directly with one or more contractor. The contracting agency can request that several, hand picked, contractors submit price proposals to them for construction of a particular project. At a previously set date, time and place, all of the proposals are submitted to the owner by the contractors. The contracting agency then takes the proposals and reviews each of them for a period of time that could be months, but is typically about 10-30 days. After review, the owner can then negotiate with all, or any number, of the contractors who submitted proposals. Usually, the owner will pick one contractor at this time and negotiate particular parts of the proposal with them. The contractor can be asked to show the owner how he obtained costs on certain parts of the cost proposal. After negotiation and discussion on the proposal, both parties typically come to a mutually beneficial price, along with all the other terms of a contract, for construction of the project.

COST-PLUS PROJECTS

Cost-plus projects are very similar to negotiated projects as far as submission of the proposal to the owner or contracting agency. Usually the owner will negotiate cost-plus projects with only one contractor. While construction of the project will be discussed in great detail, the negotiation is generally centered on what will be considered direct cost and what the margin, or plus portion, of the contract will be. As discussed in Chapter Two, cost-plus contracting can also be utilized by the contractor negotiating with a subcontractor.

FIXED PRICE PROJECTS

Fixed price projects are the most widely used style of public works construction contracts, especially with publicly funded jobs. Typically, the contracting agency will provide each potential bidder with an identical bid proposal package. The package will contain many disclaimers to be completed by the bidder along with a bid item section, bid bond section and a signature page. Each bidder must accurately complete all of this paperwork or their bid may be rendered non-responsive and not accepted by the owner. The contracting agency will set a date, time and place for the proposals to be submitted. In most cases, immediately after the time for submission, the owner will open and read the bid price of each contractor who submitted a proposal package, to those in attendance. After review of the bid documents, if the low bidding contractor has met all of the pre-bid conditions, the owner will award the contract to that company within approximately 2 - 4 weeks of the bid date. Again, as discussed in Chapter Two, no negotiations occur between the owner and the contractor at this time. Also, any pre-qualified contractor may submit a proposal on publicly financed projects of this type as opposed to the select bidders list of many privately financed jobs.

Figure 9-1 details a typical proposal page for a contract that is bid by the bid item, while Figure 9-2 shows a typical proposal page for a contract that is bid on a lump sum basis. Note that many public works construction projects that are bid on a lump sum basis can have alternates to the contract (either additive or deductive) that the contracting agency can choose to make part of the contract.

FIGURE 9-1   Typical Proposal Page - Bid Item

PROPOSAL FORM

Contracting Agency
P.O. Box 5599
Gresham, Oregon 97555

The bidder warrants that he has carefully examined the contract documents for the contract described as follows:

RECONSTRUCTION OF 14TH AVENUE

The bidder further warrants that he has examined the proposed work area independently of the indications in the contract documents and has made such investigations as are necessary to determine:

> 1) the character of the materials to be handled,
> 2) the probable interference traffic conditions,
> 3) other conditions to be encountered.

The bidder further warrants that if this bid is accepted, he will accept the contract with the Contracting Agency and will to the extent of his bid provide all things necessary for the performance of the contract, including, but not limited to, bonds, labor, materials, transportation, equipment, and anything else required to complete the work in accordance with the requirements of the contract documents.

The Contracting Agency anticipates award of this project on April 23, 20--. Following award, a pre-construction meeting will he held and a Notice to Proceed will be issued as soon as the bidder completes all agreement and bond requirements and they are approved by the Contracting Agency. It is anticipated that this Notice to Proceed will occur on or about May 7, 20--.

The bidder further promises that all work will be complete and the road fully open to traffic by July 15, 20--. The bidder agrees to pay, as liquidated damages to the Contracting Agency for any delay in completing all work, the sum of $500 per day for each day of delay. Final payment will not be made until the work is judged complete by the Contracting Agency.

The bidder submits and proposes the following unit bid prices, to wit:

RECONSTRUCTION OF 14TH AVENUE

Bid Item	Quantity	Units	Unit Price	Total Price
1) Mobilization	1	L.S.	$_____	$_____
2) Traffic Control	1	L.S.	$_____	$_____
3) Clearing & Grubbing	2	Acres	$_____	$_____
4) Demolition/Remove	1	L.S.	$_____	$_____
5) Roadway Excavation	975	C.Y.	$_____	$_____
6) Structure Excavation	90	C.Y.	$_____	$_____
7) Aggregate Base	500	C.Y.	$_____	$_____
8) Asphaltic Concrete (B)	350	Tons	$_____	$_____
9) 12" Concrete Pipe	100	L.F.	$_____	$_____
10) 24" Concrete Pipe	80	L.F.	$_____	$_____
11) Type I Inlet		1 EA.	$_____	$_____
12) 48" Dia. Manhole	12	V.F.	$_____	$_____
13) Concrete Curb	1,000	L.F.	$_____	$_____
14) Concrete Sidewalk	500	S.Y.	$_____	$_____
15) Temporary Signs	128	S.F.	$_____	$_____
16) Permanent Signs	160	S.F.	$_____	$_____
17) Chain Link Fence	1,000	L.F.	$_____	$_____

Total Proposal $_____

RECONSTRUCTION OF 14TH AVENUE

The bidder hereby acknowledges receipt of Addendum Nos. _____
to these contract documents.

Accompanying this bid is a bid security in the amount of 10% of the total bid. This bid many not be revoked for a period of 60 days after the date bids were opened. It is agreed that if the bidder does not execute and delivery the Agreement, including a satisfactory performance and payment bond for the full amount of the contract, within 10 days after the Notice of Award, said bid security may be collected as liquidated damages at the option of the Contracting Agency.

The surety company requested to issue the performance and payment bond will be _____. The bidder hereby authorizes said surety to disclose to the Contracting Agency any information concerning the bidder's ability to supply a performance and payment bond for the full contract amount.

The bidder certifies that he is a __ Resident Bidder __ Nonresident Bidder

If a Nonresident Bidder, the bidder certifies residency in _____.

Name of Bidder: _____

Address of Bidder: _____

_____

_____

Phone Number: _____ Fax Number: _____

Federal ID #: _____

CCB #: _____

Signature of Authorized Person: _____

Printed Name of Authorized Person: _____

Title of Authorized Person: _____

Date: _____

FIGURE 9-2   Typical Proposal Page - Lump Sum

PROPOSAL FORM

Contracting Agency
P.O. Box 5599
Gresham, Oregon 97555

Project: Improvements at Whitelane Park

The undersigned, having full knowledge of the specifications, drawings, and conditions, the site of the hereby proposed work, and all the conditions relating to the plans and specifications included in this bid package, hereby offers and agrees that this bid shall be irrevocable for at least forty-five (45) calendar days after the bid opening date and time, and if accepted proposes to furnish all materials, labor, transportation, service, equipment, costs and other necessary fees and do all the work in strict compliance with the terms and conditions of the Contract Documents, for the sum of:

TOTAL BASE BID – Award of the contract will be based on the Total Base Bid.

LUMP SUM: _____ Dollars
(Words)

$_____ Dollars
(Figures)

The bidder states below whether he is doing business as an individual, a partnership, or as a corporation. If a partnership, all partners are named and the person signing on behalf of the partnership states his position with the partnership. If a corporation, the bidder gives the State in which it is incorporated, whether it is licensed to do business in this State and the position of the person signing on behalf of the corporation.

FIRM NAME: _____

FIRM ADDRESS: _____

_____

_____

PHONE NUMBER: _____ FAX NUMBER: _____

FEDERAL ID #: _____

CCB #: _____

Improvements at Whitelane Park

RESIDENT BIDDER: __ YES __ NO

FORM OF ORGANIZATION: _____

IF INCORPORATED, IN WHICH STATE: _____

LICENSED IN THIS STATE: __ YES __ NO

IF PARTNERSHIP, LIST PARTNERS: _____

_____

AUTHORIZED SIGNATURE: _____

TITLE: _____

Accompanying this bid is a bid security in the form of a:

___ Bidders Bond ___ Certified Check ___ Cashiers Check

in the amount of 10% of the total bid.

All work under this contract shall be fully completed within 90 calendar days. If the contract is not fully completed within this time frame, liquidated damages will be accessed at $250 per calendar day that the project runs past the set time of completion.

Figure 9-3 details a typical non-collusion affidavit that the contractor must complete and submit with his bid on most governmental projects. The affidavit basically states that the contractor has arrived at his bid independently and has not conspired with any other contractor or company to alter any pricing of other contractors.

FIGURE 9-3 Non-Collusion Affidavit

AFFIDAVIT OF NON-COLLUSION

State of _____

County of _____

I (Type/Print Name)_____, state that I am
(Position/Title)_____ of (Name of Firm) _____
_____ and that I am authorized to make this affidavit on behalf of my firm, and
its owners, directors, and managers. I am the person responsible in my firm for the price(s) and
the amount of this bid.

I state that:

1) The price(s) and amount of this bid have been arrived at independently and without
consultation, communication or agreement with any other contractor, bidder or potential bidder,
except as disclosed on the attached appendix.

2) That neither the price(s) nor the amount of this bid, and neither the approximate price(s) nor the
approximate amount of this bid, have been disclosed to any other firm or person who is a bidder
or potential bidder, and they will not be disclosed before the bid opening.

3) No attempt has been made or will be made to induce any firm or person to refrain from bidding
on this contract, or to submit a bid higher than his bid, or to submit any intentionally high or
noncompetitive bid or other form of complementary bid.

4) The bid of my firm is made in good faith and not pursuant to any agreement or discussion with,
or inducement from, any firm or person to submit a complementary for other noncompetitive bid.

5) (Name of Firm)_____, its affiliates, subsidiaries, managers,
directors and employees are not currently under investigation by any governmental agency and
have not in the last four years been convicted of or found liable for any act prohibited by State or
Federal law in any jurisdiction, involving conspiracy or collusion with respect to bidding on any
public contract, except as described in the appendix attached.

I state that (Name of Firm)_____ understands and acknowledges
that the above representations are material and important, and will be relied on by (Contracting
Agency) in awarding the contract(s) for which this bid is submitted. I understand and my firm
understands that any misstatement in this affidavit is and shall be treated as fraudulent
concealment from (Contracting Agency) of the true facts relating to the submission of bids for this
contract.

_____
Signature

Subscribed and sworn to before me this __ day of _____, 20__

_____
Notary Public

My Commission Expires:_____

Figure 9-4 shows a typical certification of non-segregated facilities. On federally funded projects, the contractor must submit with his bid proposal this certificate which basically states that the contractor does not maintain or provide areas within his workplace that are segregated based on race, color or religion.

FIGURE 9-4 Certification of Non-Segregated Facilities (Federal Funds)

CERTIFICATION OF NONSEGREGATED FACILITIES

The bidder certifies that he does not maintain or provide for his employees any segregated facilities at any of his establishments, and that he does not permit his employees to perform their services at any location, under his control, where segregated facilities are maintained. The bidder certifies further that he will not maintain or provide for his employees any segregated facilities at any of his establishments, and that he will not permit is employees to perform their services at any location, under his control, where segregated facilities are maintained. The bidder agrees that a breech of this certification will be a violation of the Equal Opportunity clause in any contract resulting from acceptance of this bid. As used in this certification, the term "segregated facilities" means any waiting rooms, work areas, restrooms and washrooms, restaurants and other eating areas, time clocks, locker rooms and other storage or dressing areas, parking lots, drinking fountains, recreation or entertainment areas, transportation, and housing facilities provided for employees which are segregated by explicit directive or are in fact segregated on the basis of race, color, religion, or national origin, because of habit, local custom, or otherwise. The bidder agrees that (except where he has obtained identical certification form proposed subcontractors for specific time periods) he will obtain identical certification form proposed subcontractors prior to award of subcontracts exceeding $10,000 which are not exempt from the provisions of the Equal Opportunity clause and that he will retain such certification in his files.

Date: _____

Name of Bidder: _____

Address of Bidder: _____

_____

_____

Authorized Signature: _____

Title: _____

CHAPTER TEN

SETTING UP THE PROJECT AFTER AWARD

Now that your company has been awarded a construction contract, the real work starts. All of the work done prior to this point was from a speculative viewpoint, now it is time to prove that the work can be constructed within the time frame and within the costs that you estimated.

This chapter will deal with setting up and managing the project after award from a paperwork standpoint. Following are the categories in which we will breakdown the information:

The Contract and Contract Time,

Certificates of Insurance & Bonds,

Buying Out the Project: Material Suppliers & Subcontractors,

Project Cost Control,

Project Schedule,

Submittals,

Project Documentation.

THE CONTRACT AND CONTRACT TIME

Within the specifications with which your company prepared your bid proposal, there will be a time frame given for the contracting agency to award the project. Within that specified time frame (typically 30 to 60 days) the owner will review the apparent low bid to see that it meets all of the pre-bid proposal requirements. If the bid does not meet all of these requirements, the owner may re-bid the project at a later date, waive the pre-bid requirements that were not met or review the apparent second low bid for award to that contractor. Assuming that the low bid does meet all of the proposal requirements, the owner will send the apparent low bidder several copies of the contract package for signature by a company manager.

Once the contractor has the contract package there is also a time frame specified by which he has to return the signed package back to the contracting agency. If the package is not returned to contracting agency within the specified time frame (typically 10 to 14 days), the owner can deem the apparent low bidder non-responsive and move to award the contract to the apparent second low bidder. The contracting agency at that time may require that the apparent low bidder forfeit their bid bond for failure to execute the contract. This situation of forfeiture of bid bond is a very extreme case; typically the contractor can negotiate an extension of time for return of the contract package if he just cannot get it back to the contracting agency within the specified time frame. Once the signed contract package has been returned to the agency, they then sign each copy of the contract and return the fully executed copy.

Once again, there is a specified time frame after award and execution of the contract in which the contractor has to begin work at the project site. After this specified time (typically 10 to 14 days) has elapsed, the contracting agency starts counting contract days. The number of contract days (time for project completion) would have been previously stated in the bidding project specifications. Contract time is generally counted by one of the three methods detailed below:

"Specified date" contract time is exactly what it sounds like. The contracting agency states a specific date that the contract shall be substantially complete, regardless of when the contractor actually begins work, holidays or normal weather conditions for the area.

"Calendar day" contract time is simply a count of calendar days starting from the day the contractor receives the notice to proceed. The contract specifications will state the number of calendar days the contracting agency is allowing for any particular project. This type of contract time does not recognize holidays or normal weather conditions for the area, for contract time extensions.

"Working day" contract time is a count of the actual working days the contractor has spent on the project starting the day the contractor receives the notice to proceed. If some unforeseen condition, outside of the contractor's control, causes the contractor not to be able to work on operations that are on his scheduled critical path, then the day is not counted as a working day towards the contract total. Weekends and holidays are also not counted towards contract time. Again, the contract specifications will state the number of working days the contracting agency is allowing for any particular project.

Many times if the project is required to be constructed within a specified time frame, there will be monetary penalties detailed in the contract specifications for not completing the project on time. These penalties are called liquidated damages. Liquidated damages are often in the range of $100 to $1,000 per contract day, but can go up to $10,000 per contract day on large, critical projects. With this in mind, one can see that if the project was initially scheduled during the bid process to be near the amount of contract days allowed by the contracting agency, the contractor can waste no time beginning construction once contract days start counting.

Liquidated damages may also be assessed on times within the overall contract time, such as opening up a road or lane closure by a specified time each day, or on a specific date.

There have been liquidated damages specified for these types of time requirements in the range of $500 per minute. Construction contracts virtually always encounter unforeseen circumstances that adversely effect the projects schedule; therefore, it is a good idea to keep a good handle on how your schedule compares to contract days remaining on the project.

In recent years there have been many contracts written that incorporate liquidated damages for late contract completion along with monetary incentives for early contract completion. With the earthquakes in California during the early 1990's, many constructions contracts were bid, awarded and completed that had these liquidated damages vs. incentive clauses as part of the contract. This rewarded the contractors for completing their project in a very timely manner, working around the clock if necessary, to repair critical highways, bridges and other facilities.

CERTIFICATES OF INSURANCE & BONDS

Contained within the contract package, mentioned above, will be a request from the contracting agency for the contractors Certificates of Insurance and Payment/Performance Bonds. At this time your company needs to contact your agent that handles insurance and bonding to request the certificates from their office. They will generally require a copy of the contract before they will issue your certificates. Most insurance companies will either forward all copies of these certificates to your company or send the originals to the contracting agency, with copies to your office. The minimum limits of insurance and bonding will be specified in the bidding specifications supplied to your company by the contracting agency. Sometimes these certificates take slightly longer to acquire than is given by the owning agency to submit. Most owners will allow a few days extra for submission of these certificates if the contractor contacts them and explains the situation as discussed above.

BUYING OUT THE PROJECT: MATERIAL SUPPLIERS & SUBCONTRACTORS

During the bid process your company set up quote comparison sheets for categories of material suppliers and subcontractors. Now that you have a contract from the contracting agency, your company may issue purchase agreements and subcontracts to the low bidders from your quote comparison sheets.

Purchase agreements should be made for all major material suppliers. Generic purchase agreements may be purchased at most office supply retailers and Figure 10-1 shows one of these. These agreements should include such information as the following:

1.) Legal names of the buyer and seller,

2.) Detailed project name and location,

3.) Detailed description of the material to be purchased along with the unit and total prices,

4.) Any applicable taxes or discounts,

5.) Where and when the material is to be delivered,

6.) Payment terms,

7.) Language about which party is responsible for late deliveries, damaged materials and any other applicable additional provisions,

8.) Original signatures by both parties.

FIGURE 10-1   Purchase Agreement Sample

PURCHASE AGREEMENT

Federal ID #_____   Job Number _____

This agreement, made this _____ day of _____, _____ by and

between _____, hereinafter called the Seller, and

_____, hereinafter called the Contractor.

SECTION 1) The Seller agrees to furnish the Contractor the materials set forth in Section 2 hereof

necessary in the construction of _____

for _____ hereinafter called the Owner, located at

_____ in accordance with the prices and under the terms and conditions hereinafter set out. The Seller hereby agrees to be bound by the terms of the contract between the Contractor and the Owner.

SECTION 2) It is agreed that the materials to be furnished by the seller are as follows:

Quantity	Materials	Unit Price

Terms: _____ Discount if Paid in _____ days after receipt by Contractor of Seller's invoice.

SECTION 3) All materials furnished under this agreement shall be delivered F.O.B. _____with freight allowed to _____

SECTION 4) The Seller shall promptly deliver the said materials at such time and to such place as the Contractor shall from time to time specify: _____

SECTION 5) Payment for the materials furnished by the Seller under this agreement will be paid by the contractor within _____ after delivery. Payment for these materials shall not by construed as final acceptance of the materials by the Contractor or the Owner.

SECTION 6) All materials furnished under this agreement shall be first-class in every respect, shall be satisfactory to the Contractor and shall conform strictly with the plans and specifications and all modifications thereof. In the case of materials ordered by sample, the materials shall be equal in every way to the sample submitted. In the case the materials furnished do not comply with the requirements set out in this agreement, or are otherwise defective, or for any other reason, the Seller shall immediately upon notice by the Contractor remove said materials and the replace the same with proper materials that are satisfactory to the Contractor.

SECTION 7) The Seller shall send a shipping list with each delivery. Each invoice for delivery much be supported by separate purchase order issued with shipping or delivery release.

SECTION 8) It is agreed that in the case of materials to be furnished in bulk or by any other unit of measurement, the quantities hereinbefore set out in this agreement are approximate only and that this agreement is intended to cover the actual requirements. The Contractor is under no obligation to purchase or accept any such materials not actually required by it in the construction of such work by the plans, specification or modification, but the Seller shall furnish all such materials required for construction by the plans, specifications or modifications.

SECTION 9) If the Seller shall fail to furnish any of the materials set out herein with the time specified by the Contractor or in accordance with the requirements of this agreement and to the satisfaction of the Contractor, then the Contractor may at his election purchase said materials elsewhere and the Seller shall upon demand pay any excess in the cost of such materials purchased over and above the price herein specified along with any additional expenses incurred by the Contractor in connection therewith. The Seller shall not be liable under this paragraph if such default is caused by strikes, lockouts or acts of God beyond the Seller's control, but the Contractor may under these conditions and at his option terminate this agreement.

SECTION 10) The performance of each and all conditions herein by the Seller shall be a condition precedent to the payment of any monies hereunder.

SECTION 11) The Seller warrants and guarantees the materials covered by the agreement and agrees to make good at his own expense any defect in such materials which may occur or develop prior to the Contractors release from responsibility.

SECTION 12) The Seller agrees to indemnify and save the Contractor, its agents and employees harmless from any and all claims, suits and liabilities arising out of or connected with the materials furnished under this agreement.

SECTION 13) The Seller shall not sublet or assign this agreement or any part thereof, including payments due for to become due without the written consent to the Contractor.

ADDITIONAL PROVISIONS:

Contractor: _____   Seller: _____

By: _____   By: _____

Date: _____   Date: _____

Time and effort should be spent on these agreements to ensure their accuracy in the event of a future dispute. A purchase order may also be used for smaller purchases that are not critical to the project. Your superintendents and foremen will also use purchase orders during the construction process to purchase all of the miscellaneous materials and supplies that are required on a day-to-day basis. A purchase order generally includes the information listed in 1 through 5 above and a sample is shown in Figure 10-2.

FIGURE 10-2   Purchase Order Sample

ABC ASPHALT PAVING COMPANY   **PURCHASE ORDER 1001**
1234 E. Delta Avenue
Gresham, Oregon 97777   Date: _____
Office (503) 663-6633   Ship to: _____
Fax (503) 663-6634   _____

"Serving the Northwest"

TO: _____

     _____

     _____

When Ship	Ship Via	F.O.B. Point	Terms

Quantity	Description	Price	Amount

Please show this Purchase Order Number on all packages and invoices.

Authorized by: _____

Subcontracts should be made for all subcontractors on a project regardless of the size of their contract. This is due to the fact that your subcontractors are tied into the same contract that your company signs with the owner and all of those contractual requirements need to be included within the subcontract. Again, generic subcontract forms may be purchased at most office supply stores and one is shown is Figure 10-3. Subcontracts should include such items as the following:

1.) Legal names of the contractor and subcontractor,

2.) Detailed name and location of the project,

3.) Detailed description of the work to be done along with the unit and total prices (with or without sales tax),

4.) Terms of payment, including any retention held by the owner,

5.) Bonding and insurance agreements,

6.) How change orders and disputes are to be handled,

7.) What happens in the event of delays or failure to perform the work,

8.) Safety provisions,

9.) Any additional provisions that the contractor is required to adhere to by means of their contract with the owner.

FIGURE 10-3   Subcontract Sample

Form A325   **SUBCONTRACTOR AGREEMENT**

THIS AGREEMENT made and entered into by and between _____,

hereinafter called the Contractor and _____,

hereinafter called the Subcontractor, WITNESSETH:

     The Contractor, for the full, complete and faithful performance of this subcontract, agrees to pay to the Subcontractor, in accordance herewith, the sum of _____ Dollars ($ _____) payable as follows:

     In consideration therefor, the Subcontractor agrees as follows:

     1.   To furnish all supervision, labor and materials, and perform all work as described in Paragraph 3 hereof, for the consideration of _____

for _____,
hereinafter called Owner, in accordance with the Contract dated the _____ day of _____,
19____ , between the Owner and the Contractor, and the general and special conditions of said Contract, and in accordance with the drawings and specifications and addenda of said construction by _____,

Architects, all of which documents in their ENTIRETY are hereinafter referred to as the Main Contract.

     2.   To be bound by the terms of said Main Contract with the Owner (including every part of and all the general and special conditions, drawings, specifications and addenda), in any way applicable to this Subcontract, and also by the PROVISIONS PRINTED ON THE REVERSE SIDE HEREOF, which are hereby referred to and made a part of this Subcontract, Subcontractor certifies that the Main Contract has been read by him.

     3.   That the labor and materials to be furnished, and the work to be performed by the Subcontractor are as follows: _____
_____
_____
_____

     IN WITNESS WHEREOF the Contractor and Subcontractor have executed this agreement in duplicate, this _____ day of _____ , 19 ____ .

_____
Contractor

_____
Subcontractor

(PROVISIONS PRINTED ON REVERSE SIDE HEREOF ARE PART OF THIS AGREEMENT AND BINDING UPON THE PARTIES)

(Revised 4/94)

THE SUBCONTRACTOR AGREES:

(a) To assume toward the CONTRACTOR, so far as the SUBCONTRACT work is concerned, all the obligations and responsibilities which the CONTRACTOR assumed toward the OWNER by the MAIN CONTRACT which includes the general and special conditions thereof, and the plans and specifications and addenda, and all modifications thereof incorporated in the documents before their execution (which documents shall be available to the SUBCONTRACTOR). The SUBCONTRACTOR agrees not to assign or sublet said work or any portion thereof without the written consent of the CONTRACTOR.

(b) To start work when notified by the CONTRACTOR, and to complete the several portions and the whole of the work herein sublet, at such times as will enable the CONTRACTOR to fully comply with the contract with the OWNER, and to be bound by provisions in the MAIN CONTRACT with the OWNER for liquidated damages, if caused by SUBCONTRACTOR.

(c) To submit to the CONTRACTOR applications for payment at such reasonable times as to enable the CONTRACTOR to apply for and obtain payment from the OWNER, and to receive as progress payments from the CONTRACTOR the amounts allowed to the CONTRACTOR by the OWNER on account of the SUBCONTRACTOR'S work to the extent of the SUBCONTRACTOR'S interest therein. Final payment shall be made within a reasonable time after the completion and acceptance of the subcontract work unless a definite time for final payments is fixed on the face thereof.

(d) The CONTRACTOR may, without invalidating this SUBCONTRACT, order extra work or make changes by altering, adding to, or deducting from the work; the price herein being adjusted accordingly. All such work shall be executed under the conditions hereof, and of the MAIN CONTRACT, except that any claim for extension of time caused thereby must be agreed upon at the time of ordering such change.

(e) To make no claims for extras unless the same shall be fully agreed upon in writing by the CONTRACTOR prior to the performance of any such work.

(f) The CONTRACTOR has the status of an employer as defined by the Unemployment Compensation Act of the State, and all similar acts of the National Government, and including all Social Security Acts; that he will withhold from his payrolls the necessary Social Security and Unemployment reserves, and pay the same; that the CONTRACTOR shall in no way be liable as an employer to or on account of any of the employees of the SUBCONTRACTOR; that the SUBCONTRACTOR will as an employer, to the extent of any of his employees under this contract, conform to all rules and regulations of Social Security Acts and Unemployment Commissions created by said laws, and that he will furnish satisfactory evidence to the CONTRACTOR that he is conforming to said laws, rules and regulations. The SUBCONTRACTOR hereby releases and indemnifies the CONTRACTOR from any and all liabilities under said laws.

(g) That the SUBCONTRACTOR will pay any and all federal, state and municipal taxes and licenses, including sales taxes, if any, for which the SUBCONTRACTOR may be liable in connection with the labor and materials herein, or in carrying out the SUBCONTRACT, prior to final payment being made to him.

(h) To pay industrial insurance and all other payments required under Workmen's Compensation laws as the same become due, and to furnish the CONTRACTOR with evidence that the same has been paid before final payment is made on this SUBCONTRACT.

(i) That all materials delivered by or on account of the SUBCONTRACTOR and intended to be incorporated into the construction hereunder shall become the property of the OWNER as delivered; but the SUBCONTRACTOR may repossess himself of any surplus remaining at the completion of his contract. That all scaffolding, apparatus, ways, works, machinery and plans brought upon the premises by the SUBCONTRACTOR shall remain his property, but in case of default and the completion of the work by the CONTRACTOR, the latter shall be entitled to use the said scaffolding, apparatus, ways, works, machinery and plans without cost or liability for depreciation of damage by use and without prejudice to CONTRACTOR'S other rights or remedies for any damage or loss sustained by reason of said default.

(j) In the event the contract herein is upon a unit price, it is understood and agreed that any quantities and amounts mentioned are approximate only and may be more or less at the same unit price and subject to change as ordered and directed by the CONTRACTOR.

(k) To indemnify and save harmless the CONTRACTOR from and against any and all suits, claims, actions, losses, costs, penalties and damages, of whatsoever kind or nature, including attorney's fees, arising out of, in connection with, or incident to the SUBCONTRACTOR'S performance of this SUBCONTRACT. Within ten (10) days after the execution of this SUBCONTRACT agreement, to obtain and furnish to the CONTRACTOR, a corporate surety bond in the full amount of the SUBCONTRACT price. Said bond shall name the CONTRACTOR as the Obligee and shall indemnify the CONTRACTOR against loss or damage arising by reason of the failure of the SUBCONTRACTOR to perform all of the provisions of this SUBCONTRACT according to its terms and conditions, to pay for labor performed or material furnished in connection with this SUBCONTRACT. Said bond shall be furnished by a corporate surety company acceptable to the CONTRACTOR, and authorized to do business in the State in which the work is located. The premium for said bond shall be paid for by the CONTRACTOR, but only to the extent of the premium which would result from the application of the lowest available surety bond rate which is applicable to the work involved.

(l) To immediately, after receiving written notice from the CONTRACTOR, proceed to remove or take down from the grounds or buildings, all materials condemned by the CONTRACTOR whether worked or not, as unsound or improper, or as in any way failing to conform to the MAIN CONTRACT, including the general or special conditions, drawings, specifications or addenda. Failure of the CONTRACTOR to immediately condemn any work or materials as installed shall not in any way waive the CONTRACTOR'S right to object thereto at any subsequent time.

(m) To commence and at all times to carry on, perform and complete this SUBCONTRACT to the full and complete satisfaction of the CONTRACTOR, and of the Architect or OWNER. It is specifically understood and agreed that in the event the CONTRACTOR shall at any time be of the opinion that the SUBCONTRACTOR is not proceeding with diligence and in such a manner as to satisfactorily complete said work within the required time, then and in that event the CONTRACTOR shall have the right, after reasonable notice, to take over said work and to complete the same at the cost and expense of the SUBCONTRACTOR, without prejudice to the CONTRACTOR'S other rights or remedies for any loss or damage sustained.

(n) Upon completion of any unit of the work, and upon final completion thereof, to clean up all refuse and rubbish, same caused by the SUBCONTRACTOR, and to promptly remove all excess material, tools, structures, etc., which may have been brought on the premises or erected by the SUBCONTRACTOR, and in the event of the failure of the SUBCONTRACTOR so to do, the CONTRACTOR may, after reasonable notice to the SUBCONTRACTOR, clean up the premises at the cost and expense of the SUBCONTRACTOR.

© E-Z Legal Forms. Before you use this form, read it, fill in all blanks, and make whatever changes are necessary to your particular transaction. Consult a lawyer if you doubt the form's fitness for your purpose and use. E-Z Legal Forms and the retailer make no representation or warranty, express or implied, with respect to the merchantability of this form for an intended use or purpose.

The accuracy and completeness of these documents can save your company court time, confusion, and dollars in the event of a failure by a subcontractor to comply with any aspect of their subcontract. Your company, as the prime contractor, should send the subcontractors an initial package of information that includes such items as the following:

1.) The appropriate number of the subcontract itself,

2.) A letter detailing the kind of certificates of insurance that the subcontractor needs to return along with an acceptable bond form, if necessary.

3.) Details of the reporting of certified payroll for the project; what kind of forms are to be used and how often they need to be submitted.

4.) A copy of the initial schedule for the project that clearly defines when your company expects the subcontractor to be on-site and when you expect their work to be complete.

Details of any contractual labor hiring goals or requirements for the project and what your expectations are for their particular company.

All subcontractors should know, via letter or verbally, that all communication with the contracting agency or other subcontractors should go through the general contractor. The general contractor needs to be aware of all jobsite communication so they can determine any effects to products, schedule, change orders or other subcontractors.

FIGURE 10-4   Sample Subcontractor Letter

ABC ASPHALT PAVING COMPANY
1234 E. Delta Ave.
Gresham, Oregon 97777
Phone (503) 663-6633
Fax (503) 663-6634

February 29, 20--

COAST CONCRETE COMPANY
7890 W. 234th Street
Gresham, Oregon 97777

RE: N.E. 12th Street Reconstruction
City of Portland, Oregon
Concrete Subcontract

Mr. Del Walters:

ABC Asphalt Paving Company was low bid on and has been awarded the above mentioned project. Coast Concrete Company was the low concrete subcontract bid and enclosed is a subcontract for this work.

ABC Asphalt Paving Company would appreciate it if you could have an authorized manager sign all three copies of the subcontract and send them back to our office for signature by our manager. We will then forward you a fully executed copy of the subcontract for your records.

ABC Asphalt Paving Company also needs a certificate of insurance from your agent that meets the contract requirements. Please also have your agent forward us a payment and performance bond for the amount of you subcontract.

Weekly, we will need your company to submit Certified Payroll Reports to our office that correspond to the applicable labor wage rates. On a monthly basis, we will need your company to submit to our office your EEO report on the appropriate form.

If required by the contract documents, please forward all of your companies' submittals as soon as possible so that we may submit them to the owner for approval and not delay any the operations of this project. Our crews, the owners' representative and any subcontractors on site, will meet every Monday morning to conduct a toolbox safety meeting.

If you have any questions, please contact me at our home office and we look forward to working with your company.

Sam D. Maple
President
ABC ASPHALT PAVING COMPANY

Purchase agreements and subcontracts should be sent to the appropriate parties, in duplicate or triplicate, unsigned by the contractor. The supplier or subcontractor should then sign all copies of the document after they take the necessary time to review their contents. Any corrections the supplier or subcontractor makes to the documents should first be approved by the issuing contractor, and then be initialed on the documents by the supplier or subcontractor near the change. Upon return of all copies of the agreement, the contractor should review any changes made by the supplier or subcontractor and initial them on the documents if he is in agreement with the changes. The contractor should then sign the documents and return one fully executed original copy to the supplier or subcontractor, keeping the remaining copies for the project files.

During this process of buying out the project, you as the general contractor may get additional quotes from material suppliers or subcontractors. Many questionable general contractors will consider these quotes or even attempt to request more quotes from material suppliers or subcontractors beyond the quotes they inserted into their pre-bid estimates. If the general contractor did not get quotes from within a certain area of work and inserted his best guess of the cost of the material or subcontract prior to the bid, then there is no problem with trying to get an economical price after you are the apparent low bidder. But on the other hand, if you as a general contractor used the price of the low material supplier or subcontractor in your estimate for a certain area of work, then you are ethically obligated to use that company. If in fact bids for materials or subcontracts come in after the bid date, simply tell those companies that you appreciate their interest but that you can only accept bids prior to the bid date and time; furthermore, that you would gladly accept their bids on future projects. Those general contractors that conduct business through this bid shopping technique soon get a negative reputation throughout the community of material suppliers and subcontractors and they do not get the best possible quotes from these companies, if they get the quotes at all.

PROJECT COST CONTROL

The project cost control sheet will be the contractors method of tracking the actual costs spent on the project against those costs that were previously estimated. The cost control sheet should be broken down into the operations of work similar to how it was broken down in the estimate. Also, the operations should be categorized similar to the estimate, such as:

1.) Company owned equipment and outside rented equipment costs,

2.) Small tools and miscellaneous (S.T. & M.),

3.) Direct labor costs and labor accruals (Taxes/Insurance/Fringes),

4.) Permanent materials and

5.) Subcontractors

Another item that is useful to track on a cost control sheet is actual manhours used versus those previously estimated. Manhours are even more useful to analyze if the work you are constructing is labor intensive. This means that a large percentage of the base cost of your project is in labor related costs. Most public works and structural type projects are labor intensive and therefore your manhour report should be carefully analyzed on these jobs as an indicator of how the work is progressing. Typically excavation and grading projects are equipment intensive as compared to labor intensive. This means that the equipment costs are the major expense of the project that could have the most effect as to the outcome of the job financially. In the case of an equipment intensive project the cost per unit of work ($/cubic yard for example) might be the information that needs to be studied. This type of unit cost analysis is also useful in commercial construction.

Figure 10-5 is what the beginning cost control sheet might look like from our previous example of a concrete retaining wall, before any work is complete. As you can see from this example, there are two lines for costs on every operation. The first, or top line for each operation is the estimated costs. The second, or bottom line is where you would enter your actual costs. The same concept is used in the manhour report. By entering the current project information in this manner you can analyze how the job is progressing with respect to each operation. Most construction companies who use this style of cost control will update and print the reports weekly. The reports should be analyzed by the managers, superintendents or engineers of the company and discussed with the work force if any adjustments need to be made.

FIGURE 10-5   Cost Control

XYZ CONSTRUCTION, INC.
PROJECT COST CONTROL
DATE: June 11, 1997

COST CODE	DESCRIPTION	QUANTITY	UNITS	EQUIPMENT	S.T. & M.	LABOR	MATERIALS	SUBCONTR	TOTAL	THIS WEEK			TO DATE		
										M.H.	QUANT.	RATE	M.H.	QUANT.	RATE
100	MOBILIZATION														
0.01	Mobilization	1	L.S.	$800.00	$2,000.00	$625.00	$0.00	$0.00	$3,425.00	25	1	0	25	1	0
			L.S.	$0.00	$0.00	$0.00	$0.00	$0.00	$0.00	0	0	0	0	0	0
0.02	Demobilization	1	L.S.	$500.00	$250.00	$625.00	$0.00	$0.00	$1,375.00	25	1	0	25	1	0
			L.S.	$0.00	$0.00	$0.00	$0.00	$0.00	$0.00	0	0	0	0	0	0
200	RETAINING WALL														
0.01	Fabr. Forms	2500	S.F.	$200.00	$5,000.00	$2,500.00	$0.00	$0.00	$7,700.00	100	2500	25	100	2500	25
			S.F.	$0.00	$0.00	$0.00	$0.00	$0.00	$0.00	0	0	0	0	0	0
0.02	Form Footing/Wall	4500	S.F.	$1,980.00	$2,250.00	$7,620.00	$0.00	$0.00	$11,850.00	300	4500	15	300	4500	15
			S.F.	$0.00	$0.00	$0.00	$0.00	$0.00	$0.00	0	0	0	0	0	0
0.03	Place Concrete	121	C.Y.	$100.00	$150.00	$1,250.00	$7,650.00	$0.00	$9,150.00	50	121	2.42	50	121	2.42
			C.Y.	$0.00	$0.00	$0.00	$0.00	$0.00	$0.00	0	0	0	0	0	0
0.04	Wet Finish/Cure	1250	S.F.	$50.00	$125.00	$625.00	$0.00	$0.00	$800.00	25	1250	50	25	1250	50
			S.F.	$0.00	$0.00	$0.00	$0.00	$0.00	$0.00	0	0	0	0	0	0
0.05	Strip Forms	4500	S.F.	$300.00	$400.00	$4,500.00	$0.00	$0.00	$5,200.00	180	4500	25	180	4500	25
			S.F.	$0.00	$0.00	$0.00	$0.00	$0.00	$0.00	0	0	0	0	0	0
0.06	Dry Finish	4000	S.F.	$50.00	$125.00	$1,000.00	$0.00	$0.00	$1,175.00	40	4000	100	40	4000	100
			S.F.	$0.00	$0.00	$0.00	$0.00	$0.00	$0.00	0	0	0	0	0	0
300	SUBCONTRACTS														
0.01	Reinforcing Steel	7500	LBS.	$62.50	$75.00	$117.50	$1,875.00	$1,500.00	$3,630.00	5	7500	1500	5	7500	1500
			LBS.	$0.00	$0.00	$0.00	$0.00	$0.00	$0.00	0	0	0	0	0	0

There are several computer programs on the market today that tie a companies accounting procedures to the cost control of their projects. This type of system can save an immense amount of time by only entering the cost information into the computer one time. As your company begins to grow, one of these programs may be a good investment. Since these programs can vary with respect to the type of company they are designed for (heavy/highway, commercial, public works, etc.) some research needs to be done before the investment is made.

A projects cost control sheet can be a valuable resource for the contractor to track the progress of each job if the information is entered timely and accurately. From these cost controls and the below discussed project schedule the contractor can develop cost projections on what it will take to complete the remaining operations of the project. From these projections, the contractor can analyze how the project might come out financially if it were to be completed at the current production rates. Your insurance and bonding company will probably be requesting these projections of your jobs from time to time, so it is a good idea to be as accurate as possible when developing them.

Figure 10-6 shows what the projection of our concrete retaining wall would look like in relation for our cost control.

FIGURE 10-6   Retaining Wall Cost Projection

XYZ CONSTRUCTION, INC.
PROJECT COST PROJECTION
DATE: August 5, 1997

COST CODE	DESCRIPTION	QUANTITY	UNIT	EQUIPMENT	S.T. & M.	LABOR	MATERIALS	SUBCONTR	TOTAL	TO DATE M.H.	TOTAL	% COMP	M.H	PROJECTION TOTAL	DIFF.
100	MOBILIZATION														
0.01	Mobilization	1	L.S.	$800.00	$2,000.00	$625.00	$0.00	$0.00	$3,425.00	25	$3,425		25	$3,425	
		1	L.S.	$700.00	$2,500.00	$550.00	$0.00	$0.00	$3,750.00	21	$3,750	100.00%	21	$3,750	($325)
0.02	Demobilization	1	L.S.	$500.00	$250.00	$625.00	$0.00	$0.00	$1,375.00	25	$1,375		25	$1,375	
		0	L.S.	$0.00	$0.00	$0.00	$0.00	$0.00	$0.00	0	$0	0.00%	0	$1,375	$0
200	RETAINING WALL														
0.01	Fabr. Forms	2500	S.F.	$200.00	$5,000.00	$2,500.00	$0.00	$0.00	$7,700.00	100	$7,700		100	$7,700	
		1500	S.F.	$100.00	$3,750.00	$1,350.00	$0.00	$0.00	$5,200.00	55	$5,200	60.00%	91.67	$8,667	($967)
0.02	Form Footing/Wall	4500	S.F.	$1,980.00	$2,250.00	$7,620.00	$0.00	$0.00	$11,850.00	300	$11,850		300	$11,850	
		2000	S.F.	$750.00	$775.00	$3,000.00	$0.00	$0.00	$4,525.00	125	$4,525	44.40%	281.3	$10,182	$1,668
0.03	Place Concrete	121	C.Y.	$100.00	$150.00	$1,250.00	$7,650.00	$0.00	$9,150.00	50	$9,150		50	$9,150	
		55	C.Y.	$35.00	$75.00	$550.00	$3,200.00	$0.00	$3,860.00	17	$3,860	45.45%	37.4	$8,493	$657
0.04	Wet Finish/Cure	1250	S.F.	$50.00	$125.00	$625.00	$0.00	$0.00	$800.00	25	$800		25	$800	
		525	S.F.	$30.00	$100.00	$300.00	$0.00	$0.00	$430.00	10	$430	42.00%	23.8	$1,024	($224)
0.05	Strip Forms	4500	S.F.	$300.00	$400.00	$4,500.00	$0.00	$0.00	$5,200.00	180	$5,200		180	$5,200	
		1500	S.F.	$100.00	$125.00	$1,600.00	$0.00	$0.00	$1,825.00	61	$1,825	33.33%	183	$5,476	($276)
0.06	Dry Finish	4000	S.F.	$50.00	$125.00	$1,000.00	$0.00	$0.00	$1,175.00	40	$1,175		40	$1,175	
		1000	S.F.	$25.00	$50.00	$200.00	$0.00	$0.00	$275.00	8	$275	25.00%	32	$1,100	$175
300	SUBCONTRACTS														
0.01	Reinforcing Steel	7500	LBS.	$62.50	$75.00	$117.50	$1,875.00	$1,500.00	$3,630.00	5	$3,630		5	$3,630	
		4250	LBS.	$50.00	$25.00	$55.00	$1,000.00	$775.00	$1,855.00	2	$1,855	56.67%	3.5	$3,273	$357

PROJECT SCHEDULE

The project schedule should be similar to one you created during the estimating process. Now that you have the contract, a project schedule that is much more detailed should be developed along with entering actual dates in the timeline as opposed to generic weeks.

Project items should be broken down into their basic operations and have a duration attached to them that reflects all of the adjustments and new construction schemes that your company has developed since the estimate. If your company has access to computer software scheduling programs, the schedule should have even more detail.

Remember that this is the outline that your field personnel will use to construct the project as it was envisioned in the estimating process. Just as with the project cost control sheet, much consideration and accuracy should be used in developing these tools.

SUBMITTALS

Typically the contracting agency will list the exact kinds of materials to be used on the project in the "Special Provisions" of the Contract Specifications. For example the concrete for a retaining wall might be specified as the following: Concrete - Minimum 28-day compressive strength shall be 3300 PSI. Aggregate used in this mix design shall be 1" minus or 3/4" minus. Cement for this mix shall be Type II Modified.

Or exterior lumber might be specified as the following: Lumber - Incise and pressure treat all lumber with pentachorophenal (Type C Solvent). Posts shall be treated to a retention level of 0.60 and all other lumber to a retention level of 0.40. Lumber shall be Douglas Fir or Hem-Fir conforming to WCLIB Standard Grading Rules. Galvanize all hardware and nails.

The Contract Specifications will also tell the contractor how he is to submit paperwork to the owner that certifies that the material being used conforms to the listed requirements. This paperwork supplied to the contracting agency is referred to as the projects submittals.

Submittals may be required for materials supplied to the project and also on procedures and designs. For example the contracting agency might require submittal drawings for temporary bridge falsework, reinforcing steel layout or a traffic control/lane closure plan. All of these submittals should be given to the owner with enough time for the owner to review their content and return to you an approval before the applicable construction operation or material ordering begins. If the submittal information does not conform to the specifications or if there are mistakes within the drawings, the contracting agency may return your submittal for corrections and re-submission. For this reason it is always a good idea to submit your information well before the material or drawing is to be used.

The contractor collects most of the information for submittals from material suppliers or subcontractors. For example, if your company is planning to purchase the above mentioned concrete or exterior lumber, you should contact the supplier that was the low bidder from your estimates quote comparison sheets. This should have been already done when you drafted the suppliers purchase agreement and a provision should have been included in the agreement that the supplier will provide your company with submittal information for approval by the owner. It is necessary to communicate with your suppliers and subcontractors to make sure they realize when you need the documentation submitted to your office. This communication and organization prior to the start-up of the project can save your company many delay costs and confusion once construction of the project is underway.

A contractor should keep a submittal log that details everything that has transpired with respect to any particular submittal. When a large project is undertaken, there can be literally hundreds of submittals. Typically, commercial building projects will have the most submittals associated with them, while most public works construction projects have a controllable number of submittals, unless the project is very large or complex. Therefore, a contractor needs to be very organized in his control over submittals so that he may know the status on any of them when he needs to. An un-organized submittal process can lead to expensive delays on your projects. Figure 10-7 is what your submittal log might look like.

FIGURE 10-7   Submittal Log

XYZ CONSTRUCTION, INC.
SUBMITTAL LOG
Project:

SUBMITTAL NUMBER	DESCRIPTION	DATE RQRD	DATE SUBM.	DATE RCVD.	ACCEPTED REJECTED	DATE RE-SUBM.	DATE RCVD.	ACCEPTED REJECTED

PROJECT DOCUMENTATION

Organizing the paperwork flow of a project prior to the start of actual construction is very important. Once construction has begun, the contractor has many operations happening at one time and needs to be organized in order to keep control of the project. Documentation developed by a construction project may be in the range of thousands of pages depending on the size of the job; therefore, one can see that having the ability to find a particular document when it is needed can be important.

One of the first organization items upon receiving a contract is to give the project a job number that identifies it from your companies other projects. One of the easiest and most efficient ways to do this is to use the year the contract was received along the corresponding number of that job to the total number of contracts your company has received that year. For example, your first contract of 2001 would be numbered 01-01 and your third contract of 2002 would be numbered 02-03.

Project documentation for each of your individual contracts may be organized under the following categories:

ORGINAL CONTRACT - this file should contain a copy of the contract between your company and the owner. If your company is a subcontractor, the file should contain a copy of the subcontract with the prime contractor. It may be helpful to include a copy of the estimate proposal that you submitted to the prime contractor at bid time. This file should also contain copies of any certificates of insurance or bonds that have been submitted to the owner or prime contractor.

CHANGE ORDERS - this file should contain a copy of all the proposed and agreed upon change orders to the original contract. (This area of contracting will be covered more extensively in the following chapter.)

LETTERS TO THE OWNER (or Prime Contractor) - this file should contain a copy of all the letters that you send to the owner or prime contractor. It is also a good idea to sequentially number the letters for easier reference at a later date.

LETTERS FROM THE OWNER (or Prime Contractor) - this file should contain a copy of all the letters received from the owner or prime contractor. This file and the "Letters To" file are very valuable in the case of a claim later in the project. These files should contain complete documentation of all correspondence between your company and the owner or prime contractor.

PAY ESTIMATES - this file should contain a copy of all the payment requests that your company has submitted to the owner or prime contractor. At the pre-construction meeting the contracting agency will set a specific date each month in which progress pay estimates will be accepted. At this time each month the contractor and a representative of the owner will mutually agree upon quantities completed the previous month and a progress payment will be initiated. Time frames for payment should have been documented and agreed to in the original contract. Typically it will take approximately one to four weeks to receive a check from the owner. If your company is a subcontractor it could take an additional one to three weeks to get your check from the prime contractor. Once payment has been received it is advisable to attach a copy of the check to the corresponding payment request.

SUBMITTALS - this file should contain a copy of all of the submittal documentation that your company has sent the owner. The response from the owner should be attached to the original submittal. It is also a good idea to keep a separate record sheet on the status of all of your submittals that includes the date submitted, date returned, whether the submittal was stamped accepted/rejected/resubmit and the corresponding resubmission dates (if applicable) as discussed previously.

LABOR AGREEMENTS - this file should contain a copy of any labor agreements your company might have on the project. If you are operating the project without the assistance of labor organizations, then keep wage rate information in this file. This is also a good place to keep any termination slips from fired employees or any other wage/fringe benefit information.

DAILY LOGS - this file should contain a copy of your superintendents and foreman's daily logs. These daily logs are very important when you are trying to remember what happened on a particular day last week, last month or last year. These logs should contain the names of any visitors to the jobsite, a brief description of the work completed that day, any material deliveries, if subcontractors were on the jobsite, and any verbal agreements made with the owner or subcontractors or any unusual happenings at the project site. These logs can be critical in claim situations with the owner, subcontractors or material suppliers. Figure 10-8 is a sample of a jobsite daily log.

FIGURE 10-8   Daily Log Sample

DAILY CONSTRUCTION REPORT

JCB			LOCATION	
WEATHER	TEMPERATURE 8 A.M. 1 P.M.		DAY	DATE

MATERIALS RECEIVED

COMMENTS

PURCHASE ORDERS - this file should contain a copy of each purchase order written on the jobsite. This file is for the small purchases not covered by a purchase agreement as previously discussed in this chapter. Purchase agreements would be in file no. 14, Materials.

SCHEDULES - this file should contain a copy of the project schedule along with any updates that have been developed. Along with the complete project schedule it is a good idea to have your jobsite personnel develop a two week schedule, made on a weekly basis, that goes into much more detail than the overall project schedule. This will make the project supervisors think of all the small details that will come up in the next two weeks, which if overlooked, would cause delays to the project. Below is what a two-week schedule might look like using our previous example of a concrete retaining wall.

FIGURE 10-9   Two-Week Schedule

XYZ CONSTRUCTION, INC.
Two Week Schedule
Concrete Retaining Wall

ACTIVITY	MONDAY	TUESDAY	WEDNESDAY	THURSDAY	FRIDAY	MONDAY	TUESDAY	WEDNESDAY	THURSDAY	FRIDAY
Set up fenced yard	XXXXXXXX									
Mobilize office trailer		XXXX								
Deliver portable toilet		XXXX								
Deliver forming material		XXXXX								
Fabricate footing forms			XXXXXXXX							
Fine grade footing area				XXXXX						
Form Footings				XXXX	XXXXX					
Fabricate wall forms					XXXX	XXXXXXXXX				
Deliver reinforcing steel					XXXX	XXXX				
Install rebar in footings						XXXX				
Order concrete for footings					XXXX		XXXX			
Place concrete in footings							XXXX	XXXXX		
Strip footing forms								XXXXX		
Form Walls								XXXX	XXXXXXXX	XXXXX
Install rebar in wall									XXXXX	XXXX

SAFETY - this file should contain a copy of every jobsite safety meeting report. This report should include the names and signatures of all personnel who attend the meeting. Subcontractors and representatives from the contracting agency should be required to attend these meetings. Typically these safety meetings should occur once a week, unless the project is going to begin a new operation that needs to be discussed with the field personnel prior to the next weeks meeting. The superintendent or foreman who is handling the meeting should have a particular topic to discuss each week and ask for input from those in attendance. These meetings are important to the welfare of your companies personnel, owners representatives, subcontractors and the general public around the jobsite and they will only require approximately 15 minutes a week. Figure 10-10 is a sample of a jobsite safety meeting.

FIGURE 10-10   Jobsite Safety Meeting Sample

MINUTES OF CREW SAFETY MEETING

ABC Asphalt Paving Company

Date: _____

Project:

_____

Old Business:

_____

_____

_____

_____

_____

Subjects of this Meeting:

_____

_____

_____

_____

Suggestions Made:

_____

_____

_____

_____

_____

Attendees Signature   _____   _____

                        _____   _____

                        _____   _____

                        _____   _____

                        _____   _____

Foreman/Superintendent: _____

_____

COST CONTROL - this file should contain a copy of the original and weekly updated cost control as discussed previously in this chapter.

TIME CARDS - this file should contain a copy of the daily time cards filled out by each foreman and signed by all applicable employee. There should be a spot for each individual to check if they were injured that day. This check mark could be important in the event of a workers compensation claim at a later date. Also, these are the time cards that will be used to determine payroll, along with charging the crews individual time to specific cost codes for incorporation into the projects cost control sheet. A typical time card might look like the example in Figure 10-11.

FIGURE 10-11   Time Card Sample

DAILY TIME SHEET AND LABOR DISTRIBUTION REPORT

JOB NO.		DATE	
NAME		REPORT NO.	
LOCATION		SHEET NO.	

COMMENTS: LABOR CLASSIFICATION   WEATHER

TEMPERATURE
8 A.M.
1 P.M.

| OCCU-PATION | EMPLOYEE'S NAME | EMPLOY NUMBER | | HOURS | RATE | AMOUNT |

TOTALS

C E-Z CONTRACTORS FORMS FORM NO EZ 100

MATERIALS - this file should contain a copy of all the material purchase agreements for the project as discussed previously in this chapter. Also, to be included with each supplier is a copy of any correspondence with that supplier.

SUBCONTRACTORS - this file should contain a copy of all subcontracts on the project. Again, all correspondence with that subcontractor should be included in this file.

RFC's - this file should contain copies of your "requests for clarifications". Most contracting agencies will provide the contractor with a form that they like to use for RFC's. When questions regarding the plans or specifications arise, the contractor may complete one of these forms for submission to the contracting agency. This is a good way to document in writing any confusion with respect to the contract plans and specifications. If a number of RFC's are anticipated, many contractors will set up a log similar to that of the submittals.

Your company's particular project may call for more or less files than those listed above. The best way to ensure your field supervisor and home office manager has access to the same information is to have an identical filing system and documentation at both locations. A good way to keep identical information at both sites is to make a copy of all documents and distribute them to the other office on a weekly basis. This dual system may not be necessary if the project is very small or the project is a short distance away from the company's home office.

Your company should also include a cover "Letter of Transmittal" with each document sent to your suppliers, subcontractors and the contracting agency. This document will greatly assist you in making sure the information you are sending is detailed in an organized manner that can be easily tracked by anyone within your organization and the organization to whom you are sending the information.

FIGURE 10-12 Letter of Transmittal

ABC ASPHALT PAVING COMPANY
1234 E. Delta Avenue
Gresham, Oregon 97777

LETTER OF TRANSMITTAL

TO:		Date:	Job.No.
		Attention:	
		RE:	

WE ARE SENDING YOU THE FOLLOWING ITEMS:

___Attached	___Under separate cover
___Shop Drawings	___Drawings ___Specifications ___Calculations
___Copy of Letter	___Change Order ___Addenda ___ _____

Copies	Date	No.	Description

THESE ARE TRANSMITTED AS CHECKED BELOW:

___ For Approval	___ Approved as Submitted	___ For Review and Comment
___ For Your Use	___ Approved as Noted	___ For Bids Due _____
___ As Requested	___ Return for Correction	

REMARKS:

_____

_____

_____

COPIED TO: _____

SIGNED: _____

Your company will have to develop its own style and size of documentation system for the projects you construct. The above system should give you a representation of the kind of information that needs to be tracked in order to keep a well-organized project and company.

CHAPTER ELEVEN

MOBILIZING TO THE JOBSITE

At this time all of the project paperwork should be in place or very near completion. Now it is time to mobilize to the jobsite. Depending on the distance from your home office to the project, this operation can vary in complexity. Regardless of the projects distance from your home office, mobilization needs to be given much consideration or the cost and duration of this operation can quickly get out of hand.

This chapter will deal with the following mobilization operations:

Setting up the Construction Yard,

Mobilizing Equipment & Materials,

Mobilizing the Contractors Personnel,

Utility Locates.

SETTING UP THE CONSTRUCTION YARD

The mobilization operation of setting up the construction yard depends largely on two variables: project type and size.

First we will discuss the type of construction project you are building. If your job is to construct a commercial building, consideration will have to be given to site layout. You will need to develop a plan view drawing of the overall project site. Allow a buffer area for normal construction activity around the actual work area. Space will need to be designated for an office trailer along with contractors personnel and visitor parking. Space will need to be allotted for material storage along with adequate access for delivery trucks and unloading. Typically, temporary power and telephone lines will have to be installed; therefore, these activities will need to be given consideration in your construction yard layout. An example of a plan for a construction yard layout is detailed in Figure 11-1.

FIGURE 11-1   Construction Yard Layout

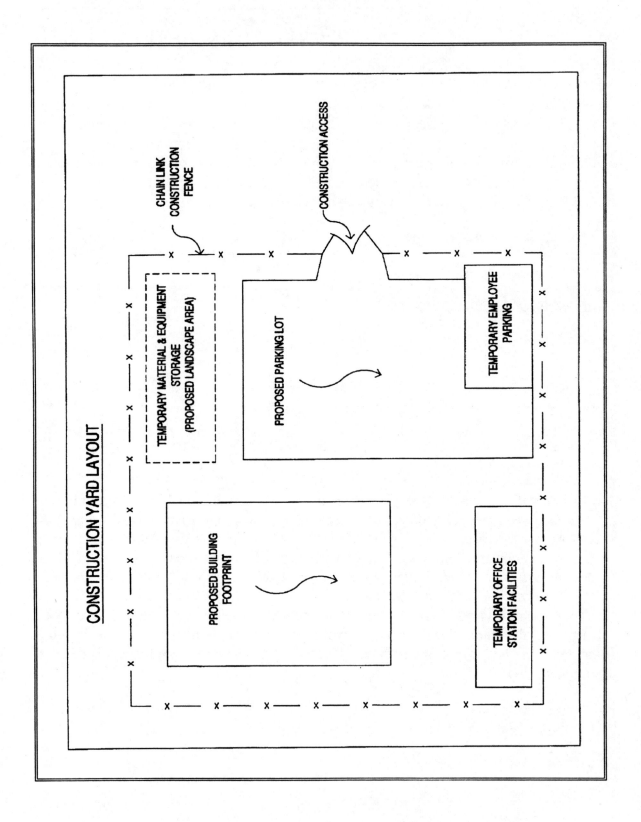

As opposed to a commercial building, if your job is to construct a portion of roadway that is three miles long, consideration will have to be given regarding where to set up the office and construction yard. The contractor will have to purchase or rent a portion of land in order to set up the yard, if this space is not provided by the contracting agency or within the limits of the project itself. Often the contractor does not have a choice with respect to the yard site due to land availability in that particular area. Many times the land that is available for your construction yard is not adjacent to the project. Items that need to be considered when choosing a construction yard for this type of work are:

1.) Where is the equipment going to be parked during non-working hours,

2.) Where is the equipment maintenance going to take place,

3.) Does the office have adequate access to the jobsite,

4.) Can telephone and power lines be installed or do you have to resort to cellular phones and generators for power,

5.) Does the yard site have adequate parking and storage space,

6.) How will the office and materials be secured during non-working hours?

Secondly, the size and duration of the project your company is constructing will have an affect on the construction yard site. If the job is large, all of the above items (along with others) will have to be accounted for. On the other hand, if the project is small and short in duration, the jobsite supervisor might work out of his vehicle instead of setting up an office. The supervisor could use a cellular telephone instead of trying to bring in an overhead or underground telephone line to the field office. On short duration projects only daily equipment maintenance will take place; therefore, there is no need for a repair/maintenance yard. Also on these small projects, mailbox retailers that are available in most cities can handle mail and small deliveries; this is only necessary if the project site is not within an easy driving distance from your home office.

As one can see, the setting up of the construction yard site needs to be a well thought out operation. If the yard space at a commercial site is not laid out properly, the contractor can experience many additional costs associated with having to make extra moves in relocating equipment and materials along with lost production. In the scenario of a small road construction project, the contractor can spend a great deal of time and money setting up an entire construction yard site when it might be more efficient to work out of a pick-up for an office. Conversely, lost production and confusion can be developed if too large or complex of a project is attempted to be managed out of a pick-up by your jobsite supervisors and major equipment maintenance is attempted along side of the road in the dark, non-working hours. These kinds of situations are just asking for trouble when a little preparation would avoid many of these problems.

MOBILIZING EQUIPMENT & MATERIALS

The obvious question that needs to be addressed before mobilizing equipment and materials to the jobsite is; "what equipment and materials does the project require?" Once that question is answered (from your estimates quote comparison sheets and construction schemes), the contractor needs to decipher when the project needs a certain piece of equipment or particular materials. This answer can be found on your previously prepared construction schedule. If your company is large enough to employ an equipment manager, this person can work out schedules between your company's individual jobs to make sure each project has the equipment they need and when they need it. This scheduling can involve both company owned and outside rented equipment. Time and thought needs to be given to the choice of buying or renting equipment. Many contractors like the low risk associated with simply renting most of their equipment, yet if your company has the ability to maintain the equipment, we believe you will be better served owning the equipment that you need on a regular basis. Owning will also be beneficial when you need a piece of equipment for sporadic, short durations and don't want to pay the mobilization and short term rental fees every time you need that particular piece of equipment. When making the decision to purchase a piece of equipment consider the following items and compare them to the costs associated with renting that piece of equipment:

1.) Amortize the purchase price considering the salvage value.

2.) Costs associated with insurance.

3.) Equipment repair and maintenance costs. When renting a piece of equipment these costs are borne by the rental agency.

4.) Any cost savings of having the equipment available to your company at all times.

5.) Working capital. Your bonding company will look at the owning of equipment in a very different way than most construction managers.

The bonding company will take 12 months worth of the principal you are paying on equipment purchases off their calculated total of your working capital. If your working capital decreases, then your bonding capacity decreases at a comparative rate. Because of this situation, the construction manager needs to give much consideration to how equipment purchases effect the bonding company's view of your working capital.

Similar to the mobilization of equipment, temporary materials that are owned by the contractor (such as forming/shoring lumber or temporary culverts) may need to be scheduled between projects. This can also be handled by your equipment manager or whomever the person is organizing the mobilization for your company.

At this time, thought needs to be given on how to make the most efficient use of trucking time for hauling the equipment and materials to the jobsite. Items that need to be considered include:

1.) Weight of the equipment or material,

2.) Size or volume of the equipment or material,

3.) Jobsite unloading capabilities,

4.) Capabilities of your company owned trucks as compared to those available from commercial trucking companies.

With some calculated scheming, the contractor can make efficient use of the resources he has available to mobilize equipment and materials to the jobsite for an economical cost.

MOBILIZING THE CONTRACTOR'S PERSONNEL

Mobilizing of the contractor's personnel to the jobsite depends mainly on one variable: project duration. This needs to be considered only if the project is more than a few hours away from the contractor's home office. In most cases, contractors will only consider moving their supervisors, engineers, and foremen along with any critical craft employees. Obviously the size of the project you are going to construct will dictate how many and which of these employees you will need at the jobsite. Most temporary or general craft positions will be filled from the workforce in the surrounding area, if the project is a distance away from your home office or area.

For a short term project (usually six months or under in duration) the contractor typically will not pay to move the employee and his family to the jobsite location. Instead the employee would be expected to find their own temporary living site, while the contractor would pay the employee subsistence. Subsistence is living allowance paid to the employee outside of their regular payroll. Subsistence amounts may vary with conditions such as the employees' position and how long they have been employed with the company, but reimbursement would typically range in today's dollars from $125 -$250 per week.

For long-term projects (typically over six months duration) the contractor will generally pay to have the employee and his family moved to the project area. This cost can be substantial if the employee has a family, household furnishings, etc.; regardless, it may be less expensive and disruptive to the employees life as compared to paying weekly subsistence and having the employee live away from their family.

All of the costs associated with mobilizing to a jobsite should be considered in the projects estimate under the indirect cost category. Once a contractor has successfully completed a few projects and by following these guidelines, the costs of setting up a construction yard and mobilizing equipment, materials and personnel to the jobsite can be closely estimated.

UTILITY LOCATES

When mobilizing to a new project site, and before actual work has started, it is necessary to call to have the existing utilities located. Most cities have a one-source notification center that the contractors can call to have utilities located within their work area. The locators will generally paint layout lines of the existing utilities in a designated work area within approximately three working days. The locators will generally specify that the location of utilities will be good for ten days. It is in most areas the responsibility of the contractor to keep these locates identifiable to your working personnel. If your company is working on a pipeline or a road, for example, that progresses linearly, break the overall work into operations that will last 7 to 10 days in duration. As your crews are about to complete one operation, call for utility locates for the next section. This scheme will keep utility locates fresh for your crews and greatly insure your companies liability against the cost or danger of damaging an existing utility.

The utility notification center is a free service to contractors that should be actively utilized by anyone who will be working near underground or overhead utilities. The notification center will contact and locate utilities such as water, natural gas, electrical, telephone and cable television. They will also tell the contractor which of the utility organizations they will contact. If there are any other utilities in the area that the contractor is aware of, then he will have to contact that utility organization himself. Although, the notification centers will typically have access to all of the utilities in the area.

This program is beneficial to everyone involved. By locating the utilities before work has begun, the contractor can perform his work in a safer manner, the utility company can avoid costly repairs, and the consumer can experience a reduced loss in any particular utility caused by damage from construction operations.

CHAPTER TWELVE

PRE-CONSTRUCTION MEETING/PARTNERING

At this time in a project, a pre-construction meeting with the owner or contracting agency will generally take place. The contractor and contracting agency will mutually agree on a time and place to meet and discuss how the work is to progress.

The contracting agency will generally lead the meeting. They will, in most cases, have prepared an outlined format for the direction of the meeting. This outline will cover all of the highlights of the construction contract that the owner wants to make sure the contractor is aware of and understands. The contractor should go to this meeting prepared with construction schedules, material submittals and any other drawings or plans that the company can get ready by the time of the meeting. The contractor should also give the contracting agency a list of names including the jobsite foreman, superintendent or managers along with what authorization each individual has as it relates to change orders, pay estimates or general jobsite decision-making. The contracting agency will generally require a list of after construction hour phone numbers from the contractor in case there is an emergency at the site in which the contracting agency has to reach a contractors representative. This preparedness, early in the project by the contractor, will show the contracting agency that your company knows exactly what they are doing and may ease them from "over inspecting" a project from the beginning. The contractor should also enter this meeting with a "partnering" attitude.

Partnering is a relatively new buzzword in the construction industry. While the buzzword is somewhat new, the attitude is not. Partnering simply means that everybody involved in a project (owner, contractor, subcontractor, architect, engineer, etc.) will ethically try to solve all disputes at the lowest authority level possible. Everyone involved should try to negotiate open-mindedly to resolve all conflicts since; in fact, they are all partners in constructing the project. The goal is to save the funding party, whether it is a governmental agency or a private investor, claim dollars and delay time associated with conflicts with the contractor. On the other hand, it is also a goal to keep the contractor profitable while providing the most economical project for the contracting agency. If the reputable contractors cannot stay profitable, then they will not be around long to bid on the future projects of that particular agency. Many agencies and contractors co-sponsor a partnering seminar, directed by an independent party that can coincide with the pre-construction meeting. Typically, everyone involved who has the authority to make decisions regarding the project, will sign a document assuring that they will approach the project with a partnering attitude and do everything within their power to keep the unnecessary paperwork, change orders and delays to an absolute minimum; therefore, providing the most economically managed project that can be realized.

Now that your company has mobilized its construction personnel, equipment and initial materials to the jobsite and has had its pre-construction meeting with the owner, physical work on the contract items can begin. During this actual construction period, the companies general or project manager can oversee the progression of work in many different ways. Some of the items that we suggest the manager take a look at during construction of a project are discussed in this chapter as follows:

Information Transfer From the Estimator to the Jobsite,

Ordering Materials and Scheduling Subcontractors,

Putting Out the Daily Fires,

Safety/Accidents,

Monthly Progress Payments,

Certified Payroll of Prevailing Wage Rates,

Accounting for Project Purchases,

Company Checking and Savings Accounts,

Jobsite Visits and Project Review Meetings.

INFORMATION TRANSFER FROM THE ESTIMATOR TO THE JOBSITE

The construction project has by now evolved from a conglomerate of ideas, speculations and calculations in the estimators (or estimating teams) mind and now needs to be transferred into reality at the jobsite. Obviously, the person who knows all of the details of the project along with the anticipated construction techniques is the person, or persons, who estimated the project.

An appropriate amount of time needs to be spent between this estimator and the person who is going to be in charge of construction at the jobsite. The estimator needs to transfer all the information he can on how he envisioned the construction of all of the projects operations. The estimator should pass on information on specific materials, equipment and manpower usage, particular construction techniques along with all of the other aspects of the estimate that the jobsite supervisor needs to know to efficiently construct the project. One can see that if the project is very large at all, that this information transfer is not going to happen in one meeting. Efficient communication along with accurate pre-construction documentation such as the cost control, schedule, submittals, etc. (as previously discussed) will assist the jobsite supervisor in building the project as envisioned during the estimating process.

ORDERING MATERIALS AND SCHEDULING SUBCONTRACTORS

For your company's jobsite supervisors, trying to get a project built from an on-site field perspective can be a hectic daily experience. Anything that the general or project manager can do to assist the field supervisor will greatly enhance the chances for a successful project.

Through proper communication with the on-site supervisor, the manager can assist the project by ordering and scheduling major materials that are to be delivered to the project. The manager can also assist in the scheduling of subcontractors onto to the jobsite when it is time for them to start their work. This ordering and scheduling of materials is often a situation that requires the manager and jobsite supervisors to have many telephone conversations in order to ensure the correct materials are delivered to the project at the proper time. This phone-tag also applies to getting subcontractors on the project at the proper time. For the manager in the office, this relay of phone conversations can be handled much easier, as compared to your supervisor in the field. Typically, the jobsite supervisor has too many field operations happening simultaneously, on a daily basis, to efficiently manage all of the material suppliers and subcontractors that will be incorporated into the project in the coming weeks. In other words, the manager typically has a broader view of the project since he is viewing it from a distance while the project supervisor will have a much more detailed perspective. In order for the manager or superintendent to be efficient in this ordering and delivering of materials and subcontractors to the project site, they need to thinking of the operations that are in the future, not the operations that are occurring right now. If the supervisor waits to think of an item that he needs when the operations has started, then a delay or re-organization of the operation will occur that costs the company time and money.

Proper communication must be upheld between the manager and the jobsite supervisor. Both must transfer information between each other for this management operation to be handled in the most efficient manner. It is also best to have only one person from your company handle this ordering and scheduling with respect to any particular supplier or subcontractor on a project. Many orders and deliveries have been bungled due to more than one person within a company trying to talk to the supplier or subcontractor, with neither person knowing what the other said. This lack of communication situation can lead to nothing but project delays and a damaged relationship with your suppliers and subcontractors.

SAFETY/ACCIDENTS

Safety is a very important aspect of the dangerous public works construction industry. Having your crew adequately trained with respect to safety can be vital to the life of any of the company's personnel. All of the labor organizations, whether it is the craft unions or the AGC, have safety and CPR classes in which the contractor may enroll his people. After successfully completing these classes the attendees will receive a wallet card to prove that that class has certified them. These cards are generally valid for one or two years, depending on the class. The company should pay for the classes since it is imperative that the contractor make sure his employees are trained with respect to safety and that they have valid safety cards.

No matter how safe the contractor's employees are on the job or how well planned the jobsite is by the project managers, accidents are unfortunately going to occur in this industry. All of the pre-job planning will severely reduce the amount of accidents and assist everyone in responding when an accident does occur on the project. One of the administrative responses that is very important to the public works construction company is the accident investigation report (Figure 12-1). The project supervisor must complete this form with absolute accuracy and timeliness to protect the liability of the company in the event of a fraudulent lawsuit against the company by anyone involved in the accident.

FIGURE 12-1   Accident Investigation Report Form

ACCIDENT INVESTIGATION REPORT FORM

Date Incident Reported: _____   Report No.: _____

WHO was involved or injured?

Name: _____   Witness (1): _____

Address: _____   Phone No.: _____

_____   Witness (2): _____

Home Phone: _____   Phone No.: _____

Length of employment: _____

WHEN did the accident occur?

Date: _____   Exact time: _____

WHERE did the accident occur?

Address: _____   Detailed Location: _____

_____   _____

WHAT happened?

_____
_____
_____
_____
_____
_____

WHY did the situation develop that led to the accident?

Training?

_____

Supervision?

_____

Maintenance?

_____

_____

ACCIDENT INVESTIGATION REPORT FORM   PAGE 2 OF 2

WHAT were the causes of the accident?

_____
_____
_____
_____
_____
_____

HOW can a similar accident be prevented?

Corrective actions already taken and by whom?

_____
_____
_____
_____
_____
_____
_____

Corrective action to be taken and by whom?

_____
_____
_____
_____
_____
_____

Prepared by: _____ Title: _____ Date: _____

Reviewed by: _____ Title: _____ Date: _____

Attach a sketch showing all measurements applicable to the accident scene. Also attach any photographs or additional pages of the descriptions.

Unfortunately, due to the public exposure and nature of the business, construction companies are often viewed as the party with the deepest pockets in the event of accident litigation. For this reason, it is imperative that the contractor have the necessary paperwork available to the jobsite personnel so they can accurately, and immediately, complete the correct forms if an accident occurs. In the event that the contractor has an employee injured on the job, it is very important to the contractor to get that employee back to work in whatever capacity the employee is capable of performing. The injured employee should return to a temporary light duty assignment as soon as the company can get a medical release from the employees' physician. This effort of getting an injured employee back to work as soon as possible will greatly reduce your workers compensation insurance rates and assist your employees in a speedy recovery. The company manager should definitely consult with their workers compensation insurance carrier with respect to this area of a "return to work" policy.

PUTTING OUT THE DAILY FIRES

Within the construction industry, unanticipated confusion seems to come up continuously. This is mainly due to the fact that it is hard, or impossible, to foresee all of the situations that can arise during the construction of a project or even a particular operation. This coupled with the many channels of communication (architect with owner, architect with contractor, owner with contractor, contractor with material suppliers, contractor with subcontractors, contractor with city or county planning divisions, contractor with labor organizations) tends to create many gaps in which critical information may slip between the cracks during the planning and construction periods.

Because the above-mentioned human inefficiencies are inevitable, the construction manager will often have to negotiate between various parties to keep the construction process moving. By keeping as calm as possible, being honest and trying to see all sides of an issue, the general or project manager can, in most cases, put out these daily fires before they get to the point where the projects schedule or financial outcome is effected. A manager's reputation, within the construction industry, is often molded around how he deals with problems and conflict. An absolute professional approach to these challenging situations will pay dividends for you and your company in the future as your positive reputation spreads throughout the industry. This does not mean that the construction manager should roll over to conflict, but simply pick you battles carefully and negotiate with the facts, in good faith. All construction managers are going to have disagreements with the contracting agencies, but the manner in which you present yourself will distinguish you from others in your field.

One of the best ways to keep these situations within your control is to know the contract and its specifications better than anyone else involved. If you know the details of the contract and its specifications you can immediately answer the questions that arise with respect to them and manipulate them to best solve the situation at hand. Many contractors have done extra, unnecessary work at their cost because the project manager or jobsite supervisor did not study and have a good handle on the contract and its specifications.

The situations that arise within the construction industry can be very challenging and frustrating, but how they are handled distinguishes the difference between a proficient and an inefficient manager or supervisor.

MONTHLY PROGRESS PAYMENTS

Since construction of the project is underway and your company is spending money on equipment, labor, and materials, it is time to begin to request payment from the contracting agency for your work. Virtually all contracting agencies will develop progress payments on a monthly basis. Each individual contracting agency has a specific cut-off date each month through which the progress payment will be made.

Sometimes the contracting agency will develop the monthly progress payment without any input from the contractor. In other cases the contractor will make up his own monthly progress estimate and submit it to the contracting agency for payment. Figure 12-2 is a sample of what a contractor's monthly progress billing might look like.

FIGURE 12-2   Monthly Progress Payment Request - Bid Item

PAYMENT REQUEST #1

PROJECT: Improvements at Whitelane Park   DATE: Sept. 2, 20--
OWNER: Contracting Agency
          P.O. Box 5599
          Gresham, Oregon 97555

Bid Item	Description	Quantity	Unit Price	Total Price
1.)	Mobilization	1 LS	$7,500.00	$7,500.00
2.)	Traffic Control	0.25 LS	$2,500.00	$625.00
3.)	Clearing & Grubbing	2 Acres	$1,250.00	$2,500.00
4.)	Demolition/Remove	1 LS	$2,000.00	$2,000.00
5.)	Roadway Excavation	400 CY	$15.00	$6,000.00
6.)	Structure Excavation	25 CY	$20.00	$500.00
7.)	Aggregate Base	0 CY	$25.00	$0.00
8.)	Asphaltic Concrete	0 Tons	$60.00	$0.00
9.)	12" Concrete Pipe	100 LF	$35.00	$3,500.00
10.)	24" Concrete Pipe	80 LF	$50.00	$4,000.00
11.)	Type 1 Inlet	1 EA	$800.00	$800.00
12.)	48" Dia. Manhole	12 VF	$200.00	$2,400.00
13.)	Concrete Curb	0 LF	$15.00	$0.00
14.)	Concrete Sidewalk	0 SY	$27.00	$0.00
15.)	Temporary Signs	128 SF	$5.00	$640.00
16.)	Permanent Signs	0 SF	$8.00	$0.00
17.)	Chain Link Fence	0 LF	$15.00	$0.00

Total This Period (8-1/8-30)	$30,465.00
Less Retainage	$1,523.25
Less Previous Payments	$0.00
TOTAL AMOUNT DUE	$28,941.75

All of the above quantities, unit prices, change orders and all other information is accurate and complete to the best of my knowledge.

Signed: _____

Title: _____

Date: _____

ABC ASPHALT PAVING COMPANY
1234 E. Delta Ave.
Gresham, Oregon 97777

The final and most efficient way to develop a progress payment is for a representative of the contracting agency and the contractor to mutually agree upon the percentage of work, or quantity per bid item, that was completed that particular month and process that amount, less retainage, for payment. Retainage is the amount the contracting agency holds out of each month's progress payment, until the job is complete. This is held for insurance that the contractor will adhere to all aspects of the projects contract and to make sure that the contractor will totally complete the project before all of the contract money is released. A typical retainage ranges from 5% -10%.

Depending on if the contract is based on bid items or a lump sum payment, the pay estimate may take on slightly different forms. If the project is a bid item job, then the pay estimate will look similar to Figure 12-2, with the quantity of each bid item calculated for the period that the pay estimate represents. Even if the contract is on a lump sum basis, the contracting agency will often require the contractor to break the project into ten or more monetary operations that will be paid on a percentage complete basis. (Figure 12-3) The type of payment request form differs from contracting agency to contracting agency, with many accepting forms similar to Figures 12-2/12-3, while many use other standard forms available. The contractor simply needs to read this section of the contract specifications to make sure he uses an acceptable form so that no payment delays are caused by the use of an unacceptable payment request form.

FIGURE 12-3 Monthly Progress Payment Request - Lump Sum

PAYMENT REQUEST #1

PROJECT: Improvements at Whitelane Park   DATE: Sept. 2, 20--
OWNER: Contracting Agency
           P.O. Box 5599
           Gresham, Oregon 97555

Bid Item	Description	Quantity	Unit Price	Total Price
1.)	Mobilization	1 LS	$10,000.00	$10,000.00
2.)	Clearing & Grubbing	1 LS	$2,500.00	$2,500.00
3.)	Demolition	1 LS	$2,500.00	$2,500.00
4.)	Drainage	0.5 LS	$5,000.00	$2,500.00
5.)	Water Lines	0.5 LS	$3,500.00	$1,750.00
6.)	Excavation	0.75 LS	$4,500.00	$3,375.00
7.)	Aggregate Base	0.5 LS	$2,750.00	$1,375.00
8.)	Concrete Work	0.25 LS	$10,000.00	$2,500.00
9.)	Playground	0 LS	$30,000.00	$0.00
10.)	Site Furnishings	0 LS	$8,500.00	$0.00

Total This Period (8-1/8-30)	$26,500.00
Less Retainage	$1,325.00
Less Previous Payments	$0.00
TOTAL AMOUNT DUE	$25,175.00

All of the above quantities, unit prices, change orders and all other information is accurate and complete to the best of my knowledge.

ABC ASPHALT PAVING COMPANY   Signed: _____
1234 E. Delta Ave.
Gresham, Oregon 97777   Title: _____

                               Date: _____

In general, it will take anywhere from 10 to 30 days for the contracting agency to process a payment request and get the contractor a check. Although in some cases it can take as long as 45-60 days to get a check from the owner. A call to the contracting agency is often in order if your company has not received a progress payment within 30 days of your payment request. In many cases your contract specifications will incorporate an interest factor that your company can accrue if the contracting agency does not make payment within the specified time period. Payment delays can be multiplied if your company is a subcontractor in a case where the prime contractor does not get paid for 30 days or more from submission of the payment request, leading to your company having to wait 45 days or more to receive payment.

CERTIFIED PAYROLL OF PREVAILIING WAGES

Prevailing Wage Rates are the minimum wages, including fringe benefits, that must be paid to all workers employed in the construction, reconstruction, major renovation or painting of any public works in excess of a specified total contract amount. Different rates are established for specific trades and specific geographical areas. Copies of these rates will be incorporated into the bid specifications when the advertisement for a public works contract is issued. The rates in effect at the time the bid specifications are advertised and the contract is executed, are those that apply for the duration of the project. An employee may be paid less than the prevailing wage rate if the worker is registered in a bona fide apprenticeship program of the U.S. Department of Labor, Bureau of Apprenticeship and Training. Also, office/clerical employees and supervisory employees who are supervisory only and do not perform any hands-on labor are not required to be paid the prevailing wage rate.

Contractors and subcontractors who construct public works projects are required to keep records necessary for determining if prevailing wage rates were paid. These records must include the Payroll and Certified Statement Form (WH-38), which need to be submitted to the contracting agency on a weekly basis, as well as the following:

1.) The name, address, and social security number of each employee,

2.) The work classification(s) of each employee,

3.) The rate(s) of wages and fringe benefits paid to each employee,

4.) The rate(s) of fringe benefit payments made in lieu of those required to be provided to each employee,

5.) Total daily and weekly compensation paid to each employee,

6.) Daily and weekly hours worked (including overtime) by each employee; apprenticeship and training agreements,

7.) And payroll and other records pertaining to the employment of the contractor's personnel on a public works contract.

FIGURE 12-4 Certified Payroll

CERTIFIED PAYROLL REPORT

ABC Asphalt Paving Company 1234 E. Delta Avenue Gresham, Oregon 97777	Prime Contractor ☐ Subcontractor ☐	Project Name: Project Number: Project Location:	Payroll Number: For Week Ending:

Name, Address, and Social Security Number of Employee	# of Exemp	Work Classification	OT ST	Day and Date S	S	M	T	W	TH	F	Total Hours	Rate Of Pay	Gross Amount Earned	Deductions FICA	With-Holding Tax	Oregon With-Holding	Workers Compens. Assessm.	Other	Total Deduct.	Net Wages Paid For Week
				Hours Worked Each Day																
			O																	
			S																	
			O																	
			S																	
			O																	
			S																	
			O																	
			S																	
			O																	
			S																	
			O																	
			S																	
			O																	
			S																	
			O																	
			S																	
			O																	
			S																	

Date:

I (Name of Payroll Person), (Title) do hereby state:

1.) That I pay or supervise the payment of the persons employed by Andersen Pacific Contractors, Inc. on the _____ and that during the payroll period commencing on the _____ day of _____ and ending the _____ day of _____ all persons employed on said project have been paid the full weekly wages earned, that no rebates have been or will be made either directly or indirectly to or on behalf of ABC Asphalt Paving Company from the full weekly wages earned by any person and that no deductions have been made either directly or indirectly from the full wages earned by any person, other than permissible deductions as defined in Regulations, Part 3 (29 CFR Subtitle A) issued by the Secretary of Labor under the Copeland Act, as amended (48 Stat. 948, 63 Stat. 108, 72 Stat. 357; 40 U.S.C. 276c), and as described below:

2.) That any payrolls otherwise under this contract required to be submitted for the above period are correct and complete; that the wage rates for laborers or mechanics contained therein are not less than the applicable wage rates contained in any wage determination incorporated into the contract; that the classifications set forth therein for each laborer or mechanic conform with the work performed.

3.) That any apprentices employed in the above period are duly registered in a bona fide apprenticeship program registered with the State apprenticeship agency recognized by the Bureau of Apprenticeship and Training, United States Department of Labor, or if no such recognized agency exists in a State, are registered with the Bureau of Apprenticeship and Training, United States Department of Labor.

4.) That:

(a) WHERE FRINGE BENEFITS ARE PAID TO APPROVED PLANS, FUNDS, OR PROGRAMS

[ ] In addition to basic hourly wage rates paid to each laborer or mechanic listed in the above payroll, payments of fringe benefits as listed in the contract have been or will be made to appropriate programs for the benefit of such employees, except as noted in Section 4 (c) below.

(b) WHERE FRINGE BENEFITS ARE PAID IN CASH

[ ] Each laborer or mechanic listed in the above referenced payroll has been paid, as indicated on the payroll, an amount not less than the sum of the applicable basic hourly wage rate plus the amount of the required fringe benefits as listed in the contract, except as noted in Section 4 (c) below.

(c) EXCEPTIONS

Exception (Craft)	Explanation

(Name of Payroll)
(Title)
ABC Asphalt Paving Company

The willful falsification of any of the above statements may be subject the contractor or subcontractor to civil or criminal prosecution. See Section 1001 of Title 18 and Section 231 of Title 31 of the United States Code.

ACCOUNTING FOR PROJECT PURCHASES

When a purchase is made for any particular project a copy of the purchase order and packing slip (receipt that is supplied at the time of purchase by the supplier) should be turned into the main office on a weekly basis. As previously discussed, your companies job number and the appropriate cost account should be clearly labeled on each of these purchase orders and packing slips. These purchase orders and packing slips should be organized in alphabetical files until the invoice for the purchase is received from the supplier. When the invoices arrive at your office, the purchase order/packing slip from your files should be matched up with the individual invoices, checked to make sure the invoice matches the purchase order and placed in a file of invoices to be paid. Once the invoice has been paid, a copy of the purchase order, packing slip and paid invoice should be organized alphabetically in an individual file for each project.

Each of your material supply companies will have different terms on which your account is to be settled. Most companies have terms that are net 30 days. This simply means that the invoice you receive from them needs to be paid off within 30 days of the invoice date. If the invoice is not paid off within the period stated, they will have interest charges accrue on the balance until the account is paid off. These terms and interest charges are agreed upon at the time you set up your credit account with each supplier. It is a good idea to keep your credit accounts clean with each of your suppliers. If you do not pay off your accounts in a timely manner, the suppliers have no choice but to enforce the interest charges because they have to pay for the materials that you have taken out on credit from their inventory. If your accounts are consistently behind the payment terms, or sometimes late only once, the material suppliers will often drop your line of credit with them and place your account on a C.O.D. (cash on delivery) basis. This type of money management by your company can affect the reputation of your company in a detrimental manner. If your company keeps its accounts up to date you can often times get better service out of your material suppliers, which can greatly help your project coordination and profitability.

Many material suppliers, especially rock and concrete suppliers, offer early payment discounts. The terms for these early payment discounts are often near a 2% reduction if paid by the tenth of the following month. This author firmly believes in taking every early payment discount available. This not only helps your company's reputation with the material suppliers, but also makes good economic sense. If your money is in a money market savings account it may earn 4% on an annual basis, which is approximately 0.33% per month. Therefore, a 2% per month discount on your material purchases saves your company substantially more money than it would make in your money market or savings account.

COMPANY CHECKING AND SAVINGS ACCOUNTS

When a start up public works construction company (or any business for that matter) begins business, they will generally only have a checking account in place with their chosen banking institution. As the company grows, with some luck and efficient management, the amount of money in their checking account should grow also. Corporations cannot have an interest bearing checking account; therefore, at the point when there is a steady sum of money in the checking account, the company should consider setting up a money market savings account with your bank (or a similar style of account that your bank can offer). This savings account should be interest bearing so that your steady, or growing, sum of money that you keep in the bank can go to work for you.

All of the checks that you receive from the owners of the project that you are constructing should be deposited into this savings account. When payroll, material invoices, or subcontract requests for payment are due to be paid, enough money can then be transferred, with a simple telephone call to your bank, out of your interest bearing savings account into your non-interest bearing checking account. Most companies do this money transfer on a weekly basis in order to maximize the amount of time that money can be in the interest bearing savings account. There is a federal limit of 7 money transfers that can occur per month.

JOBSITE VISITS AND PROJECT REVIEWS

Typically, the general or project manager will have daily communication with each of his project supervisors. If the project is large enough, the project manager might even have his office on the project site with a general manager at the home office, over him. Nevertheless, whoever is managing the job from the home office should visit each project on a weekly basis in addition to the daily communication. If the project site is not within driving distance, the manager could visit the jobsite on a bi-weekly basis. As a manager you will have to determine the amount of time you need to spend at a particular jobsite to feel comfortable with its progress; this can also be a function of the experience and past performance of your jobsite supervisor.

Jobsite visits are important because a person can get a much better feel for what is actually happening on a project by physically being there as compared to only getting information from your jobsite supervisor over the telephone. Jobsite visits can also reaffirm with the jobsite crew your personal, along with the company's, commitment to the success of the project. If the manager never makes it to the construction site, the employees working on the project can get the feeling that the company either doesn't care about the project or that they are being left stranded, without any management assistance.

When the manager is on the jobsite, this is also a good opportunity to have an overall project review meeting with the job supervisors. This review can incorporate some, or all of the following construction jobsite meeting outline:

1. Review with the supervisor his weekly, and two-week schedule and how it relates to the overall project schedule. Discuss corrective action if necessary.

2. Review the project cost control, along with the manhour report, as it relates to the different construction operations.

3. Review subcontractor performance and what corrective action to take, if necessary.

4. Review the overall project quality.

5. Review the projects adherence to the company safety plan.

6. Review the projects labor, equipment, and tool requirements for upcoming operations.

7. Review any problems with the daily project documentation such as time cards, purchase orders, daily reports, billings and notice of potential claims.

8. Review with the supervisor how the company is getting paid by the contracting agency for the current and future operations. It is imperative that the manager and the supervisor know exactly how the company is getting paid for each operation. For example, say the contract calls for your company to construct a concrete slab on top of a layer of baserock with the bid items appearing as below:

Bid Item	Quantity	Unit Price	Total Price
Baserock	15 C.Y.	$ 40.00	$ 600.00
Concrete	135 S.Y.	$ 21.00	$2,835.00

In the case of your company being paid on a unit price basis for actual quantity constructed, one typically does not want to install less baserock and more concrete. This scenario would result in a financial loss due to the fact that your concrete bid item is on an area pay basis, not a volume pay basis. Therefore, if you fill part of your baserock volume with concrete, not only are you losing some baserock payment, but you are filling that volume with expensive (as compared to baserock) concrete that will not be paid for by the contracting agency because concrete is paid on surface area, not volume. Understanding the basis of how each bid item within the contract is paid, and adjusting the construction based on this understanding, can greatly enhance the profitability of the project.

These review meetings can be beneficial for both the manager and supervisor since it gives them both a chance to sit down, eye to eye, and discuss the best path for the project to take while learning from the good and bad events that have taken place since the previous meeting. A general jobsite review can also take place with the contracting agency representatives. We suggest that once a week the appropriate managers meet at the jobsite with the contracting agency representatives to discuss the projects schedule and any problems that have been encountered or that are foreseen. It is best to analyze the potential problems before they become a costly conflict for both parties.

While the manager is on the project site, it is a good time for him to take photographs of the operations taking place and any situation that might come up at a later date. Your jobsite supervisor should keep a photograph diary of the project that is detailed beyond that of the manager. If during any operation of the project, an item is uncovered or is about to be covered up that could be questioned by the contracting agency at a later date, a picture should be taken as proof of what actually happened. The use of frequently taken jobsite pictures has critically helped many contractors win claims later in the project that, without the photographic evidence, might not have been won. When it comes down to a conflict or a claim situation, often the contractor or the contracting agencies representative cannot remember (for whatever reason) exactly how an operation was constructed or what a particular area looked like before construction started. Although oral representations may differ and create no positive proof in either direction, well-timed photographs may prove to be the pivotal evidence you need to win the claim.

INTRODUCTION TO PUBLIC WORKS CONSTRUCTION INSPECTION

General. The information in this chapter is designed as an introduction to public works construction inspection. Included are general policies and procedures of a good public works construction inspection and contracting agency as well as information to orient the Construction Inspector to his duties and responsibilities. This chapter is written mainly from the perspective of the contracting agency. This perspective will greatly assist the construction contractor in his day to day dealings with the contracting agency.

Contracting Agency Inspectors must work constantly to achieve a high standard of excellence in the administration and quality control inspection of public works improvement contracts. The accomplishment and stature of the Contracting Agency must inevitably stem from the individual commitment and performance of each employee and each individual has a responsibility to perform in such a manner that personal goals and the Contracting Agency goals are not in conflict.

The Contracting Agency organization must deal effectively with a relatively difficult control and communication problem. Inspectors are widely dispersed to the various project sites and spend the entire workday there. Isolated from immediate supervisory resources and control much of the time, the Inspector must make literally hundreds of individual judgments affecting the quality of construction. Communication is vital under these conditions and the Inspector and the Contracting Agency Management must utilize all available resources and opportunities to discuss and resolve problems as they arise.

The Inspector must have the perspective of knowing the role or function and purposes of other organization elements and groups of the Agency and the relationship they have to his own immediate area of participation and responsibility. With such understanding the Inspector can utilize their support in the process of accomplishment the work of the Contracting Agency.

Each Inspector must be motivated to increase personal knowledge and skills and to be informed regarding the latest construction materials and techniques.

Ideally, the Construction Inspector at the jobsite is a leader who gains the confidence and respect of the people he is dealing with by demonstrating his knowledge and ability. He does not depend solely on the rights and powers vested in him by the Contracting Agency because he knows that such an exclusive and arbitrary approach is not likely to produce the result he and the Contracting Agency are seeking.

He has to have a thorough knowledge of the project or phase of work to which he is assigned.

He must know sound construction methods and the latest techniques for their implementation.

He has to be experienced in all standard inspection and testing procedures and coordinate his work with the contractor or his representative so that the combined effort will produce the specified quantity and quality required by the specifications.

He must be timely, but not hasty to condemn. However, once he is aware of work that endangers the quality, he is to be firm in his insistence or corrective action. He must keep in mind that an order to "tear it out" is sometimes necessary, but realizes such action often is questionable benefit to the public, which very often pays the cost in delay, inconvenience and eventually in money represented by rising construction costs on future work. The most important attribute the inspector needs to have is an honest partnering attitude towards the contractor and the contracting agency. This partnering approach on the part of the contracting agency and the contractor creates the best product and service to all parties involved, especially the tax paying public.

Legal Aspects of Construction Inspection. State and local laws require detailed and continuous inspection of all work paid for with public funds or which is performed in a public field of contractor-public relationship and gradually developed as a safeguard against the natural tendency of individuals or groups to divert portions of the benefits to themselves.

The Construction Inspector is the representative of the Contracting Agency at the site of the work who is empowered to enforce the provisions of the contract or permit. He is authorized to reject materials and workmanship not in conformity with the contract or permit requirements.

Characteristics and Training of the Inspector. The Inspector must be mature, confident, patient, meticulous in carrying out his duties, and a person of integrity who also possesses good judgment. He should have had practical experience in engineering construction and possess and understanding of the principles involved, as well as having a thorough knowledge of the policies, procedures and specifications applicable to his work.

To be successful, the Inspector must have a character and personality of such strength as to merit the respect of those with whom he works. He must be firm but fair in his decisions and follow through to insure that he obtains compliance with his instructions. He should be understanding of the contractor's problems and willing to cooperate at all times to secure acceptable work at the least cost without compromise of the plans and specifications. He should gain the respect of his associates through his knowledge of the work, impartial decisions, exercise of good judgment, his conduct, and exemplary performance.

The Inspector must be alert and observant. He must maintain a spotless service record and conduct himself in such a manner that will reflect credit upon the Contracting Agency. A proper sense of proportion will enable him to give greater attention to the more important matters.

INSPECTION IN CONSTRUCTION

Inspection is the vital link between design and its fulfillment in the field. The inspector is too often underrated in the engineering organization and looked upon as a necessary evil by contracting forces. However, the function is essential in the contract system; inspectors and their problems deserve more understanding and upgrading from both sides.

If all inspectors possessed ideal qualifications, if they had experience coming from years in engineering and contracting, if they had wisdom adequate for a supreme court judge, if they were masters in the art of human relations, there would be little reason for the usual misunderstandings.

It is a stubborn fact that an inspector has authority to point out deviations from specifications, but does not have corresponding authority to approve changes, however minor. This leads the contractor, or his harassed general superintendent, to complain that the inspector can always say "No," but is never able to say "Yes." The function of the inspector begins and ends with seeing that field operations produce results called for in the plans.

Occasionally an inspector appears to take delight in using his authority in situations that exhibit pure cussedness. To illustrate: Several years ago during and editorial trip over a job with the construction engineer we found the pouring of the invert in a by-pass tunnel temporarily shut down and the foreman fuming over the action of the inspector. The work stoppage resulted from the inspector finding boot tracks made by a finisher on the concrete surface and stopping the pour until they were removed. This was concrete to be covered by rock ballast for the railroad and at the end of construction the tunnel was to be permanently plugged. Such action of the inspector would appear to be lacking in reasonable judgement although it must have been in accord with the letter of specifications. However, there is always the other side to such an incident. Possibly this inspector had been pushed around by the contractor's supervisory staff until he was waiting in desperation for some opportunity to retaliate and assert his authority.

If he has the qualities of firmness with patience, and judgment with a desire to be correct but practical, he will advise his staff to appreciate the function on the construction team. And if the contractor advises his staff to appreciate the function of the inspector, even as they use their own methods for accomplishing the results he must secure, then the other half of the team is in harmony. This combination will result in securing the completion of the contract with minimum friction, maximum speed, and complete effectiveness.

AIDING THE YOUNG OR INEXPERIENCED INSPECTOR

There has been such an expansion in the field of highway construction in recent years that the experience level of inspector personnel has been considerably lowered. Experienced field engineers and inspectors have been elevated to supervisory positions, while new and younger men have been hired to perform inspection. These new inspectors, in general, have been sincere and have performed diligently. There, has however, been a noticeable deterioration in the quality of inspection, primarily because of the relative inexperience of these new men. This is reflected by more frequent complaints by contractors that "those inspectors don't know their business ..." or "... we would sure like to have a good, tough inspector who knew what he was talking about ..." Engineering administrators and supervisors have noticed more and more work being accepted by inspectors that is of borderline or even substandard quality.

In all fairness, it must be said that the tempo of construction has greatly increased, making inspection more difficult. In addition, contractors have been finding it more difficult to hire experienced crews, resulting in the need for greater effort to maintain the quality of workmanship. This, of course, requires more experience and ability on the part of the inspector.

Even though the vast majority of highway construction is performed under contract, wherein one party agrees to perform certain work to a given standard in return for an agreed remuneration by the other party, it has been necessary for one party to check on the other to see that the contractual agreements are met. This has been the traditional function of the inspector. His duty is to see that the field operations produce the results called for by the plans and specifications. In this capacity, he has the responsibility to point out deviations from the specifications but not the authority to approve minor changes, even for improvement.

It is generally agreed that the inspector nearly always looks upon his work seriously and tries to do the best job possible. It is an extremely rare occurrence when the inspector is not interested in seeing the work performed correctly. He sometimes loses sight of the over-all picture because of his concentrated interest, but on the other hand, his duties usually do not include an appreciation of the over-all picture. He will usually do a good job within the limits of his knowledge and ability.

The engineering administrator and, often, the engineering supervisor look upon the job of inspection as routine. They are usually aware of its importance but prone the quality of inspection that exists. Quite frequently, they feel they are unable to do otherwise because of the press of their many duties. They instruct the inspector to inspect a specific item of work and then accept his results. In some cases, if the inspector has never been exposed to the particular item of work, the supervisor will give him a short verbal lesson on what to do or turn him over to an "old timer" with one or two jobs under his belt. Supervisory personnel generally bewail the need for inspector training but produce innumerable excuses why such a program cannot be instituted in their particular organization.

At the opposite extreme is the new inspector who, for the first time in his life, is thrust into a position of authority. Occasionally the individual is greatly over-impressed with his own importance and position and begins to throw his weight around unnecessarily. In doing so, the new inspector probably thinks he is behaving in the correct manner. The trouble with this approach, however, is that there are probably several things concerning the work which he doesn't know, thus making him ludicrous. And, his demeanor is such that the people he is dealing with get the impression of supreme arrogance. This man frequently makes the mistake of so behaving that he assumes superintendence of the work, putting him in the untenable position of judging the quality of results reached by means that he dictated. It is this type of inspector who draws the more fire from the contractor.

Requirements for a Good Inspector

The inspector invariably works under the strict requirements of contract plans and specifications in addition to the further instructions of his supervisor. He must deal with a myriad of minor and major problems and details. In doing this, the ideal inspector must have desire, sagacity, firmness, good judgment, knowledge and courtesy. His relations with the contractor's crew must be friendly and firm without familiarity, and must be conducted with the skill of a diplomat.

Knowledge. First, it is of paramount importance that the inspector has a knowledge of the work he is inspecting. This knowledge should include information concerning the materials, as well as considerable acquaintance with the equipment and procedures. There is a school of thought that the less the inspector knows about a specific item, the better and more objectively he can enforce the letter of the specifications. I believe this is fallacious. The more knowledge the inspector has, the better prepared he will be to discharge his duties.

Common Sense. The second commodity that is invaluable to a good inspector is an abundance of common sense. While it cannot be substituted for knowledge and specifications, it is the means of applying one to the other, and of interpreting the specifications to properly enforce their intent. It may grow on knowledge but it cannot be learned out of a book.

Observation. The inspector must perform his function by observation of what is going on about him. Thus, another basic requirement is the ability to see what he looks at. "Seeing" in this context includes observation with the eyes as well as considered thought about the image observed. It is amazing how an inspector can observe an incorrect condition and not realize it as such. This situation is aggravated by lack of knowledge or common sense and, most seriously, by mental laziness. Too many inspectors just don't bother to think about what they are looking at.

Physical Tools. Besides the personal requirements of the inspector, there are tools he must use to perform his function. These include the general run measuring devices as well as necessary testing equipment. Perhaps the most important tools are a notebook and pencil.

The importance of job records listed in the inspector's notebook cannot be over-emphasized. It may be that the information recorded will never be needed and never be reviewed, but, if it is ever needed, it will be needed badly. The notebook should contain every bit of information possible concerning the work being inspected. Such related information as weather conditions, the time and place any incident occurred, breakdown of equipment, length of work stoppages, number of men and type of equipment affected by work stoppage, any unusual incident or condition, even a change in color of a material should be recorded. If the item seems unusually important, it should be recorded and analyzed in sufficient detail to make it fully understandable at some later date. The notebook information will become a reference for future performance of the work, a certain reference in the event of legal action or litigation by any interested party, and possibly most important, it may contain a clue for a future investigator, in the event the job fails. There is nothing too trivial to be included in the inspector's notebook, and the very act of recording will help him to learn and remember.

Courtesy. A major part of the inspector's job is to inform the contractor when unsatisfactory conditions exist or when the specifications are not being met. This is accepted and expected by the contractor, yet is the source of most of the poor relations that can develop on the job between the contractor and the inspector. Since valid criticism or objections by the inspector are expected by the contractor, it cannot be this factor that strains relations; it must be, not what is said, so much as the way it is said that is important. Of course, an aggravating manner of speech is not limited to inspectors. But when poor relations develop between the inspector and the contractor, the work suffers as well as everyone connected with it. A little common courtesy will go a long way.

Improving the Inspector

In light of the requirements for a good inspector and the situations that develop on the job, what can be done to improve inspectors? It can be seen that many of the things that might need improvement depend entirely on the individual. While some assistance may be possible with such things as the employment of common sense, the ability to observe, and the development of personality, the main effort in these things must come from the inspector himself. The main area of assistance is in increasing knowledge through education, better specifications, and training in the use of the tools of inspection.

Better Plans and Specifications

Specifications are the means of communication between the designer and the constructor in achieving the completed structure. The inspector, as well as the contractor, is stuck with the specifications. Both the inspector and the contractor must thoroughly understand the specifications, and they are legally and morally bound by them. Specifications play an important part in the inspector's behavior. These are the rules by which he must referee the game of construction. Their clarity and simplicity play a big part in his performance. In the matter of plans and specifications, the designer can do much to reduce the burden on the inspector. Simple, concise wording of specifications makes their intent clear and minimizes the amount of interpretation needed. The same can be said of the plans. The designer should have in mind some of the problems of construction when the plans are drawn so that they can readily be transposed into the desired structure. These two things alone can greatly simplify the inspector's work, providing instructions do not become so sparse that the inspector must make design decisions in the field that should have been settled long before. It is interesting to note that good plans and specifications will frequently reduce the bid price because the contractor knows better what he must do and, therefore, how he can do it more efficiently.

Aid in Acquiring Knowledge

The primary assistance that can be given the inspector is by increasing his knowledge. This education must be a continuous thing because new people are constantly entering the field and because new developments are constantly appearing. In many items of inspection, the individual will have contact with the item only infrequently and needs a periodic refreshing of his knowledge.

There are several ways for inspectors to increase their knowledge, some of which can be taken advantage of concurrently. For technical education concerning materials, theory, etc., there are courses offered by various colleges. These courses give the inspector a better background of information regarding the materials and theories he is dealing with and allow him to perform his function more intelligently. Formal schooling alone, however, is not enough and must be used in conjunction with other methods of education.

Probably the best place for inspectors to increase their knowledge is on the job. Here the very things they are trying to learn are happening. It is the ideal place to learn. However, it is of great importance to have a qualified individual available to give explanations of things to be learned. Only in this manner can the new or inexperienced inspector learn and, at the same time, avoid the pitfalls of inexperience. It is true he can probably learn without a tutor, but it will take much longer and he will be making the very mistakes it is desirable to avoid. Furthermore, with explanations of why certain things occur or why it is best to use a certain method, the lesson is retained better and the reasoning can be adapted and applied in future instances.

This on-the-job training can be set up in many ways, from a completely informal, hit-or-miss method to a carefully planned and integrated program. Unfortunately, if anything prevails, it is the former. Planned and integrated programs are practically non-existent. The administrator or supervisor often says such programs are impossible in his organization for many reasons. I wonder where this administrator thinks the trained inspector is going to come from or what organization can afford to plan some training on the job. It is my belief that if training and education are desired for inspectors by any engineering supervisor, then that supervisor must instigate it. His efforts will be almost useless, however, unless he is honestly interested and wants to provide planned training on an intelligent, sustained basis. The only weakness may be that the supervisor has no one capable or qualified (including himself) to train the inexperienced man. But nothing will be accomplished.

The literature is another very important source of knowledge. For many years, technical papers have been presented on a wide variety of subjects dealing with highway engineering or construction. In addition, most construction trade magazines make a continuing practice of printing articles concerning proper methods. Articles such as these can be invaluable to the inspector as well as the contractor's crew as a reference source for any particular item or work. Supplementing these sources are many of manufacturers of materials and equipment who publish literature dealing with their particular line. Trade associations increase the availability of technical literature. All these sources should be drawn upon for information in any training program. Since it is difficult for the individual to determine what to read and where to find it, a real contribution would be the development of a list of suggested reading for inspectors, to be published yearly.

Training aids such as movies, film strips and models have been produced by many of the larger contracting organizations. These can be put to good use in any training program. For example, the California Division of Highways has produced several film strips dealing with different types of construction and aimed at informing the inspector. Many trade associations have also produced movies dealing with their product that can be of great help to the new inspector. Most of these organizations are willing to make training aids available, on request, to any agency desiring them.

Handling the Inspector

One of the intangibles in increasing the knowledge of the inspector is his desire for self-improvement. Plans and training are of little value if the object of their attention is disinterested. If desire is lacking in the inexperienced inspector, he should either be shifted to some other line of endeavor or an attempt should be made to instill the desire.

The inspector is a very important person in our system of contracting for construction work. At the same time, mainly because of tradition and economy, the inspector-designate is usually the newest, youngest and most inexperienced engineer on the job. He is thrust into a position of considerable authority, frequently with little preparation. He can be helped if his supervisor makes clear the inspector's position as well as giving him some technical instruction. The psychology of personnel management dictates that the supervisor encourage and support the subordinate. This is doubly true in the case of the inspector. In order to build up his confidence, as well as his technical knowledge and personality, he must be equitably supported by his supervisor. If it becomes necessary to overrule one of his decisions, the reasons should be explained in such a way as not to humiliate him. The inexperienced inspector should be tactfully and fairly handled by the supervisor in order to develop fully his potential.

Cooperation by the Contractor

The people who are most in contact with the inspector are contractor personnel. They do much influence and mold the inspector, especially the new one. It is also true that most contractors' supervisory personnel will do their best to **please** the inspector even though this involves a slight change operation. Unfortunately, there are few contractors who try to **help** the inspector. Yet, it would seem that some assistance by the contractor might be in order since he is one of the parties most interested in the inspector's work. Still more unfortunate, there are some contractors with their representatives who actually seem to harass the inspector. Their philosophy sees to be that the most advantageous position for them is to get the inspector on the defensive and keep him there. Assistance by the contractor and his representatives might take the form of explaining the reasons for doing certain items of work in a particular way. This assistance, of course, becomes valuable only if the one giving it knows what he is talking about.

Possibly the greatest assistance that could be rendered by the contractor would be a deliberate effort to appreciate the function and the problems of the inspector. At the same time, the inspector will assist himself, as well as the contractor, if he will try to understand the problems of the contractor. The inspector is primarily interested in quality, while the contractor is primarily interested in quantity. Under no condition should the contractor expect, nor the inspector permit, a reduction in quality in the interests of quantity, although within specified limits of quality all efforts for maximum quantity should be encouraged. Right here, in fact, is the crux of the understanding that is needed by both parties.

Summary

In summary, some of the things that are needed to assist the inexperienced inspector are:

(1) Desire for self-improvement on the part of the inspector.

(2) Planned training programs on a sustained basis to increase knowledge. The supervisor must interest himself in the program to make it successful.

(3) The taking into account by designers and specification writers of the construction problems connected with their projects.

(4) Application of common sense by the inspector. This requires a little logical thinking on his part and involves the very important ingredient of careful observation.

(5) Emphasis on record keeping, especially a notebook or a diary. Besides providing job record, the act of writing down things helps the inspector learn and remember.

(6) The studied practice of common courtesy by the inspector.

(7) Increasing knowledge by taking full advantage of information and educational aids offered by manufacturers and trade associations.

(8) Continued perusal of available literature in conjunction with work on the job.

(9) A program developed and supported by some contractors' organizations aimed at discussing with inspector and their supervisors the contractors' problems and viewpoints.

Many of the things discussed in this presentation may seem completely idealistic. Undoubtedly, many of the sophisticated will ridicule and sneer at certain concepts. I submit, however, that honest effort along the lines discussed will be necessary not only to aid the inexperienced inspector but to elevate his position to a level commensurate with its importance.

PROJECT INSPECTION GUIDELINES

General. At the outset of anew assignment as project inspector, the Public Works Construction Inspector must be prepared for his leadership role with knowledge and background material concerning the work to be accomplished, as a foundation to establish and effective working relationship with contractor, other agencies, engineering personnel and to meet with representatives of the public who are interested in the project.

During construction, he must evaluate the contractor's operations and production with respect to quality and progress, as well as for construction safety and public safety and convenience. He must anticipate and recognize problem areas, exercise judgment, take action and make decisions literally hundreds of times before the project is completed.

If he has organized and administered his responsibilities to control the work in a logical, mature and efficient manner, his reward will be a high quality public improvement that will function and endure as the designer intended.

Project Inspector's General Check List. The following information, in the form of a "check list", is designed to guide the Construction Inspector in his overall organization and approach to duties on any project assigned to him.

PROJECT INSPECTOR'S GENERAL CHECKLIST

(a) **Review Job Plans, Specifications and all other Contract Documents.**

   1) Plan Notes.

   2) Standard Plans, Reference Specifications.

   3) Traffic Control Requirements.

   4) Special Phasing or Sequence.

5) Unusual Methods and Materials.

6) Utility Conditions.

7) Encroachments, Obstructions, Removals.

8) Soil and Boring Data

9) Shop Drawings and Other Submittals Required.

10) Permit Requirements from Other Agencies.

(b) **Check Job Envelope for Supplemental Information**

1) Utility Notice and Report of Utility Meeting.

2) Correspondence.

3) Progress Schedule.

4) Record and Reports.

5) Grade Sheets.

(c) **Relate Requirements to Site Conditions.**

1) Check Adequacy of Survey Staking

2) Note Adjoining Property "Conditions."

3) Note Vehicular Traffic and Pedestrian Problems.

4) Utility Interferences.

5) Drainage Conditions.

6) Work Space, Storage and "Stock Piling."

(d) **Review Project with Supervisor**

1) Apparent Problem Areas.

2) Interpretation of "Gray" Areas.

3) Apparent Plan Errors or Omissions.

4) Public Relations.

5) Contractor's Organization.

6) Clarification of Inspection Procedures (unfamiliar and unusual materials and methods).

7) Engineering Liaison.

8) Inspection Supplies and Equipment.

9) Street Maintenance Liaison.

(e) **Review Project with Contractor.**

1) Contractor's Organization.

2) Subcontractors.

3) Important Job Conditions.

4) Construction Safety.

5) Public Safety.

6) Construction Methods and Procedures, Sequence of Construction.

(f) **Exercise Controls During Construction.**

1) Coordination and Communication.

2) Construction Inspection Procedures.

3) Sampling and Testing.

4) Safety and Convenience Procedure.

(g) **Maintain Accurate and Complete Records and Reports.**

1) Log Job Progress and Status. (Schedule).

2) Report Special Conditions and Events.

3) Log Important Request, Notifications.

4) Maintain Orderly Filing System.

5) Measure Completed Work to Verify Monthly Pay Requests.

INSPECTION POLICY

General.   Inspection of public works construction is a control exercised by a governmental agency over the materials, methods and workmanship used by contractors in the performance of their work. The purpose of inspection is to ensure compliance with the plans, specifications and other requirements of contracts, purchase orders and permits for public works construction, including compliance with the pertinent provisions of Federal government. It is essential that the Inspector read all the contract documents, and re-read them from time to time to insure complete familiarity with all provisions and requirements.

Inspectors shall not prescribe or interfere with the contractor's methods of performing work. However, if, in the opinion of the Inspector, the methods of the contractor will not meet the requirements of the contract, purchase order or permit, the contractor shall be warned and, if the contractor's methods will produce a hazard to life, health or property, or will result in defective work which would be impractical to correct or replace subsequently, the Inspector shall stop the portion of the work involved and immediately notify his supervisor. The Inspector shall cooperate fully with each contractor and assist him in all practical ways to complete the work economically, expeditiously and satisfactorily.

Relations with the Contractor.   While an Inspector does not generally have the authority to allow deviations from essential contract requirements in trivial construction details or technicalities. The Inspector should not interfere with the contractor's methods of performing the work. Advise, but do not try to force him arbitrarily to a certain course of procedure where the specifications permit more than one method. Such interference may operate to release the contractor in whole or in part from the responsibility he has assumed under the contract to obtain specific results. If, however, a contractor's methods are obviously improper, inadequate, unsafe, or likely to result in damage or future expense to the Contracting Agency, it should be called to the contractor's attention at once. If prompt objection is not made to unsatisfactory work, it will be difficult to prove later that it was not satisfactory.

Orders to the contractor should be in writing, or later confirmed in writing, so that instructions will be clear and no misunderstanding develop over controversial issues. Particular care should be taken that no instructions are given which could be construed as assuming superintendence of the work. Poor judgment in this respect could result in claims against the Contracting Agency.

Instructions or formal orders should be given by the Inspector directly to the contractor, his superintendent or foreman only. The contractor must provide a competent foreman who is more than a "pusher." A foreman must be able to read plans and perform the necessary layout for the work and is expected to **direct** the activities and operations of the workers in accomplishing the work. All of these activities are properly the contractor's business and the foreman's responsibility and not that of the Inspector.

Relations with the contractor and his employees should be agreeably maintained. Surliness or an overbearing attitude or abusive language on the part of the Inspector is uncalled for. The Inspector should enforce his decisions through his personality, and his judgment should be fair and impartial. His knowledge of the work and the codes under which it is to be accomplished should be so thorough that he will achieve respect and compliance. He should not give any directions which are not justified by the contract documents, but should insist on compliance with such directions as he does give. A reputation of being slack or easy, though it is quickly, is difficult to overcome.

An order suspending any part of the work betrays a serious condition on the project. Public relations as well as economic loss may be involved. Consequently, such orders are never to be given lightly or in a spirit of punishment. However, situations will occasionally occur when orders suspending work must be issued. Certainly the work must be built to line and grade. Dimensions and quality must be as specified, and the finished work must be acceptable within specification tolerances.

Relations with Other Agencies. The closest co-operation should be maintained at all times with all other levels of government with which the Contracting Agency conducts business. Any suggestions or criticism of any of the functions or personnel of another governmental unit shall not be tendered directly to such unit but shall be reduced to writing and forwarded through the Inspector's supervisor.

The Inspector's Public Image. Inspection personnel are expected to conduct themselves at all times in such a manner as to reflect credit upon themselves and the Contracting Agency they represent. The Public Works Construction Inspector is expected to be pleasant, courteous and business-like in meeting the public. Above all, his conduct must be governed by common sense. To the public, he represents the Agency. The people hold him responsible for accomplishing his work in a manner which will afford the greatest public benefit and the least public inconvenience. It is important to be helpful and considerate in answering questions asked by the public, and if the employee cannot definitely answer their questions, he should refer them to the proper authority or department for accurate information.

PUBLIC INFORMATION

General. Information furnished by the Inspector to the general public should be restricted to construction details for the specific project involved. Inquiries from representatives of the press requesting data for publication and to representatives of Chambers of Commerce and similar groups requesting data for reports to their organizations, should be referred by the Inspector to supervisory personnel for reply.

The Inspector is cautioned not to furnish any information on the following matters:

(a) Matters pertaining to claims or lawsuits involving the Agency. Such information shall be released only by the legal counsel for the Agency.

(b) Estimates made by the Agency for future work to be done under contract.

(c) Information as to policies, procedures and official actions of the Agency, and as to property and rights in property to be acquired by the Agency.

(d) Results of tests of competitive materials and products. Employees are not discredit any contractor, product or manufacturer.

JOBSITE COMMUNICATIONS AND INSTRUCTIONS

General. Important communications and instructions should be issued in writing. If a contractor, permittee or other agency performing work under the supervision or control of the Contracting Agency, fails to comply with or violates any contract provision, ordinance or lawful instruction, or causes any unsafe condition, written instruction are mandatory.

Verbal Instructions. Appropriate verbal instruction relative to non-compliance, safety violations, etc., shall be given at once by the Inspector or Supervisor to the representative of the contractor or permittee on the job. The text of these instructions shall be noted by the Inspector in the Job Report.

Job Memorandum. The Job Memorandum is intended for the purpose of permitting the Inspector to issue written instructions regarding routing matters or confirming verbal instructions to a contractor or permittee where a Notice of Violation or Non-Compliance may be inappropriate or where the circumstances are not of such importance as to warrant the issuance of the latter form.

An example of the use of the Job Memorandum would be to issue to a foreman instructions not to backfill a sewer trench until the Inspector returns from a second job assignment at a specified time to check the make-up of the pipe joints. While this particular situation will not always require the use of this form, it will prove useful in the case of a more unreliable foreman who demonstrates that he often misunderstands instructions or who later claims no such instructions were ever issued by the Inspector.

Notice of Non-Compliance. Verbal instructions regarding non-compliance shall be confirmed the same day by a Notice of Non-Compliance, except in those instances where instructions are compiled with at once. When this notice is issued to a party on the job who is not the permittee or prime contractor, the notice will be addressed to both the party on the job and the permittee or prime contractor.

Job Inspectors are authorized to issue Notices regarding violations or non-compliance with plans, specifications and legal requirements of the project, and to stop work on any portion of the job if the contractor's methods cause unsafe conditions or will result in defective work which would be impractical to correct or to replace subsequently, while permitting other (conforming) portions of the work to continue. Notices stopping all work shall be issued only by or at the direction of the Supervisor having jurisdiction over the work, or higher authority. Issuance of a Notice stopping all or nay part of the work shall be reported by telephone immediately to the Main Office of the Contracting Agency and the appropriate District Office.

RECORDS AND REPORTS

General. Keeping accurate records and reports is a very important function of the Field Inspector. These records are necessary for a number of reasons. Some of the most common reasons and their use as references are as follows:

(1)   Time and material accountability, including quantities for periodic progress payments and extra work under cost plus change order procedures.

(2) Verify actions and decisions of the Inspector, contractor or engineer.

(3) Establish job status and site conditions in the even of an accident or liability claim.

(4) Verify time charges and justify inspection activities to the permittee who questions inspection fees.

(5) Clarify the continuity of a project (working days, delays, activities) when the contract time is in dispute.

(6) Prepare responses to inquiries and complaints.

(7) Evidence in legal actions.

(8) When there is a change of inspection personnel; progress or status of project to orient the newly assigned personnel.

These uses should be kept in mind and job records should be prepared accordingly.

Job Reporting Procedure. The record of activities on a construction job should be reported in the exact sequence that they take place.

The basic daily reporting medium, the Construction Inspector's Daily Report commonly referred to as the "Job Log." It is a continuing report of job progress and provides for the use of as much space as is necessary to adequately report each day's progress and activities. Each day's report begins on the line following the previous day's report and each page is numbered consecutively.

Each daily entry should be brief but at the same time be complete, clear and factual and include all work accomplished by the contractor as well as pertinent related information. In other words, the inspector should think "who did what, where, when, how and how much." Entries shall be made daily to avoid errors or omissions, and include the number of hours charged against the job and the Inspector's **legible** signature. Abbreviations are desirable as long as their meanings are not confusing and have a common acceptance.

On Permit projects, the Daily Reports are the only continuing job reporting medium utilized, except for the Inspector's personal diary or notes and "as-built" information recorded on the project plans.

On most large projects, especially those requiring more than on Inspector, the Project Inspector in charge will keep a daily Job Log and other assigned inspectors will assist him in keeping the Job Log current. He will use the Job Log to record all project activity.

On contract projects, the Job Log should include a daily record of all men and equipment working on the job. This information is not included when a special reporting form is utilized for this purpose.

Job Reporting Checklist. Following is a general checklist of entry items applicable to all jobs:

DAILY REPORT CHECKLIST – GENERAL

(1)   All entries shall be printed in black ink or typed.

(2)   The first report in any series should begin with the job title, job number, contractor's (and subcontractor's on first working day) name, address and phone number; and the superintendent or foreman's name and job office phone number.

(3)   Be brief, but include all work and activities and related information.

(4)   Entries are clear, accurate and legible.

(5)   Total regular inspection hours worked and signature for each daily entry.

(6)   Overtime hours noted by the initials "OT" after the number of hours.

(7)   Entries made the same day the work is performed to avoid errors or omissions.

(8)   For each daily entry, include the pre-printed number of any Job Memorandum or Notice of Non-Compliance issued on the job and underline in red.

(9)   If work being done is Change Order work, record the Change Order number and description of the work as part of the entry on the job report.

(10) Record any verbal instructions or authority form the Design Engineer on the job report on the day received, including the Engineer's name.

(11) All job related incidents must be noted on the job reports, such as men and equipment working, traffic accidents, damage to existing improvements or utilities, injuries, etc.

(12) On permit work, when inspection costs are charged to the permittee, and the contractor (after requesting inspection) does not show up on the job site, a Job Memorandum must be issued to the permittee and a note accounting for the time charge (usually two hours) entered on the job report.

(13) Job progress must be reported in terms of quantity, distances, stations and weight, as they are appropriate and applicable. Reporting must account for all bid item quantities including when, where an what was constructed by exact limits so as to establish an accurate audit trail.

(14) Mention important visitors to the project and the nature of their business.

OTHER SIGNIFICANT ITEMS TO BE REPORTED

For all types of construction, the following items are to be considered and reported where appropriate:

(1) Factors adversely affecting progress of the work, such as delay in utility work completion, delivery of materials and equipment, unforeseen conditions, strikes, plan changes, poor contractor management, severe weather and resulting soil conditions, etc.

(2) Unsatisfactory work performed by the contractor and corrective actions proposed or taken.

(3) Conditions that may require changes or extra work, or generate controversy or claims. The proposed methods of handling the situations should be described. Any indications by the contractor of his intention to file a claim should be reported along with pertinent job report.

(4) Unusual or difficult engineering, construction or traffic problems involved and their solution.

(5) Unusual conditions regarding safety. Precautionary measures taken with respect to protecting construction workers, the traveling public and abutting property from injury or damage as a result of the construction operations.

(6) Right-of-way, public utility and public transportation problems.

(7) Quality of the work produced.

(8) Provisions for movement of traffic, access to property detours and signing.

(9) Causes of retarded progress and delays. Contract time, percent of work completed and time extensions granted.

(10) Unusual material and equipment brought on project or removed form projects when this is considered a significant effect in maintaining satisfactory progress.

(11) Documentation of actions taken and justification therefore.

(12) Field sampling, testing and laboratory test results, particularly failures and resolution.

(13) Developments regarding problems or undesirable conditions discussed in one inspection report should be followed up in a subsequent report indicating final solution or disposition.

(14) When shutdown periods occur, the dates of suspension and resumption must be included in the project records.

(15) Observations and conclusions concerning the overall review of construction operations with particular emphasis on the actual construction features.

Surface and Grading Project Reporting. Daily reports should be used for all common surface improvements, such as: rough grade, fine grade, curb and gutter, paving and sidewalk.

Typical entries for grading projects should include:

(1) Limits of slide or alluvial material removed.

(2) Name of soils Engineer or Geologist who checked area (if available).

(3) Methods and equipment used. Soil type and lift thickness.

(4) Size and limits of subdrains laid.

(5) Limits of fills placed.

(6) Compaction tests and results reported by the laboratory.

(7) All failures should be circled in red and be accounted for by a subsequent entry reporting "retest passed" or other resolution.

Typical entries for street subgrade jobs should include:

(1) Limits of roughgrade, cuts and compacted fill.

(2) Limits of roughgrade checked. Tests by the laboratory and results.

(3) Base material source, number of samples and test results must be noted.

(4) Limits and thickness of base material placed and compacted.

(5) Location and results of compaction tests taken on base material. (failures resolved as indicated in grading, Item 6).

(6) Fine grade checked and approved for paving.

Typical entries for curb, gutter, walk and driveways should include:

(1) Name of subcontractor (if any) if not entered on first card.

(2) Station to station limits of forms placed when concrete is not placed the same day.

(3) Station to station limits of concrete placed, type of concrete and additives if any, number of cubic placed and source of the concrete.

(4) Type and size of curb and gutter.

(5) Width and thickness of walk; width and thickness of driveways together with the distance from curb face to top of driveway slope.

(6) Number of concrete test cylinders taken.

(7) Station locations of blockouts for catch basins, curb drains.

Typical paving entries should include:

(1) Name of paving subcontractor if not shown on first card.

(2) Source of material.

(3) Method of laying, type, thickness, base or wearing surface and tons of asphalt paving material laid (or cubic yards of concrete placed), limits such as station to station and width and square feet if payment is made on this basis.

Sewers and Storm Drain Project Reporting. Storm drain main lines are usually designated by a letter such as Line "B". Laterals are usually designated by the letter of the main line and a number such as Line "B-1 or B-2". A catch basin can be identified by the lateral associated with it. If main lines and laterals are not designated by letter or number, then laterals should be located by station and catch basins should be located by station and/or by distance to nearest intersection.

Typical entries for sewers and storm drains should include:

(1) Station to station limits of removals, excavation, main line pipe laid (including size) and for storm drains, the D-load, bedding, backfill and water densifications by jetting or by other consolidation methods.

(2) Station of sewer house connections excavated or laid, including backfill and jetting or other operations performed. Include numbers of any house connection change reports.

(3) Manholes or other structures completed (or partially completed) including backfill and backfill densification.

(4) Limits of all special compaction completed.

(5) When curved sewer lines have been balled, make note on job report and underline in red.

(6) Any sampling and field tests performed as well as the test results reported.

(7) When lines has been air pressure tested, make note on job report and underline in red.

(8) Location of concrete collars (storm drains).

(9) Make not on job card when "Y" sheets are sent to main office.

Instructions are also included for the "Request for Change of House Connection Sewer," which is required whenever it is necessary to omit, add or relocate a house connection more then five feet.

Street Lighting and Traffic Signal Projects Reporting. Daily progress of construction should be recorded on the job reports as on other phases of construction, such as: removals, limits of conduit laid, backfilled or "jacked," cable pulled, pull boxes set, concrete for bases placed and equipment installed.

Other items that must reported included:

(1) When working on existing circuits, record any streetlight "safety clearances" obtained with date, time, duration and reason for clearance.

(2) For equipment and material entries include types and other descriptive information such as manufacturer and model number.

(3) All receipts for salvage equipment should be recorded on job report.

(4) When traffic signal shut-down takes place, record location, date, time, duration and substitute traffic control utilized during shut-down.

Contract Time. With rare exceptions, the contract time for agency projects is specified in working days.

The Contracting Agency further specifies the assessment of liquidated damages when the contractor fails to complete the work within the contract time. Therefore, the accuracy the substantiation of work progress and delays must be thoroughly and accurately documented. Records, reports and communications with the contractor during the course of construction are of the utmost importance in accomplishing the enforcement of the contract provisions in this regard.

From the information reported by the Project Inspector and reviewed by the District Supervisor the contractor is apprised by mail of the number of days charged and days remaining on the contract by means of a form entitled "Recapitulation of Contract Time."

After notification of award of a contract and prior to start of any work, the contractor submits his proposed construction progress schedule to the Contracting Agency. The schedule, in the form of a tabulation, chart or graph (or critical path diagram, when required) must be in sufficient detail to show the chronological relationship of all activities and the procurement of materials. The construction schedule must reflect the completion of all work under the contract within the specified time. If the contractor wishes to make a major change in his operations after beginning construction, he must submit a revised construction schedule in advance of the revised operations.

As soon as it becomes evident that the actual rate of progress will not be sufficient to complete the project as scheduled and within the contract time, the Project Supervisor must initiate written correspondence to the contractor to advise him of this likelihood and urge him to expedite the work. Written correspondence should follow in progressive steps, i.e. utilizing field correspondence first (Job Memorandum, Notice of Non-Compliance); and if necessary, the Supervisor should prepare a letter draft for issuance by the Division Chief. The ultimate action by Contracting Agency management is to recommend that the contractor be declared in default of the contract.

PROGRESS PAYMENT REPORTING

General. Monthly progress payments are made to the contractor on all contract projects and on assessment act projects when general funds finance a portion or all of the project. The quantities of work accomplished are estimated and reported monthly by the Project Inspector on a progress payment estimate form which he submits to his Supervisor for review and approval.

The normal closing date for this report is the fifteenth of each month (or a different date when requested by the contractor). A report is not required if no progress is made and the Project Coordinating Section in the Main Office is notified.

Cost Plus Change Orders. The cost of extra work, for which unit prices or stipulated prices cannot be applied, is usually negotiated. When a negotiated agreement cannot be reached, the work is done under "Cost Plus" change order procedures established by the Standard Specifications.

It is imperative that complete and accurate records be maintained to substantiate the labor, material and equipment used on the cost plus change order work. To facilitate accuracy and agreement with the contractor, the following detailed procedure has been established for such work:

PROCEDURE FOR EXTRA WORK (COST PLUS) CHANGE ORDERS

Unless otherwise specified in the contract documents, the following accounting procedure shall be in effect whenever a change order for extra work requires that the cost of such work be based on the accumulation of costs as provided for by the provisions of the Standard Specifications.

The project Inspector shall prepare a "Daily Report for Cost Plus Changes," in triplicate, for each day that work is performed and chargeable to the change order. The form "Daily Report/Invoice for Cost Plus Changes" shall be utilized for this purpose by checking the box "Daily Report." Daily reports shall be numbered consecutively in the upper right hand corner and the last report shall include the word "Final." The daily report shall account only for the time and quantities of labor, equipment and materials used.

At the close of each working day, the Project Inspector shall review the daily report entries with the contractor's representative, obtain his signature and sign the document. In the event that a disagreement develops regarding an item, either the Inspector or the contractor's representative shall establish a record of the disagreement by entering appropriate notes on the form prior to signing the document. The Inspector's Supervisor shall attempt to resolve any disagreements noted on the daily report with the contractor prior to submittal of the invoice.

Distribution of the "Daily Report"

Original—Project Engineer (via Supervisor)

cc: Contractor's Representative

cc: Job Envelope or File

Following completion of the extra work, the contractor shall submit to the Engineer the "Invoice for Cost Plus Changes". The contractor may utilize the Agency's form for cost plus changes, or submit his own form if acceptable to the Engineer. The invoice for cost plus changes submitted shall state the cost for each item agreed to and listed on the "Daily Reports." After verification, the Engineer shall negotiate the final costs with the contractor and prepare a change order for the extra work done.

In the administration and control of extra work change orders, the Inspector must be alert and accurate to the extent that he approves only those charges which are necessary to complete the change order work.

The following instructions are to be observed by the Inspector engaged in cost plus change order work:

(1) He must have a thorough understanding of what the change order requires. If the limits of the work are not clearly defined or understood, he cannot be accurate with his accounting.

(2) When the construction operations for contract work and the cost plus change order work are being pursued at the same time and they are closely associated or integrated with other work, special care must be exercised in accounting for the charges against the change order.

(3) He must be thoroughly acquainted with the Standard Specification controls governing the cost of such work.

(4) He must be prompt in recognizing and resolving any situations involving unnecessary or inefficient use of labor, equipment or material.

(5) The daily report must be fully descriptive. Labor must be accounted for by trade, name, classifications and specialty when applicable. Equipment must be identified by specific type, model, size and accessories where differences in these items will affect the rental rate. (See the following guidelines). It must also be clearly indicated as to whether the equipment is rented bare (in which case the operator could be listed under labor) or operated and maintained. Description of material items must be specific with units and quantities clearly indicated. For example, the class must be shown for concrete, and asphalt base materials must be listed by type. Miscellaneous items of work may be involved (such as dump charges) which should be accounted for on the daily report by number of loads and dump location with dump fee receipts attached to the report.

(6) The Inspector must also report daily construction activity on the cost plus change order work in the job log.

COST PLUS CHANGE ORDERS GUIDELINES FOR LISTING EQUIPMENT

Air Compressor. All types rated in accordance with the manufacturer's rated capacity in cubic feet per minute at 100 pounds per square inch gage pressure. Indicate hose length over 50 feet.

Jack Hammers. Rated by tool weight in pounds (other air tools not rated by pounds).

Asphalt Paving Machines. Rated by manufacturers and model, includes all attachments and accessories.

Road Brooms. Towed rental rates by broom widths. Street-sweeper type, self-propelled, rated in accordance with the hopper capacity in cubic yards.

Compactors (Impact and Vibratory Types). Hand-guided, gasoline powered, including all attachments and accessories. Rated by load weight or vibratory impact weight.

Light Plants and Generators. Rated in accordance with the manufacturer's continuous rating in kilowatts.

Barricades (Lighted and Unlighted), Delineators. Includes all servicing. Rate per unit per day. (List types).

Loaders, Crawler. Includes all attachments and accessories except for clam action buckets and auxiliary backhoe units. Rated by manufacturer and model.

Loaders, Rubber Tired. Same as crawler type.

Motor Graders. Rated by manufacturer and model.

Pumping Units. Manufacturer's rated capacity.

Rollers, Tamping and Grid. Rated by number of drums and drum dimensions (diameter and length).

Rollers, Street (All Types). Rated by manufacturer and model.

Rollers, Vibratory (Towed and Hand Guided). Rated in accordance with type and net horsepower of the engine mounted on the roller.

Rollers, Vibratory (Self-Propelled Types). Rated by manufacturer and model.

Saws, Concrete Cutting. Rated by the net engine horse-power of the gasoline engine. Indicate lineal footage and depth of cut.

Hydraulic Cranes and Excavators (All Types). Rated by manufacturer and model. Includes all attachments and accessories.

Shovels, Power Crawler or Truck Mounted. Rated by manufacturer, model type and use (shovel, backhoe, driving piles with leads, crane, clamshell, dragline or for driving piles without leads).

Tractors, Crawlers. Listed by manufacturer and model. Includes all attachments and accessories such as power control units and push blocks (when needed) but does not include bulldozer and ripper units. Indicate accessories used.

Tractors, Rubber Tired Small Industrial and Farm Types. Rated by manufacturer and model. Indicate accessories such as loader, bulldozer, backhoe.

Tractors, Rubber Tired, Heavy Construction Type. Same as crawler tractors.

Trenching Machines. Rated by manufacturer and model. Includes all attachments and accessories.

Dump Trucks (On Highway Type). Rated in accordance with the total number of axles in the vehicle train. Includes all end dump, side dump and bottom dump, including all attachments and accessories.

Welding Machines, Arc. Rated by manufacturer's output rating expressed in amperes. Includes all attachments, accessories, helmets, rod holders and cable. (Diesel, gasoline or electric powered).

Cost Breakdown, Lump Sum or Bid Item Projects. In order to provide a measure of uniformity in estimating progress payment amounts, the following guidelines are furnished. These values (expressed in percent) can be used by the Supervisor in reviewing cost breakdowns of lump sum projects or lump sum bid items; and by the Inspector in the field to break down unit bid items.

PIPELINE CONSTRUCTION		PERCENTAGE		
Open Cut (Trenching)	Storm Drain	Excavation and pipe laying	80*	
		Backfill and jet (or mech. Comp.)	10	
		Perm. Resurfacing and cleanup	10	
		*Where total project is lump sum. Reduce this item by 10% to provide for structures and other work.		
	Sewer (Unlined Pipe)	Excavation and pipe laying	80	
		Backfill and jet.	10	
		Perm. Resurfacing and cleanup	10	
	Sewer (Lined Pipe)	Excavation and pipe laying	75	
		Joint make-up and liner plate welding	10	
		Backfill and jet.	5	
		Perm. Resurfacing and cleanup	10	
Tunnel	Storm Drain	Tunnel excavation	50	
		Shafts	10	
		Pipe laying	20	
		Concrete and earth backfill	15	
		Resurfacing and cleanup	5	
			Unlined	Lined
	Sewer	Tunnel excavation	50	50
		Shafts	10	10
		Pipe laying	20	25
		Liner plate welding	--	5
		Concrete and earth backfill	15	5
		Resurfacing and cleanup	5	5

PAVEMENT CONSTRUCTION

Pavement	Asphalt Concrete	Subgrade for select material base	20**
		Select material base	Per sq. ft.
		Asphalt concrete	Per ton
	Concrete	Subgrade for select material base	20**
		Select material base	Per sq. ft.
		Concrete	Per ton
		**To pay for subgrade for SMB, use 20% of the price per sq. ft. for SMB.	

STREET LIGHTING CONSTRUCTION (LUMP SUM CONTRACTS)

Street Lighting	Removals and relocations***	
	Service points	5
	Underground work	35
	Bases	5
	Conductors installed	10
	Electroliers erected and connected	35
	Resurfacing, caps and cleanup	10
	***Estimate the percentage for this item and reduce all other items to compensate.	

There will be instances where the nature or complexity of a particular project will result in the need to modify these values to more properly reflect appropriate payment to the contractor. The Project Inspector is authorized to make such adjustments subject to review by his Supervisor.

SAFETY AND CONVENIENCE

General. A primary responsibility of the Contracting Agency Inspector is to have a working knowledge of the controlling regulations, codes and directives dealing with public convenience, public safety and construction safety. He must have the ability to apply this knowledge to the construction operations to which he is assigned. In the area of safety, there can be no hesitancy on the part of the Inspector to take immediate action to reduce or eliminate a hazard or an unsafe practice. He must make a conscientious and diligent effort to eliminate any conditions which, in his judgment, would be hazardous to the workers, the public or to himself.

It is the responsibility of the Inspector to follow safe-practice rules, render every possible aid to the contractor in providing safe operations, and report all unsafe conditions or practices to the proper authority. The contractor's supervisory force must insist on employees observing and obeying every rule, regulation and order as is necessary to the safe conduct of the work, and take such action as is necessary to obtain compliance.

In 1970 Congress passed the Williams-Steiger Occupational Safety and Health Act which resulted in the creation of the Occupational Safety and Health Administration (OSHA) within the Department of Labor. OSHA safety rules and regulations take precedence over any less stringent, conflicting, or overlapping State or local safety regulations.

The federal law provides for enforcement agreements with individual states that adopt regulations that are as effective as the OSHA regulations and which have a safety enforcement organization.

Approximately half of the 50 states have entered into agreements with the federal government which permit local state enforcement.

Although the matter of safety at the jobsite is the contractor's legal responsibility, the Inspector may well encourage safe working practices by pointing out possible sources of danger. The Inspector must be safety-conscious and not hesitate to promote the safety of the job and the public.

If, in the informed opinion of the Inspector, the precautions taken by the contractor are found to be insufficient or inadequate in providing job or public safety at any time during the life of the contract, he must inform the contractor to take additional precautions. When the contractor has failed to take action on safety violations, after being advised of the unsafe condition, it is the duty of the Inspector to notify Agency management in order that the enforcing OSHA agency may be notified.

The Inspector can demonstrate interest in safety by establishing a firm, positive attitude toward the prevention of accidents. His knowledge of the construction safety orders is essential if violations arise.

Agency Personnel Safety

General. Employees of the Contracting Agency at the site of construction work shall be safety conscious at the times and shall not work under any conditions that are in apparent violation of the OSHA regulations. It is mandatory that all Contracting Agency Inspectors conform to the performance standards of all safety regulations. In this regard, the Inspector, as an employee, has the same rights and is afforded the same protection as any other public employee within the State. These are:

(1) Any employee or his representative may call the OSHA enforcing agency and report any unsafe working condition. The employee need not identify himself. If he does identify himself, his identity is kept confidential.

(2) An employee may refuse to work under unsafe conditions.

(3) Any employee has the right to observe, monitor or measure employee exposure to hazards and has the right of access to accurate records of employee exposure to potentially toxic materials or harmful physical agents.

(4) The employee may not be harassed or disciplined in any way for the legitimate calls reporting unsafe conditions or for refusing to work under unsafe conditions.

Safety Rules for Inspectors. The following are the general employee responsibilities as prescribed in OSHA law. No person shall do any of the following:

(1) Remove, displace, damage, destroy or carry off any safety device, safeguard, notice or warning, furnished for use in any employment or place of employment.

(2) Interfere in any way with the use thereof by any person.

(3) Interfere with the use of any method or process adopted for the protection of any employee, including oneself, in such employment or place of employment.

(4) Fail or neglect to do every other thing reasonably necessary to protect the life and safety of employees.

Almost all injuries and accidents are caused by someone ignoring safety orders, not following safety practices, taking a chance or failing to correct dangerous conditions. Safety orders only establish standards of work safety. They are the framework in which work can be accomplished safely. The Inspector's complete safety on the job depends on his own efforts to be safe. All unsafe acts and work practices must be avoided for the benefit of all concerned.

To prevent accidents, the Inspector must have a great desire not to have them. Attitude is the keystone of accident prevention. The individuals with proper attitudes have few accidents, whereas the careless have many. The Inspector must be able to recognize accident hazards and to eliminate them. Rules are just as necessary in working, as they are in competitive sports. Safe practice rules are needed so that every employee can work as a part of the team, without fear of injury. The following set of rules should be complied with in order to prevent accidents and to enforce safe practices:

SAFETY RULES

(1) Be in good physical condition before starting work, with your alertness and ability unaffected by illness or lack of rest, which causes fatigue and decreased efficiency. Problems can lead to accidents if your mind isn't on the job.

(2) If you become ill when at work, report to your Supervisor or Dispatcher for replacement so that you may receive proper medical attention.

(3) Wear the right work clothes and shoes for the job. Your clothing should allow freedom of action and should not hang loosely. Wear durable, hard soled shoes that fit well. Tennis shoes and other inappropriate footwear, shoes with thin or badly worn soles , or loosely fitting shoes are dangerous and shall not be worn. International Orange vests shall be worn when high visibility of personnel is advantageous.

(4) Wear your Agency-issued high impact plastic hard hat with Agency decal only on all construction jobs that are posted or otherwise designated as "hard hat areas." Check periodically for damage to the shell and suspension system.

(5) Wear safety goggles when near welding operations or near hazardous liquids or other materials which may spatter and impair your vision. Think about what you see, where to look and what to look for.

(6) Be sure the equipment you are using, or working near, is in safe operating condition, grounded, properly operated and contributes to a safe working atmosphere.

(7) Keep as clean as possible to prevent skin trouble when working with chemicals, oils, paints or cleaners. Wash thoroughly after handling anything that might be poisonous or injurious, especially before eating. Report and treat all injuries, not matter how small they may seem. Prevent serious infection by receiving first-aid treatment.

(8) Never act impulsively. Think about what you are going to do before you do it. Consider the hazards and take adequate precautions. Correct any unsafe conditions you can; report all others to your Supervisor. Always expect the unexpected.

(9) Don't attempt to handle more than you can control. Do your work the right way and safe way; taking short cuts is often dangerous. Work at a speed which is know t be safe, watch where you're walking, and never run.

(10) Use handrails on stairs or on elevated places. Never jump from platforms, scaffolds, loading docks or other elevations.

(11) Your job in fire prevention is to keep things that start fires away from things that burn. If you notice a fire hazard, see that it is corrected. Observe "no smoking" regulations where posted. You should become familiar with the operation and use of the various types of fire extinguishers provided and their locations.

(12) Always use safe practices and follow instructions. Help make the entire job safe. Watch out for the safety of other workers and help new employees learn safe work practices.

(13) Be your brother's keeper. Consider what you do in terms of the hazards it may create for others. Never leave a booby trap for the next person who may come by.

(14) Obey all traffic regulations while driving vehicles both on and off the job. Be courteous to other motorists. When not driving, be a safe pedestrian. Stay alert and don't jaywalk.

(15) The rules of safety you use at work are just as important for you and your while you are at home. For safety 24 hours a day, teach and practice safety in your own home. Safety is a year 'round job, always in season.

(16) Don't' enter manholes, underground vaults, chambers, tanks, silos, or other similar places that receive little ventilation, unless it has been determined that it is safe to enter.

(17) Plan your work ahead to prevent accidents, and take part in regular accident prevention programs. Preplanning of safety measures to meet know construction activity hazards will prevent accidents and promote efficient and economical construction. By utilizing proper attitude, basic skill, good habits thorough knowledge, fair judgment, along with mental and physical fitness, the benefits derived will be worth the effort involved and in the final analysis will prove that safety makes sense.

The Inspector's cooperation is necessary for the protection of himself and others. It is important that he follow all safety rules, take no unnecessary chances, use all safeguards and safety equipment provided and make safety a part of his job. Accidents and fires hurt all of us in many ways. The Inspector and his family suffer if he is injured; all employees lose, because accidents are wasteful. This means lost production, higher operating costs and inefficiency. The Inspector must do his part by giving full support to all safety rules.

Inspector's Protective Equipment. The Contracting Agency Inspector is expected to wear suitable clothing and protective gear to meet the needs of his employment. Specialized gear will be supplied from Contracting Agency headquarters as needed.

For his protection, the Inspector shall wear the hard hat issued by the Agency at all construction jobs where he is subjected to the hazard of falling and flying material. This will also serve to identify the Inspector on the job and set a good example for other people working on the project.

Protective vests of International Orange must be work for high visibility when the Inspector is in close proximity to traffic or moving equipment. It is better to be seen by a vehicle than to be hit by one.

It is particularly important that the Inspector insure that closed or confined spaces (such as tunnels, pipelines, tanks, manholes and other underground structures) are safe before entering or allowing others to enter. Most Contracting Agencies provide portable equipment for the detection of oxygen deficiency, lower explosive limit and for toxic concentrations of methane, hydrogen sulfide and other chemical substances. It is essential that the Inspector learn when and how to use such equipment. If it is not readily available from his Contracting Agency, he should contact the local fire department for assistance.

In the demolition of existing structures or facilities, as well as with new construction in an open area, toxic waste materials may be encountered. The Inspector should immediately request professional investigation of the site if he suspects chemical contamination, uncovers strange odors, discovers asbestos-like materials or encounters any other condition that he considers potentially hazardous.

Construction Operations Safety

General. The enforcement of regulations for the protection and safety of construction workers is the responsibility of the Federal or State Agency delegated by law to enforce occupational safety and health rules.

Employees of the Contracting Agency are cautioned that their participation in the contractor's safety program is to assist in the recognition of safe and unsafe practices. As a general rule, the contractor alone is responsible for the safe conduct of the contract and Contracting Agency employees are to avoid giving instructions to the contractor that might be construed as directing the work.

In his absence, the Contractor may delegate responsibility for safety for his employees and the public to a superintendent, foreman, and leadman. They should have sufficient knowledge and experience to be aware of hazards in and surrounding the work area.

Under the law they are considered to be "competent persons" and can be held personally liable for failing to identify and correct hazardous conditions.

The law defines a "competent person" as one who is capable of identifying predictable hazards in the surroundings; or working conditions which are unsanitary, hazardous or dangerous to employees or the public. A competent person has the authority to take prompt corrective action to eliminate any hazardous conditions.

It the Project Inspector permits all of the Contractor's supervisory personnel to absent themselves from the project at the same time, he risks unintentional assumption of being the competent person until they return.

Construction Safety Enforcement. The following general instructions are intended to provide guidelines to inspection personnel:

(1)   All inspection personnel, on projects administered by the Contracting Agency, shall exercise a conscientious and diligent effort to eliminate any conditions which, in their opinion, appear to be hazardous to the contractor's employees, the public or themselves.

(2)   Construction project safety shall be governed by the pertinent requirements of the various OSHA agency publications as they apply, including the: Construction Safety Orders, Electrical Safety Orders, General Industry Safety Orders and Tunnel Safety Orders; and special requirements as directed by the contract and Agency policy such as the *Work Area Traffic Control Handbook.*

(3)   At the beginning of a project, the Inspector shall discuss the subject of safety with the contractor to establish the interest and concern of the Contracting Agency.

(4)   The Inspector shall have a timely discussion with the contractor and his Supervisor regarding safety problems which can be anticipated during construction.

(5)   Any conditions, practice or act which develops during construction, which in the Inspector's judgment is a potential hazard, must be called to the contractor's attention at once.

(6)   Any unsafe condition, practice or act should be clearly identified and if the contractor responds promptly and the resulting condition appears to be safe, no further action is required.

(7)   When the contractor fails to take satisfactory corrective action promptly, the Inspector shall issue a Notice of Non-Compliance to the contractor and request OSHA to make and investigation. The Notice of Non-Compliance shall state that the unsafe work area shall be vacated and the unsafe condition corrected. If

the contractor fails to remove the employees from the unsafe work area, the Inspector shall consult his supervisor about notifying OSHA. If the supervisor cannot be reached immediately, the Inspector may make the decision to notify OSHA.

The Notice of Non-Compliance shall contain the following information:

- Description and location of unsafe work area to be vacated.

- Nature of the unsafe condition.

- Section number of the Safety Orders violated.

- When verbal and written warnings were issued and to whom.

- Date and time Notice of Non-Compliance was issued and to whom.

(8) Every request for investigation shall be noted in the job record, including the:

(a) Time of call;

(b) Name of OSHA authority representative contacted;

(c) Name of OSHA authority field investigator, the hour and date of his arrival on the job and a summary of his report.

(9) Judgment should be exercised to avoid unnecessary involvement of OSHA in minor violations which have little urgency. The Inspector should consult with his Supervisor when there is doubt regarding the action to be taken.

(10) Deaths and serious injuries on the project, either to the public or the contractor's employees, shall be reported to Contracting Agency headquarters by phone and a "Job Safety record Accident Report" shall be prepared by the Inspector an forwarded to the Contracting Agency headquarters as soon as possible.

Excavation and Trenches. In order to prevent deaths and injuries from cave-ins, the Inspector should satisfy himself that the contractor is familiar with the required safety standards and shoring methods to prevent cave-ins and that the employer has taken all necessary precautions to protect his workers. The protection of workers must be judged at least as effective as that provided for by the Construction Safety Orders.

Many states, such as California, have laws providing that all employers must have Division of Occupational Safety and Health permits to make excavations or trenches five feet (1.5 m) or deeper.

Agency contracts valued at more than $25,000 or which call for excavation or trenching five feet (1.5 m) or deeper, should require that detailed plans of trench shoring systems be submitted by the contractor for Agency review.

Sloping the sides of trenches to avoid the need for shoring now depends upon the soil type. The maximum allowable slopes (horizontal distance to vertical depth H/V) are:

Stable Rock		Vertical 90°
Type A Soil*	¾:1	53°
Type B Soil	1:1	45°
Type C Soil	1½:1	34°

*A short-term maximum allowable slope of ½:1 is permitted for excavations in Type A soil that are 12 feet or less in depth and will be open less than 24 hours.

There are a number of excavation configurations, including: benching, combinations of benching and sloping, trenching- shoring- sloping, etc. Consult the *Construction Standards for Excavations* promulgated by OSHA for minimum standards required. All shoring systems, including trench shields and boxes, must now be designed by a registered professional engineer.

The soil classifications described by OSHA for use in connection with excavations are:

Type A. Clay, silty clay, sandy clay, clay loam and cemented soils such as caliche and hardpan. It is not Type A if it is fissured, subject to vibration, is layered or has been previously disturbed.

Type B. Granular, cohesionless soil such as: angular gravel similar to crushed rock, silt, loam, sandy loam, silty clay loam, sandy clay loam, fissured Type A soil and unstable dry rock strata.

Type C. Granular soils such as: gravel, sand, loamy sand, submerged soil or from which water is seeping, and submerged unstable rock.

For projects requiring a Division of Occupational Safety and Health permit for excavations or trenches, the contractor is required to designate a "competent person" to make daily inspections for the excavations, adjacent areas, and the shoring systems for evidence of possible cave-in, failure of the shoring, hazardous atmospheres, or other hazards. A Competent Person/Trench Excavation Certification Form shall be completed and forwarded to the Contracting Agency Headquarters. The top portion of the form above the dashed line shall be filled out by the Inspector. The remainder of the form shall be filled out and signed by both the contractor's representative and the competent person before the end of the first day of work and prior to any workers entering the trench or excavation.

Tunnels, Pipelines and Confined Spaces. Hazardous conditions may exist or be created in a variety of situations when work is performed in confined areas. The Inspector should be alert to recognize these situations in order to insure that the contractor takes the necessary safety precautions.

The first safety consideration in tunnel construction is protection from moving ground. Most of the tunnel construction done under contract to the Contracting Agency requires approval of the contractors' proposed shoring plans by the Agency. It is the Inspector's duty to enforce strict conformance with all the requirements of these approved plans. As tunnel construction progresses, mechanical ventilation must be provided in order to provide sufficient clean air and avoid the accumulation of toxic gasses. Also, a means of quick communication with personnel outside the tunnel is vital to the protection of personnel.

"Confined spaces," as defined by OSHA Construction Safety Orders, includes the interior of storm drains, sewers, utility pipelines, manholes and any other such structure that is similarly surrounded by confining surfaces which could permit the accumulation of dangerous gasses or vapors. Most or the safe practice provisions for tunnel construction also apply to confined spaces.

A "permit-required confined space" is defined in the General Industry Safety Orders as having one or more of the following hazards: (1) atmospheric; (2) engulfment; (3) entrapment configuration; or (4) other recognized hazard. A host employer, including the Contracting Agency, must notify the contractor of the existence of permit-confined spaces. All existing sewer facilities shall be considered permit-required confined spaces, and the contractor is required to implement a "permit space program" prior to performing any such work. Other types of permit-required confined spaces shall be as identified in the specifications. For detailed information, refer to the General Industry Safety Orders.

Tests for the presence of hazardous gasses on oxygen deficiency shall be made with an approved sensing device immediately prior to a worker's entering a confined space, and continuously while a worker is inside to insure a safe atmosphere. Although this is primarily the contractor's responsibility, the Contracting Agency provides such testing devices for use by its own personnel and the Inspector needs only to contact his Supervisor whenever the equipment is needed. In general, the Contracting Agency equipment should be used on all projects as a back-up system, to check on the accuracy of the test equipment supplied by the contractor.

Sources of ignition, including smoking, should be prohibited in any confined space. If the possibility exists that confined spaces tested and found not to be hazardous may become hazardous as construction proceeds, an approved safety harness with a life line attached should be utilized with at least one worker standing by on the outside, ready to give assistance in case of emergency.

Potentially hazardous confined spaces where inspection work is performed by Contracting Agency personnel, such as the larger diameter sewer pipe lines, particularly where there is a live sewer atmosphere or plastic liner work, should be checked regularly to detect any deterioration from a safe environment.

Explosives. When the nature of the work requires that a contractor use explosives, the contractor must first obtain a permit from the proper authority. Copies of such permits must be kept on the project at all times. Such permits usually require a qualified blasting operator and special blasting inspection from the permit authority.

Accident Reporting. Inasmuch as accidents and personal injuries may involve complicated litigation in connection with workers' compensation or damage actions, complete and accurate reports are of the utmost importance. The Inspector must file accident reports in accordance with Contracting Agency reporting procedure.

Jobsite Maintenance. Good housekeeping and sanitary provisions are important to the safety of to worker and the public. The Standard Specifications should be referred to for specific requirements.

The contractor is responsible for public and private property which may be endangered by his operations and shall take every reasonable precaution to avoid damage.

Throughout all phases of construction, the rubbish and debris on a project shall be held to the absolute minimum and confined to organized disposal and storage areas. In the interest of a safe working environment, materials and equipment shall be removed from the worksite as soon as they are no longer needed. Excess excavated material should be removed from the site as soon as practical.

Dust nuisance is to be held to a minimum by cleaning, sweeping and sprinkling with water or other means as necessary. The use of water for sprinkling must be controlled so as not to generate other problems such as mud and slippery conditions.

The contractor's equipment and construction operations shall not contribute excessively to air pollution by discharging smoke, exhaust fumes, dust or other contaminants into the air in such quantities as to exceed the limits legally imposed by any local control authority (such as the local Air Pollution Control District). Likewise, severe noise pollution for protracted periods should be avoided or minimized as much as possible. Any contemplated operations of this nature, such as on-site pavement crushing, should be reported to the Supervisor.

Adequate drainage shall be maintained at all times. Existing gutters, ditches or other drainage devices are to be kept clear of spoil or debris. Blockage of normal drainage is not permitted, except in time of emergency for a few hours during the day to protect the work. The Inspector is cautioned to insure that a temporary dam does not create a nuisance or hazard to the public. The contractor shall not drain water or other liquid into an existing sewer line, and only clear water may be channeled into the storm drain system.

Sewers through the worksite shall not be disrupted. However, if sewer lines are accidentally damaged, the immediate temporary repair shall involve the resumption of sewage flow by conveyance in closed conduits until disposed of in a functioning sanitary sewer system.

Construction materials are normally not to be stored in public streets for more than five days after unloading unless specifically located within a pre-designated construction area. Care should be exercised in the placement of materials to minimize traffic hazards or damage to existing plants, trees, shrubs or ornamental objects to which the property owner attaches significant value.

Vehicular access to residential driveways shall be maintained to the property line, except when necessary for construction activity for short, reasonable periods of time during the day. Safe and adequate pedestrian zones and public transportation stops, as well as reasonable pedestrian crossings of the work at frequent intervals, are to be maintained by the contractor. Whenever possible, existing sidewalks are to remain free of obstructions.

Hazardous Materials. OSHA requires that an employer have a written Hazard Communications Program at the worksite to ensure that employees are provided with information regarding hazardous materials in the workplace. Whenever employees of different employers are on the same jobsite, it is considered a "multi-employer worksite." Multi-employer worksites are common in the construction industry.

At multi-employer worksites, the General Industry Safety Orders requires that employers inform any employers sharing the same worker area of the hazardous substances to which their employees may be exposed while performing their work and any suggestions for appropriate protective measures. The contract shall comply with the following:

(a) The contractor shall keep copies of Material Safety Data sheets at the jobsite for hazardous materials used in the work and make them available to the Inspector upon request.

(b) The contractor is required by the Standard Specifications to comply with all federal, state, and local safety laws, including hazard communications and the use of protective equipment.

(c) The contractor's employees are to confine their activities to their specific job area.

(d) The contractor shall inform the Inspector of the presence of any hazardous materials used by its employees.

Also, the contractor shall be informed of hazardous materials used by the Contracting Agency in the areas in which the contractor's employees will be working. If requested, Contracting Agency Materials Safety Data Sheet information shall be made available to the contractor's management personnel.

Hazardous Waste. The Project Engineer is responsible for the removal and disposal of hazardous material waste encountered at the jobsite which is not identified on the plans or in the specifications. The following procedure shall be followed by the Inspector when any substance suspected to be hazardous is encountered:

1. The Inspector shall instruct the contractor to stop the work in the area of the suspected hazardous material.

2. Instruct the contractor to cordon off the area affected by the suspected hazardous material and secure the area from entry by any unauthorized personnel.

3. Inform your Supervisor and the Project Engineer as soon as possible of the circumstances of the discovery and any action taken.

4. Confer with your Supervisor and the Project Engineer as needed for any further instructions.

5. Direct the contractor to continue construction activities outside the area influenced by the suspected hazardous material as appropriate under the circumstances.

6. Monitor, communicate, and document all activities affecting the execution of the contract associated with the discovery of suspected hazardous material.

7. The Project Engineer shall be responsible for preparing the manifest and arranging for the disposal of any hazardous material waste.

The following includes some examples of situations typically encountered during construction that may indicate the presence of hazardous material:

- Any abandoned tank uncovered during excavation or drilling.

- Soil contaminated by surface spills or from leaking above ground or underground storage tanks. Indications of soil contamination include odors, discoloration, and moist areas.

- Contamination caused by previous dumping of hazardous substances.

- Groundwater or surface water contamination from surface spills or leaking above ground or underground storage tanks.

- Buried drums containing hazardous or unknown substances.

- Any unusual odor or smell.

- Polychlorinated byphenyl (PCB) contaminated material from transformers.

- Asbestos materials such as piping, building insulation, old ceiling or floor tile, and roofing materials.

- Railroad ballast which may be contaminated with heavy metals, heavy petroleum hydrocarbons and/or PCB's.

Traffic Safety and Convenience

General. Agency projects on highways, major and secondary streets nearly always have special traffic control requirements, such as street or lane closures, detours and hours of work. The *Work Area Traffic Control Handbook* (a reference document in may Contracting Agency contracts) specifies general requirements and controls.

The *Work Area Traffic Control Handbook* sets forth basic principles and standards in order to provide safe and effective work areas and to warn, control, protect and expedite vehicular and pedestrian traffic through the construction project.

The responsibility of safe and proper handling of traffic rests with the contractor. The Inspector shall see that the contractor provides access for traffic as required by the contract and shall direct the contractor to correct any potentially dangerous situation that exists. If necessary, the Inspector will instruct him in writing to take action to protect and warn the traveling public.

Street Closures. Contracting Agency management is normally authorized to close residential streets for certain periods, when necessary for construction work, subject to the following limitations:

(a)   Authority is limited to residential streets normally carrying light vehicular traffic traveling only to residences within the blocks immediately adjacent to the construction area.

(b)   Closure is confined to daylight working hours.

(c)   Closure must be approved by the Division Chief or higher authority, after consultation and notification to Utility, Police, Fire and Traffic Agencies.

(d)   The Division Chief is responsible to assure that the contractor has provided for adequate advance notification to all residents affected by the closed area.

(e)   Suitable warning barricades and lights shall be placed. Refer to the *Work Area Traffic Control Handbook* for guidance in providing a safe and effective work area.

Requests from the contractor to close or restrict traffic on any major, secondary or other street (including residential streets), subject to through traffic, must be received in writing sufficiently in advance to permit adequate time for investigation and report to Contracting Agency authority. No such closing or restriction of traffic may be permitted prior to approval. The Contracting Agency Public Information Office will be responsible for notifying the appropriate elected officials of all street closings by the Agency.

In general, the Traffic Agency will post and maintain all necessary detour signs to lead traffic through the detour, without cost to the contractor. The contractor shall place and maintain all barricades, lights and signs necessary to protect the public from hazards within his area of operations.

Barricades and Striping. The preponderance of the work inspected by the Contracting Agency is performed in public streets. Consequently, nearly every project creates some increased degree of hazard to the traveling public and the imminence of moving vehicles in the streets increases the hazards to the workers and the Inspector on the project. It is imperative that the degree of interference with the normal traffic flow be kept to an absolute minimum and that the working area is adequately defined with barricades and lights to warn the traveling public of the work area and to afford a satisfactory degree of protection to the workers.

The *Work Area Traffic Control Handbook* shall be used for guidance when barricades, warning devices or signs are required. Any temporary pavement marking required will be done by the Traffic Agency. Necessary removal of existing marking and the later removal of temporary marking to restore the permanent marking is the responsibility of the contractor.

Temporary "No Parking". At the request of the contractor, the Traffic Agency will post and remove temporary "No Parking" signs. The contractor is charged for this service.

Traffic Emergency Service. When the contractor fails or neglects to adequately barricade and light his work area, and does not respond to instruction to correct these conditions, the Contracting Agency Inspector arranges for Contracting Agency forces to perform the necessary emergency work and the contractor will be billed by the Contracting Agency for the cost incurred.

CONSTRUCTION SURVEY STAKING

General. Unless otherwise specified, all construction staking on contract projects is provided by the Agency Survey Division. Construction survey staking services by private engineering firms or licensed surveyors on all permits administered by the Contracting Agency must conform in all respects to the quality and practice required of the Contracting Agency Survey Division. All grade sheets must be prepared on a grade sheet similar to the Contracting Agency grade sheet form.

The contractor is required to notify the Survey Division prior to starting work, in order that necessary measures may be taken to insure the preservation of survey monuments and bench marks. At the beginning of the job, the Inspector should verify with the Survey Division that this has been done.

Inspectors shall instruct the contractor to protect all survey stakes, witness markers, reference points and survey data painted on existing improvement.

Normally, stakes will be set for rough grade, curbs, headers, sewers, storm drains and structures. Stakes may be set on an offset with a station and a corresponding cut or fill to finish grade (or flow line) indicated on a grade sheet. The grade sheet may be issued on a standard form or be a copy of the Surveyor's completed field notes. In the case of structures, such as bridges, these stakes may serve as controls for checking the formwork prior to placing concrete. All other stakes will be set to finished grade with top colored with blue crayon (commonly referred to as "blue tops"). It is the option of the Agency as to whether grade and line are provided by blue tops or by marking the cuts and fills on pavement (or the stakes) or by referring to stakes, drill holes, chisel cuts, etc., on a grade sheet. If the Inspector is on the job when the blue tops are set, he will check all hubs (blue top reference stakes) set by the contractor before permitting the blue tops to be disturbed. To avoid damage these hubs should be set a minimum of four inches (10 cm) below subgrade. Blue tops are not to be driven down for use as a hub or other grade reference for the contractor due to the extreme possibility of error in setting and also the blue color could mistakenly indicate finished grade.

It is never permissible for a contractor to set stakes for the elevations shown on the plan unless he himself, is or employs, a registered engineer or a licensed land surveyor. However, he may set such auxiliary stakes for his own purposes as he desires and he is required to se "guineas" (usually small stakes) or intermediate grades and to transfer grades from offset stakes.

Survey Service Requests. All requests for survey service must be made by the contractor. Extra survey service, for replacing lost or disturbed stakes, or for the contractor's convenience, may result in an extra charge against the contractor. When the Inspector is doubtful that the lines or grade are to plan, due to abuse of the reference points, of for any other reason, or the line or grade does not appear to check, he may request survey to recheck or to provide additional control points. Care should be exercised to keep these survey requests to a minimum.

Earthwork Stakes. Rough grade stakes will generally be set parallel to and on an offset from the operation being performed. The interval between stakes will be 50 feet (15 m) or less if the project is less then 500 feet (50 m) long. An attempt is made to use a consistent and convenient offset, but this is not always possible due to interferences. A lath, serving as a witness marker, will be set adjacent to the stake marked to show identifying stations and offset distances. Along with the stakes and laths, a grade sheet is issued to the contractor and Inspector containing essential information such as: type of stake set, station, offset distance and cut or fill to a specific plan location.

For slopes, where heavy cuts or fill are to be constructed, slope stakes and offset reference stakes are furnished to the contractor together with a rough grade sheet to permit him to utilize heavy equipment to economically approach the final grades to within approximately 30 mm (.10 of a foot).

Witness marker laths identifying these stakes show the difference in elevation between the slope stake and reference stake; distance between slope stake and reference stake; station of slope stake and reference stake; cut or fill and distance from slope stake to toe of cut or shoulder of fill; and slope ratio.

In addition to the inevitability of these slope stakes being lost during construction and inaccuracies that develop as the contractor transfers these grades, another set of stakes and grade sheets are provided for final grading operations. The location and information for these stakes will vary depending on the requirements, such as proximity to final grades, proposed drainage bench and other improvements.

Street Improvement Stakes. After rough grading is accomplished, a survey party will set curb, gutter (or header) stakes and issue grade sheets. Except when specifically noted the grade sheets, all elevations shown refer to the top of curb (at face) or header. The contractor sometimes mistakenly uses such elevations for flow line data and the Inspector is cautioned to check the grade sheet against the construction to insure that the grade sheet is being properly interpreted and the plan requirements are being met. In case of a varying curb face, cuts or fills will be given both to the top of the curb and to the flowline. Stakes for curb construction or for monolithic curb and gutter are always set on a convenient offset (usually five feet [1.5 m]) for ease of construction. Unless otherwise indicate, the offset distance refers to the top front face (street side) of the curb. A tack in the top surface of the stake is the exact point from which to measure. Intervals between stakes are generally 25 feet (7.5 m), but this distance will decrease if the overall length of the project is less than 500 feet (150 m) or the rate of grade is under one-half of one percent. A witness lath will be set adjacent to each stake.

The Inspector must be on the alert to see that driveway depressions on the curb are not overlooked. The centerline of driveways may not be staked but the centerline station will be indicated on the grade sheet and, except for some permit work, will also be shown on the project plans.

Flow line elevations will be necessary if concrete gutter is to be placed against existing curb. The survey party provides this by chisel cutting an inverted "T" or by triangle symbols painted on the curb face. These are usually set at a constant vertical offset above the flow line [usually three inches (8 cm)]. This dimension above the flow line will be painted on the curb face at suitable intervals.

After the concrete curb and gutter, or gutter only, has been placed and the concrete has set sufficiently to prevent damage to the surface, the flow line should be checked for proper drainage by running water down the gutter.

Subgrade stakes may set by the survey party under certain conditions. These are referred to as "red tops" because the tops are colored with red crayon to distinguish them from finish grade stakes which are colored blue. When blue top stakes are set, the contractor must calculate the subgrade and set hubs besides the blue tops and far enough below subgrade to prevent them from being disturbed by the subsequent grading operations. These hubs must be checked by the Inspector for accuracy and dimension below blue tops so there will be no questions as to their elevation.

When the roadway measures less than 40 feet (12 m) in width between curbs, stakes will generally be set only where elevations are shown on the plans. If the crown section varies between centerline grade changes, additional stakes will be set on the centerline of the roadway.

For roadways whose widths are more than 40 feet (12 m) between curbs, stakes will be set on centerline at specified intervals, in addition to plan elevations.

Stakes for "T" sections are not normally set by the Survey Division. Necessary stakes along the "T" section must be set by the contractor and checked by the Inspector. Consult the standard plan for sheet crown sections for information required to perform the calculations necessary to check this operation.

In intersection "blue tops" will be set for all plan elevations. After the base courses of asphalt pavement are laid [on major streets, pavements thickness over four inches (10 cm)] the fills to finish surface of the wearing course will be painted on the base course to indicate the fill to finished grade. If the base course is to be laid in more than one lift, the Inspector may request that fills to finished grade be painted on the surface of each lift. The fill data provided by the Survey Division for paving will be at the same intervals and locations stated above for blue tops set for fine grading operations. Fill data required along "T" sections must be set with a string line by the contractor and verified by the Inspector.

In alleys, rough grade stakes will be provided on one side of the alley only where the cut or fill is extensive, except in those cases when the grades for the opposite side of the alley are substantially different. Generally these stakes will not be "tacked."

Where the cut or fill is not extensive, or after extensive cut or fill operations, tacked grade stakes or reference points, with a suitable offset, will be provided on both sides of the alley between the property lines at each end of the alley between the property lines at each end of the alley and at specified intervals, depending on the alley length and flatness of grade. A grade sheet will be issued indicating at cuts or fills for both top of header grades and the flow line of the longitudinal gutter.

If alley intersections are to be constructed, the curb returns will be staked at the time the stakes for headers are set. If the project includes the improvement of adjacent streets, the curb returns for alley intersections will be staked along with the adjoining curbs.

When requested by the contractor, after rough grade has been completed, a survey party will set blue tops for the longitudinal gutter flow line.

Many street lighting systems are construed as a portion of the work to be performed by the contractor as part of a complete street improvement project. If this is the case, the electrolier bases are usually located after the curb has been constructed. It is a simple matter and will expedite construction for the contractor to establish the location of the electroliers (and their stations) by utilizing the curb stakes. The difference in stationing between the electrolier and the nearest curb stake is used to obtain the required measurement between them.

When the street improvements exist, electrolier locations will be indicated by a "Y" painted on the curb. Where no curbs exist, electrolier stakes will be referenced to the proposed curb.

Pipelines, Utilities and Other Substructure Staking. Under normal conditions, mainline sewer and storm drain pipe will be laid from offset stakes. The offset distance will be determined by the contractor based on the equipment he intends to use, depth of cut, type of soil encountered, and trench width. Line and grade must be transferred from these offset stakes by the contractor and checked by the Inspector. These stakes will be set at specified intervals, generally 25 feet (7.5 m) or less. Generally, at least three consecutive stakes should be used to establish line and grade, either in the trench or on the surface, to detect staking errors prior to laying pipe.

Other offset stakes along the mainline will be set at the following locations: existing joins, BC's and EC's of curves and inlets, outlets, and stubs of manholes or structures. If the manhole stub is not on the sewer line produced, a stake will be set 10 feet (3 m) or more from the manhole and on the line of the stub. Offset and dimensional stakes are also used to locate other structures such as: lampholes, clean out structures and special structures. A grade sheet will be issued describing the kind of stakes, offset dimension or line indication, cut to flow line or special elevation, station and other special information.

For large diameter pipe [generally 60 inches (1.5 m) and over], sloped excavations, or cast-in-place structures, "blue tops" for line and grade may be used in the trench in addition to the offset stakes which would then be used for excavation only. Blue tops become necessary because of the high risk of inaccuracy in transferring grades from long offset distances.

As a rule, stakes for house connections are set one foot (30 cm) or more beyond the end of the pipe of the house connection line produced. Offset stakes may be set if the house connection is unusually long or is not on a straight line or straight grade. If a general note on the job plans calls for a uniform depth at property line then the stakes are set only for location. If a special depth is called put for any house connection, then both depth and location will be indicated. This and any other special information will appear on the grade sheet.

Staking for catch basins for storm drain systems are frequently included with the curb stakes. Usually on stake is set on the curb stake line opposite the center of the catch basin. For catch basins over seven feet (2.1 m) long, stakes, are also set opposite the ends of the proposed structures. If the curb stake line is too close to permit excavation for this catch basin without losing the stakes, the stakes are set on a larger offset dimension. This will be noted on the marker lath and the grade sheet.

If there is not existing curb and no curb is to be constructed, strakes for small catch basins are set five feet (1.5 m) back of the future curb on the centerline, and five feet (1.5 m) or more on each side of the centerline of the catch basin in order that the catch basin may be constructed to be on line with the future curb. For catch basins over seven feet (2.1 m) long, stakes are set opposite the ends of the structure in addition to the centerline stake, If no stations are shown on the plans, the stakes will be identified with letters.

A third condition is frequently encountered: a curb may already exist which must be removed for catch basin construction. In this case, drill holes are set in the top surface of the curb at an appropriate distance each side of the center of the catch basin. These points are identified and referenced to the grade sheet by letters. Always review the appropriate standard plans and projects plans thoroughly. The center stake may locate the center of the basin, the center of the opening or the width depending on which type of basin is to be built.

Normally, when the mainline pipe and the catch basin are already in place, a lateral pipe extending from the catch basin to the mainline transition or junction structure which is less than 25 feet (7.5 m) in length, is not staked unless the grade is less than half of one percent or other is a major line or grade change point required by plan. Catch basin outlet elevations are determined either from the project plans or the standard plan. The project plans will also indicate the elevation difference between the mainline and the connection lateral inlet flowlines. The flowlines of the lateral pipes are normally on a straight grade with their connecting structures.

In most cases, the utility companies will use the contractors rough grade stakes for any relocation of services that are required. However, when approved, a separate set of stakes for utility purposes may be set before construction of the improvement begins. This staking generally follows pipeline staking procedure.

A series of offset stakes, with cuts to specified elevations, will be provided to locate the shaft for tunnel operations. Controls to establish line and grade from the shaft into the tunnel may vary depending on the contractor's excavation methods. One popular system is for the survey party to drive nails and tins on line into timber shoring on both sides of the shaft, and above the bottom elevation of the intended crown shoring of the tunnel to be driven. The nails and tins must be out of the way of contractor's equipment. Line and grade for initial tunneling operations can be easily transferred into the tunnel by the contractor using plumb bobs hanging from these controls. As the tunnel is driven, periodic checks will be made by the survey crew and additional reference points established. The kind of points again may vary with the type of equipment and methods used by the contractor. One system is for the survey party to drive "spads" (hood type nails) on line in the soffit of the tunnel shoring. This provides a means of hanging a plumb bob and visually producing a line from some point previously set. A cut from the spad to a specified elevation is also given or nail and tin driven into the tunnel shoring on one side of the proposed pipe at springline. After the tunnel excavation has been completed, the bedding placed (when required), and the pipe is ready to be laid, blue tops must be set by the survey crew.

When jacked casings are to be used in lieu of tunneling, the jacking pits may differ from the tunnel shafts, (principally in length) in order to accommodate the jacking equipment, etc. Prior to jacking operations, the equipment and first pipe may be checked by the survey party to insure correct alignment. Since the casing will be moving, adequate permanent reference points cannot be set, therefore, it will be necessary that periodic checks of the casing alignment be made by a survey party to insure that the required line and grade are being achieved.

Bridges, Buildings and Retaining Walls. Retaining wall and bulkhead stakes are usually set on the upperside and at an offset distance great enough to allow for excavation. The offset distance will refer to a plumb face or other definite plane of the wall or bulkhead that is constant throughout its length. The Inspector must be alert to the calculations that may be needed for necessary checks, such as batter, etc. Stakes will be set opposite the ends and BC's and EC's, grade changes, angle points, changes of cross section in the wall or bulkhead, and at specified intervals in between. A grade sheet will be issued indicating the offset and will usually give two grades from each stake, one for the footing and the other for the top of the wall or bulkhead.

After grading has been completed, the original stakes may be unusable for construction due to the height of the wall or extensive offset distance. In such cases, footing stakes may also be set with offsets and grades to specific locations.

In the construction of major bridges and structures, many of the measurements are beyond the scope of the Inspector to witness and check using hand instruments and he must depend on the surveyor. On some projects, a survey crew is retained on a full time basis; on others, only periodic services are required.

Stakes are first set beyond the lines of excavation on the prolongation of the control lines of each abutment, pier or bent, which will be needed during construction. The contractor should decide on an offset that will be convenient and will not encroach upon his construction activities. After the stakes for construction are set, the Inspector and the contractor must jointly review and understand to what location each offset and cut or fill stake is referred. If piles are required, the location of the pole line is established and a temporary bench mark is set at each end of the structure at pile cut-off elevation. In some cases, pile stakes may be combined with foundation stakes. Foundation stakes may be set to subgrade if the grade has been made. Piling cut-offs may not be marked if it is feasible for the contractor to use the foundation stakes for this purpose. Temporary bench marks may be placed at some even foot (cm) above the foundation to facilitate the contractor's work. If starter walls are to be constructed as part of the foundation, it may be necessary for the survey crew to make a line and grade check of these forms prior to placing concrete.

If starter walls are not utilized, it may be necessary to have a survey crew locate the bottom of the pier or abutment wall prior to forming (or after forms are set-up) to check the top and bottom for line and elevation. A convenient elevation marked on the back of the form can be of value to the contractor and Inspector in setting and checking variable height rebar, embedded items and pour strips. Other items that may be critical should be located and checked by a survey party prior to placing concrete. These could include: bearing plates, expansion joints, grade changes and wall openings. On some projects, when warranted, the surveyor will make frequent checks during concrete placement to assure that the forms remain plumb, in line, on grade, and that no shift or displacement is taking place. The walls or other components may again be checked after concrete has been placed to verify that the plan requirements have been met.

Falsework bent lines and temporary bench marks should be staked by the surveyor for the contractor's use in building the falsework. The specifications usually require the contractor to submit detailed shop drawings of the falsework for approval. It is known that each joint in the falsework will settle when the concrete is placed; also, there will be a settlement of the bridge due to its own load when the falsework is removed. The falsework to its own load when the falsework is removed. The falsework and dead load settlement and the design camber must be taken into account to meet the plan elevations when the work is completed. Therefore, elevation checks on the bridge deck forms and girder locations are usually beyond the scope of the Inspector and will require the services of a survey party.

Other controls that may require special survey service are: the perimeter formwork, reinforcing steel layout, anchor bolts for railings and other equipment and streetlight locations. After the reinforcing steel is in place, elevations of forms should be read and the necessary adjustments made to meet the required grade and camber. The inspector must be sure to have the screeds checked to insure that the finished surface required by the plan elevations will be met.

When necessary, grade sheets will be issued in the field by the surveyor.

The above outline is a general guide to acquaint the Inspector with the information and checks that may be necessary from the survey party in staking bridges and other structures. Every project presents unique dimensional checking problems and each solution will depend on the design and existing field conditions. The Inspector, contractor and surveyor should maintain close communication so that the stakes and information supplied by the surveyor is adequate for the contractor's construction needs and for the purpose of verification by the Inspector so that the surveying services are not wasted.

PRESERVATION OF IMPROVEMENTS AND UTILITIES

General. Most public improvement contract projects are located in developed areas, either business or residential. Such areas always have public and private improvements and utilities which must be considered in design and construction of the improvement.

Contracting Agency projects provide that the contractor is generally responsible for preserving public and private property and utilities along and adjacent to the roadway insofar as they are endangered by his operations. The contractor must take proper precautions in performing the work to avoid damage to property. The contractor is obligated to repair to rebuild damaged property, or to make good any damage by other means in an acceptable manner to the affected party.

Contract plans for the project should show the type and location of the improvements within and adjacent to the construction area. In addition, the contract documents will provide for the treatment of such improvements (removal, replacement, relocation). However, the inaccuracy of available underground location records or the construction of improvements after the project plans were prepared, or other reasons, may lead to problems not provided for in the contract.

The Contracting Agency Inspector is responsible to assist the contractor as much as possible to avoid damage and further to coordinate the expeditious relocation of utilities which the construction operations require. In addition, the Inspector must see that all damage that does occur is properly repaired. One of the first duties of the Contracting Agency Inspector assigned to a new project is to walk the job with the construction plans and compare actual field conditions with those noted on the plans, making special notes on the plans where conflicts exist and where other problem areas are obvious which may result in damage to existing improvement during construction. Consultation with the contractor or design engineer to resolve potential problems prior to construction will expedite progress and avoid unnecessary expense and potential litigation at a later date.

Usually after award of the contract and before any work is done by the contractor, the Agency conducts a preview inspection of the project in considerable detail. The purpose is to insure that all damage done by the contractor during construction operations is repaired and to insure that the Agency, the contractor and the public are protected from false claims and litigation.

The preview consists of a video (or photographic) and audio (tape recorder) description of the condition of the jobsite and adjacent improvements such as walls, sidewalks, driveways, pavement, buildings, fences and landscape features. Disputes regarding the prior condition of adjacent improvements can usually be resolved by replaying the video tape on a video cassette recorder (VCR).

Encroachment and Salvage. Many times privately owned improvements need to be removed or altered because they encroach into the public right-of-way and interfere with the planned improvements. Written notice ahead of construction is given by the Contracting Agency to the property owner to remove the encroachment.

As required by the contract provisions, the Inspector should see that the contractor also gives notice for those encroachments that remain at the time of construction. The contractor may want to assist the property owner to salvage or relocate such items as a public relations gesture although he is under no obligation to do so after giving reasonable notice.

Utility Protection and Relocation. The importance of protecting existing service utilities from damage in the construction area cannot be too heavily stressed because of the possible hazard to life and property and disruption of other services should damage occur.

The Contracting Agency searches all known substructure records and furnishes location descriptions (usually on the project plans) of all utility substructures (except service connections) which may affect or be affected by the construction work required. Information concerning the removal, relocation, abandonment or installation of new utilities is furnished to all prospective bidders. After award of the contract, it is the contractor's responsibility to either call the area Underground Alert (USA) Service or request the utility owners identified in the bid documents to mark or otherwise indicate the location of their facilities (including service connections to private property).

Although the project plans or specifications may require the contractor to relocate or reconstruct certain utilities as a portion of the work required, such relocation or reconstruction is usually performed by the owner of the utility at the request of the awarding Agency. Such relocations are ordered when it is known that the existing utility will interfere with construction of the project. When feasible, the owners responsible for such reconstruction will complete the necessary work before commencement of work by the contractor. When utility interferences are considered extensive, a pre-construction meeting is called by the Contracting Agency prior to the actual start of construction to plan the necessary coordination of utility protection and relocation during construction. Such meeting s will be attended by the contractor and his interested subcontractors, the design engineer and utility representatives having installations within the project site and the Project Inspector.

Should any potential hazard exist as a result of accidental damage, the Inspector should immediately instruct the contractor to evaluate all personnel from the vicinity of the damage and to prevent anybody in the vicinity from entering the hazardous area until emergency crews from the affected utility agency have made necessary repairs and given a clearance that the potential hazard has been eliminated.

After the initial safety precautions have been taken, the contractor or the Contracting Agency Inspector shall immediately notify the owner of the damaged utility.

There is almost no limit to the variety of utility interferences which are likely to occur during construction, especially in older, heavily built-up areas for which records may not exist or are so inaccurate with respect to new improvements as to be practically worthless. It thus behooves both the contractor and the Inspector to be on the alert and cooperate with each other to the fullest in order to minimize the great hazard to life and property which can ensue from accidental damage to many hidden utility substructures.

Substructure Interference Reporting. Because jobsite interferences frequently result in claims for additional compensation by the contractor or utility owner which are often difficult to resolve and may end in court for final resolution, the accuracy and completeness of the Inspector's job records are of the utmost importance in establishing the facts surrounding the circumstances of the actual damage or extent of any claim.

Protection and Use of Sewers and Storm Drains. All mainline sewers and storm drains will be identified on the plans and their proximity to the proposed improvements will be indicated. The Contracting Agency Inspector must see that the contractor takes all reasonable precautions to protect these installations. Any major damage to such mainline installations should be reported immediately to the appropriate maintenance organization so that crews cam be dispatched to alleviate any emergency situation which might exist or be imminent. All existing sewer lines (mainline and house connections) shall remain in service and if damaged shall be maintained by temporary means to permit continued operation until permanent approved repairs can be made.

Unless specifically provided for in the contract documents, the Contracting Agency Inspector shall not permit storm water or any material other than sewage to be deposited in the existing sewer system. The contractor shall provide adequate desilting of water before it is deposited in the storm drain system.

Catch basins shall remain in service during project construction except as provided for in the contract documents during remodeling or reconstruction phases of the work. It is noted, however, that such reconstruction should not be scheduled during the rainy season or when disruption of existing storm drainage facilities would create a hazardous condition.

WARRANTIES AND DISCLAIMERS

General. When a Contracting Agency furnishes a contractor a set of contract documents, they are generally held in law to have an implied warranty to the effect that the plans and specifications are both accurate and suitable for construction the project. The accuracy of such warranty refers to the Contracting Agency's factual representations about the details and nature of the project. Should these representations subsequently be proven inaccurate or be seriously misrepresented, the contractor may recover his added costs due to a breach of warranty. An example of such a breach would be depicting deep soil borings on the drawings with no indication of rock substrata, while in actual field conditions, bedrock was encountered at a depth of a few feet.

Generally, the Contracting Agency cannot evade its implied warranty by including boiler plate disclaimers such as a statement that all warranties of the plans and specifications are disavowed and thus rendered ineffective should a disparity arise during construction of the project. Neither will the Agency evade responsibility by including a statement in the contract documents that the contractor must inspect the work site and examine all the specification and plan details to verify their accuracy. Any attempt to convey responsibility for the suitability of the plans and specifications to the contractor are usually disregarded in legal actions. The rationale substantiating this position is that the bidding process would become virtually impossible to conform to if intelligent contractors could not rely on the accuracy and suitability of the Agency's contract documents.

Contractors are aware that some discrepancies and inaccuracies occur in plans and specifications for most large projects. They understand that these will be taken care of by interpretation or by change orders at the appropriate time; and if contract time or cost is affected, the Contracting Agency will make a proper adjustment. However, when obvious or patent defects appear in the contract documents, the contractor cannot recover damages for knowledge which should have been obvious to him at the time he signed the contract. Both the contractor and the Contracting Agency are held to a standard of reasonable care, and responsibility is imposed on both parties. This doctrine is based on what an intelligent contractor knows as well as what he should have known in his field of endeavor.

With respect to a contract clause requiring the contractor to visit the worksite to familiarize himself with the site conditions, a prudent contractor would be expected to make a physical inspection of the surface of the site, and from such an inspection and subsurface data available, draw conclusions regarding its impact on his proposed methods of construction. He would not be obligated to undertake expensive and time consuming pre-bid subsurface investigation.

However, there is a doctrine that acts against an Contracting Agency that attempts to avoid responsibility by deliberate silence or omission in the contract documents. The Contracting Agency is obligated to disclose all facts in its possession which could materially affect the contractor's bid. Concealment of such facts by the Contracting Agency will subject it to liability for any subsequent claims for damages. These damages not only relate to the direct costs such as labor, materials and equipment, but for other more intangible costs such as delay costs, lost efficiency, cost escalations and extended overhead.

DIFFERING SITE CONDITIONS

General. Differing or changed site conditions are subsurface or latent physical conditions differing materially from those represented in the contract; or unknown physical conditions of an unusual nature differing materially from those ordinarily encountered and generally recognized as inherent in the type of work being performed.

A prudent contractor will include a sum of money in his bid to provide for possible unexpected costs or contingencies which might be encountered in the project. Should no problems arise, this contingency amount becomes a bonus to the contractor. Modern specifications attempt to minimize such contingency amounts by providing a "differing site conditions" clause. This is intended to induce bidders to reduce such contingency costs in their bids and to depend on the differing site conditions clause to negotiate with the Contracting Agency during the construction period when differing or unforeseen difficulties are encountered. In theory, the contractor is treated in an equitable manner and the Contracting Agency receives lower bids over a period of time, having reduced contractor contingency provisions in its bids.

PROJECT DELAYS

General. Delays are classified as excusable but non-compensable, excusable and compensable, and inexcusable as well as non-compensable. Excusable but non-compensable delays are provided for in the contract documents, and after negotiation, time extensions are granted without any compensation to the contractor. Examples of this type of delay are: labor disputes, fires, adverse weather conditions, etc., which conditions are beyond the control of the Agency or the contractor.

Excusable and compensable delays occur when a valid change in the work is required and equitable settlement to the contractor is made in time extension and costs. Inexcusable and non-compensable delays result from a contractor failure to prosecute or properly sequence the work, or his abandonment of the project. No time extension or compensation is granted by the Contracting Agency in such circumstance.

In some instances, work may be delayed for reasons attributable to both the Contracting Agency and the contractor. This is referred to as a "concurrent delay," and in such event, neither the Contracting Agency nor the contractor is entitled to recover damages for the period of delay.

The inherent dependency of the contractor on such items as weather, labor disputes and availability of materials or equipment, often results in delays in the scheduled progress of the work. Due to the domino or ripple effect on other phases or elements of the work, serious delays may completely disrupt job progress, resulting in inefficiency in deploying labor and equipment. Deviation from the most logical sequence of work may cause the project to be delayed until unsuitable weather develops, thereby extending a delay period.

In all cases, the Inspector should keep accurate records of the disposition of men and equipment during any delay. He should realistically assess the impact of rain or its effects. Often certain craftsmen (such as electricians) can be moved from outdoor to indoor work without experiencing any delay from wet weather. Conversely, cement finishers cannot work in the rain and not for days after a rain until subgrade is dry and firm. In the latter instance, the contractor would be expected to remove standing water from forms to expedite preparations for the earliest possible placement of the concrete after the rain.

It should be noted that in some jurisdictions where wet or cold weather is commonly experienced, local Contracting Agency specifications may not recognize such conditions as "adverse weather," and delays from such conditions may not qualify for time extension or compensation.

SUSPENSION OF THE WORK

General. The issuance of an order suspending all or part of the work can have a dramatic effect on the additional costs incurred by the Agency and the contractor. Such an order must be specified as the whether the entire project is to be suspended or that a clearly defined nonconforming portion of the work only is to be suspended. Suspension of the work for the convenience of the Agency should be utilized only when no other alternatives are viable or available. Most orders to suspend work are of an emergency nature involving safety or potentially hazardous situations. However, some orders to suspend work emanate from changes found to be necessary in the work or from unforeseen site conditions.

Constructive Suspension of the Work. When the Agency does not issue an affirmative order suspending the work, but the facts indicate that such an order should have been issued, the contractor may be entitled to an equitable adjustment in time and compensation under the doctrine of construction suspension of the work.

To establish entitlement, the contractor must demonstrate that the Contracting Agency did something (or failed to do something) to accommodate the forward progress of the work that a prudent owner, faced with the same issues, would have done in administration of that contract. The costs to which the contractor may be entitled include the increased direct and indirect costs caused by the delay, disruption or suspension, but does not include profit.

Failure of the Contracting Agency to respond in a timely manner to requests for information or to process shop drawings, can act in such a manner as to delay fabrication and delivery of equipment to the extent that the project is delayed. This concept can also be reversed when the contractor fails to submit or resubmit shop drawings in a timely manner.

CLAIMS AND DISPUTES

While the plans and other contract documents establish the standards for materials and construction details for the project, they constitute the minimum acceptable to the Contracting Agency. On the other hand, these specified materials and details are the most the contractor has to furnish to comply. This usually results in a narrow band of contractor performance hovering between what he has to provide and the least the Contracting Agency will accept. Thus the contractor and Contracting Agency personnel are automatically placed in an adversary position at the outset, with the constant potential for dispute.

Such disputes most often arise from ambiguities in the contract documents or differing interpretations of plan or specification details. While the Inspector has little or no authority to change a contract document, or to impose his interpretation as a final decision, he is responsible to bring such problems to the attention of those who are responsible to clarify or resolve the matter.

Most disputes are quickly resolved, some by agreement as to the intent, and some by change order when additional cost or time is involved. When agreement cannot be achieved, some disputes result in claims, which if not resolved, can lead to arbitration or litigation.

Therefore, it is essential that disputes be quickly resolved to avoid stoppage of the work, or other costs to the contractor. If the contractor is correct in his position, he is entitled to any cost or time extension associated with the dispute. In the interests of the Contracting Agency, such costs must be avoided or minimized. When engaging the services of a consultant or construction manager, most large public agencies give considerable attention to the extent of experience professed by the applicants in claims avoidance.

Claims avoidance is best achieved by the Inspector (and contractor) planning ahead, looking for problems or conflicts that could arise and result in contractor delays or disputes. By this means, there is usually time to resolve the problem before the work reaches the stage that a delay will result.

Of course there will be times when both the Contracting Agency and the contractor cannot reach agreement, and the contractor will be ordered to proceed (without prejudice) according to the Contracting Agency's interpretation, with the contractor subsequently filing a claim. The contractor will submit documentation to substantiate his claim for additional time and payment for his costs. It becomes immediately apparent how important the Inspector's records are in documenting the field conditions leading up to the incident that initiated the claim, as well as his records of the labor, materials, duration or other factors involved in the disputed portion of the work as it progresses.

This should emphasize that complete and accurate inspection records are essential in substantiating (or refuting) contractor claims, particularly when the contractor belatedly files a claim for extra compensation for something that he claims happened in an earlier phase of the work. Good records and progress photographs (or video coverage) of the work become useful tools for assisting in resolving most disputes and claims during the course of the work and are invaluable in the event of subsequent litigation.

CONSTRUCTION SCHEDULING

General. When contractors bid a project, they expect to complete the project in the least possible time at the least cost. They traditionally approach the bidding process by listing all the cost items involved such as labor, materials and equipment, then add a percentage for overhead, profit and contingencies.

One of the cost factors best controlled by the contractor is the construction time. By his estimate of the extent of the work and the size of the work force he intends to use, the contractor can arrive at a close approximation of the time required to complete the work. Since time and labor costs are directly related, the contractor can prepare his bid accordingly. If he should by good management of time and work force, and by utilizing innovative methods, complete the work in a shorter period, the resulting cost savings accrue to him, thereby increasing his profit margin. Reducing the time for completing the work means less interest charges on cash invested during construction, less supervision and overhead expense, earlier availability of equipment for use in other work, and earlier release of funds traditionally retained by the owner.

All of this emphasizes the importance of developing a realistic schedule to assist in controlling construction, which is one of the critical elements of consideration during the bidding process. Prior to World War II, most construction contracts were scheduled using bar charts or similar devices to graphically depict major construction operations. As the size and scope of projects increased, particularly during the war and immediately thereafter, it became obvious that a better system was needed to control the vastly larger projects. Early systems such as the Gantt chart, the Line Of Balance (LOB) and Integrated Project Management (IPM) were principally graphic. Procurement and construction projects for the U.S. Navy late in the war resulted in the development of a line and arrow diagram system to depict activity and the time relationships. This system was called the Program Evaluation and Review Technique (PERT).

While this system worked well for Navy procurement contracts, improvements were developed in the private sector, principally in the chemical manufacturing industry and the emerging computer industry, evolving into the critical path method of scheduling known as CPM. This method of diagram scheduling (called networking) has many attractive features. It can be programmed into a computer to take the drudgery out of processing data or schedule revisions. Also, the CPM can graphically depict the time and cost elements of each activity and subactivity, show the impact of early or late start dates, develop logical sequencing of events, and most important, define the critical path. The critical path is the least time in which a project can be logically completed.

Any factor, such as delay or the elimination of a portion of the project, will almost always impact on the critical path to shorten or lengthen it. Thus the diagram can be used to accurately estimate the effect of changes in the project with respect to cost and time.

Initial Schedule Development. The first step the contractor must take in preparing an estimate for bidding a project is to make a time schedule to fit into the project time constraints and set up a tentative plan for doing the work. This schedule should show all of the items affecting the progress of the work including the impact that the weather (seasonal) could impose. Such factors as: when delivery of critical items can be realistically anticipated; or what physical restraints may be imposed by the contract, such as keeping major traffic arteries clear near major shopping centers during the holiday shopping season, must be accurately evaluated. From these and other controlling factors, production rates for the major items of work are decided and the number, type and size of the contractor's plant, labor force and equipment needed to complete the work within the contract time, is determined.

From this schedule, the contractor should be aware of the indefinite, hazardous or other features that could affect his cost or time estimates. He can determine the total man-hours of labor and total machine-hours for major equipment required in doing the work, peak labor requirements, and controlling delivery dates for important items of equipment or materials. In addition, it can show him his cash requirements based on scheduled income and expenditures during the contract period. This is referred to as "cash flow."

Bar Charts. The most common form of scheduling, particularly for smaller projects, is the use of bar charts. They are almost always used, even in larger projects, to initially phase the major elements of the work to keep them within the contract time. Since they are graphic, they reflect logical sequencing of work, and indicate the obvious need to accelerate certain items of work, or to adjust lead times for procurement.

Bar charts are best when they are kept relatively simple, reflecting only the major items of work or administrative processing. An example of a bar chart for the construction of a $300,000, steel-framed warehouse with an overhead crane is shown in Figure 1.

FIGURE 13-1 – Bar Chart Construction Schedule

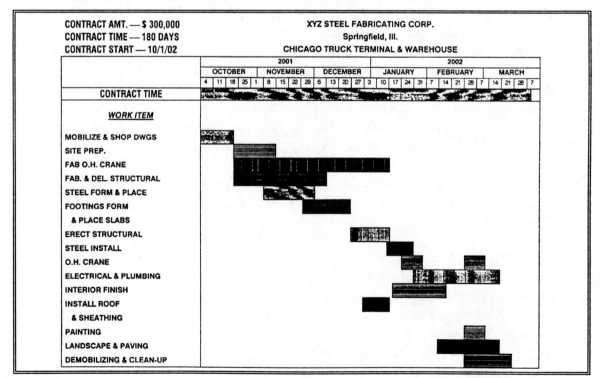

In graphing construction sequencing and showing construction time required for each major item of work, bar charts are of invaluable assistance in developing PERT or CPM diagrams. Even with a CPM schedule available, construction superintendents often use a bar chart in the field to measure and record actual progress against planned progress.

Critical Path Method of Scheduling (CPM). This tool of management is very useful for larger or complicated projects and is required by many governmental agencies on some types of construction. CPM is based upon planning and job analysis that goes far beyond that needed to bid a project. Besides knowing the detailed breakdown of the work into its elementary operations and graphing the sequential relationships, the planners must know the duration of each operation, lead time for procurement, shop drawing preparation and approval times, fabricating and delivery times for procured items and methods that the contractor intends to utilize.

Activities are generally represented on the network by arrows or nodes connected by sequence lines. Analysis may be manual or by electronic computer for establishing realistic time relationships, selecting these operations whose completion times would be responsible for establishing overall project duration, determining the impact of changes and the operations affected (including the effect on project duration), establishing realistic sequencing of the work, and determining the status of work in progress with relation to the number of days ahead or behind schedule.

Work items that are performed concurrently often have float time in them. Float is the difference between the time it actually takes to do the item of work and the time in which it must be completed.

"Free float" is the maximum time of slippage of an activity that can be tolerated without affecting the completion date of any other activity, assuming all activities start at the earliest possible times. "Total float" is the maximum slippage of an activity that can be tolerated without affecting the completion date of the overall project.

Each node represents the completion of the preceding activity and the start of the following activity. By the sequence of operations that will require the most time to complete the project utilizing normal work forces and equipment, the critical path is determined, which establishes the duration of the project. The critical path is indicated by darker(heavier) arrows or by double lines. The critical path is commonly shown on a horizontal line, but may be shown on contiguous arrows any place in the network. There is no float on the critical path.

Generally, CPM networks are evaluated to determine early and late starts with respect to each activity and its completion time. The early start dates are the earliest any activity can start, each in its own proper sequence. The late start dates are worked backwards from the specified project completion date and indicate the latest date that each activity can begin and still complete the project time.

When the cost-loaded activities in the diagram are plotted against contract time and scheduled progress, a set of curves for the early starts and late starts result in a set of curves called the "banana curves". These curves are used to predict the cash flow requirements for the contractor and the owner, and can be useful in assessing the contractor's performance. When the contractor's progress is plotted on the banana curves, it becomes obvious whether or not the is meeting his schedule. Should his progress curve fall below the late start curve, it is an indication that the contractor may be in trouble and he may need to accelerate the work to avoid exceeding the contract time. A speed-up of the work would presumably be accomplished only by more outlay for labor the equipment since the cost-loading of the normal network is presumed to be minimum.

A computer is frequently used to analyze the normal times of each activity on the critical path an compute the maximum time schedule. Then the activity is selected on the critical path which offers the least cost increase in relation to any time decrease, after which a new schedule is computed with the activity "crashed" (taken from the late start date). Under this procedure, the normal start date is considered to be the maximum time. Since these attempts are made to reduce it, the crashed time can represent a substantial reduction of time and produce a new critical path. The computer continues to "buy time" along the latest critical path as cheaply as possible. Once a final schedule has been decided, the contractor and owner are provided with sufficient information to insure adequate control over time and costs.

CPM schedules must be up-dated (usually monthly) to reflect the contractor's progress (or lack of it), change order work, delays or nay other factors that may impact the schedule. The cost-loaded network, when compared to the work actually completed, gives the Inspector the best means of measuring progress for payment purposes on large and complex projects.

CONSTRUCTIVE CHANGES IN THE WORK

Any conduct by an Contracting Agency representative authorized to issue change orders, but which are not construed as formal change orders and have the effect of requiring the contractor to perform work different from that prescribed b the original terms of the contract, constitutes a constructive change order. This could entitle the contractor to equitable compensation and time extension.

An example of a constructive change occurs when it originates from a difference of opinion as to the interpretation of the specifications or plans. An Contracting agency representative refuses to issue a formal change order for work in question because it was in his opinion, already required under the terms of the contract. Should he later prove to be incorrect, the result is a constructive change order.

Proof that a constructive change has occurred lies with the contractor. The Inspector must be aware of the consequences of his interpretations of contract documents and avoid actions which would inadvertently trigger a constructive change.

CONSTRUCTIVE ACCELERATION

In the absence of an acceleration order, a constructive acceleration order would exist if (1) the contractor is entitled to an excusable delay under the contract documents, (2) the contractor has notified the Contracting Agency of a delay and requested an extension of time for an excusable delay which is improperly denied by the Contracting Agency, (3) the Contracting Agency issues an order or implies an acceleration order to meet existing contract dates, (4) the contractor notifies the Agency that he considers and informal order to alter his schedule to be a constructive change, and (5) the order to accelerate incurs additional costs related to constructive change order.

It should be emphasized that the Inspector must assess all time delays and cause valid extensions of time to be issued based on the contract documents. Failure to do so in a timely manner can result in sizable acceleration claims presented by the contractor.

CONSTRUCTIVE SUSPENSION OF THE WORK

When the Contracting Agency does not issue an affirmative order suspending the work, but the facts indicate such an order should have been issued, the contractor may be entitled to and equitable adjustment in time and compensation under the doctrine and constructive suspension of the work.

To establish entitlement, the contractor must demonstrate that the Contracting Agency did something (of failed to do something) to accommodated the forward progress of the work that a prudent owner, faced with the same issues, would have done administration of that contract. The costs to which the contractor may be entitled include the increased direct and indirect costs caused by the delay, disruption or suspension, but does not include profit.

Failure of the Contracting Agency to respond in a timely manner to requests for information, or to process shop drawings, can act in such a manner as to delay fabrication and delivery of equipment to the extent that the project is delayed. This concept can also be reversed where the contractor fails to submit or resubmit shop drawings in a timely manner.

CHAPTER FOURTEEN

CHANGE ORDERS TO THE CONTRACT

By this time the project construction is well underway and monthly progress payments are being executed. Everything is going fine until one of your construction operations uncover an unknown, or unforeseen, obstacle that causes the operation to extend beyond the estimated time and dollars. If the obstacle needs to be dealt with to complete the contract work, and it was not addressed in the project plans and specifications by the contracting agency before bid time, then the contractor should try to initiate a change order to the contract.

Any substantial item of work that your company believes to be outside the contracted work needs to be brought to the attention of the contracting agency as soon as possible. If the work progresses, and the conflict is brought up at a later date, the contractor can often lose the opportunity to be monetarily reimbursed for the extra work along with any extension of contract time. Many slight changes to the projects operations are simply a function of the construction business. The contractor should carefully determine if the change to a contractual operation is monetarily worth his time and effort to recover money or contract time from the contracting agency. Some contractors submit notices of potential claim on every slight change that happens at the jobsite, virtually every day. It is often better for the contractor to choose his notices of potential claim more carefully, saving them for the changes that substantially effect the project; so that when he really needs a change order, he hasn't cried wolf too many times. This is a very gray area and each individual contractor will have to find his own comfort level, with respect to claims. Just keep in mind that a partnering attitude with the contracting agency will in most cases go a long way in creating and keeping a financially successful construction company with an excellent reputation within the industry.

This chapter deals with these change orders to the original contract in the following categories:

Quantity Overruns/Underruns,

Notice of Potential Claim,

Force Account Settlement,

Fixed Price Settlement,

Claims.

QUANTITY OVERRUNS/UNDERRUNS

The contract that your company signed with the owner should have established quantities with which you based your estimate and proposal upon. If it appears that one or more of these quantities are substantially in error, then your company might have the right to negotiate a new unit price for those bid items or simply get paid your bid unit price for an overrun quantity.

Most contracting agencies have a set percentage that the bid item quantity must go over or under, in order for the unit price to be re-negotiated. These percentages typically range from 15% to 25 %. If the over or under run in quantity is less that the set percentage, then the contractor will simply get paid the actual constructed quantity at the bid unit price. Some contracting agencies will allow the contractor to re-negotiate a particular unit price, even if the percentage of quantity change does not meet the requirement, if he can prove that the unit price of the item has substantially changed as compared to the bid unit price. This should be approached on a case-by-case basis, but it cannot hurt to professionally talk to the owners representative about any particular situation in which you sincerely believe that your company has sustained substantial financial damage due to a factor outside of the contract.

In the situation of a quantity underrun that is beyond the percentage allowable by the contracting agency, the contractor might be able to prove that the unit price for the actual quantity constructed is higher than that of the unit price estimated in the original proposal. For example, the contractor might have a bid item that is to construct a concrete retaining wall that is paid by the contracting agency on unit price basis, by the cubic yard. To construct the retaining wall the contractor might have to fabricate a specific amount of concrete forms to be used in several locations on a project. The cost of fabricating these forms would be spread over the cost of all the concrete to be placed and paid for by the owner. If the owner decides that part of the concrete bid item does not need to be constructed or there was simply an error in the original concrete quantity, then obviously the cost of fabricating the forms will be spread over less concrete quantity; therefore, raising the unit price. Another scenario, using a quantity underrun, might be that a contract had a specific quantity of drainage pipe to be placed at designated locations on the jobsite. At a later date the drainage system is re-designed or simply had an error in the original quantity beyond that allowed by the contract specifications. Lets also say that the re-design or quantity error resulted in pipe runs that are shorter in length and a farther distance apart on the project than those originally bid upon. This would create more moves and set-ups for the construction crew; again raising the unit price of pipe actually being placed. In both of the quantity underrun situations mentioned above, the contractor should be able to prove to the contracting agency how the underrun raised the unit price of a particular bid item and consequently should have that unit price adjusted via a contract change order. Typically, underruns in a bid item quantity will result in an increased unit price.

In the event of quantity overrun by more than the percentage allowed by the contract specifications, the unit price of the bid item may be either raised or reduced. Whether the item unit price will be raised, reduced, or remain the same depends on the particular situation. In the scenario above where the fabrication of concrete forms is spread over the total quantity of concrete being placed, the owner might add a substantial amount of small, isolated concrete structures. Although your company might be able to use the same previously fabricated forms on the concrete structures, the costs associated with moving and setting up many more times outweighs the savings in spreading out the fabrication costs over more concrete quantity. In this case the contractor may be able to prove a higher unit price on the additional concrete quantity. On the other hand, and with respect to the above drainage pipe operation, the contracting agency might add a substantial amount of pipe quantity to the contract. The pipe runs might be much longer and shallower than those shown originally on the contract plans. The owner might request a reduced unit price proposal on the extra pipe runs, due to the lower installation costs involved in longer, shallower runs. Most contracting agencies will not request a reduced unit price from the contractor, but simply pay the extra quantity at the established unit price. The contractor should go into these negotiations prepared although, because it is within the owners contractual right to re-negotiate the item if the change in quantity is beyond the set contractual percentage.

Quantity overruns and underruns can have a major effect on the financial outcome of your projects. Much consideration must be given to these quantity changes on how they effect the operation in question, and other operations, along with the effect on the flow and duration of the entire project.

NOTICE OF POTENTIAL CLAIM

A notice of potential claim is simply a written and verbal notice to the contracting agency that your construction operations have encountered a situation that appears to be outside the limits of the original contract. Most contract specifications state that a notice of potential claim must be submitted to the owner within a certain time period (typically ten days) from the date of discovering the potential conflict, for the contractor to have the right to submit a claim for reimbursement at a later date. Even if the contract does not require a notice of potential claim, it is a gesture of good faith to provide this notice to the contracting agency. It is also a good idea to make sure the owners representative knows that this notice is simply to protect your right to get reimbursed later, if the change does in fact cost your company substantial time and money.

The notice of potential claim letter (as shown in Figure 14-1) should include the reasons why the contractor believes such work is outside the contract and an approximation of what costs might be involved. In these instances it will help your case to quote contractual specifications to justify the changes that are occurring on the project; avoid speculation or what you might think is common sense. In publicly funded projects, the contracting agencies representatives cannot approve changes to the contract unless they are specifically addressed within the standard, or special, specifications. In this situation, these representatives have others reviewing everything that happens on the project from a paperwork standpoint; therefore, all change orders must be well documented and founded within the contract.

Often, even if the agencies representative believes the contractor deserves a change order on a specific item, he may not be able to approve it if he does not have an avenue with which to base it on within the projects contract. Well-organized potential claim paperwork will greatly assist you at the end of the project when the details of the events have faded in your memory and in the memory of the contracting agencies representatives.

FIGURE 14-1 Notice of Potential Claim Letter

ABC ASPHALT PAVING COMPANY
1234 E. Delta Ave.
Gresham, Oregon 97777
Phone (503) 663-6633
Fax (503) 663-6634

February 29, 20--

CONTRACTING AGENCY
4567 E. 129th Street
Gresham, Oregon 97777

9901.2

RE: N.E. 12th Street Reconstruction
Contracting Agency
Contract No.: 98345

Mr. Hal Johnson (Project Manager):

With respect to the above mentioned project, ABC Asphalt Paving Company has encountered a situation with the 12" Concrete Sewer Pipe at approximate Sta. 1+69.50 that we believe is outside the scope of this contract. ABC Asphalt Paving Company has encountered some buried car bodies that appear to be part of an old dumpsite. Obviously, the trench work for this mentioned sewer pipe is substantially different than that of the contract plans and specification.

ABC Asphalt Paving Company at this time is giving a Notice Of Potential Claim per Section X.YYY of the Standard Specifications.

We at ABC Asphalt Paving Company will work with (Contracting Agency) to resolve this matter in the best way possible. If you have any immediate suggestions or need further information, please contact me at our home office.

Sincerely,

Sam D. Maple
President
ABC ASPHALT PAVING COMPANY

FORCE ACCOUNT SETTLEMENT

Once the contractor and the contracting agency have reached the point where they agree that a particular portion of work is outside the original contract, then a contract change order will be written.

One way to determine the amount to be paid within the change order is a force account settlement. A force account settlement is very similar to a cost-plus contract, as discussed in previous chapters. The contractor and the owner track the amount of manhours, equipment hours and materials purchased that are directly associated with the extra work. Typically, each contracting agency will have a specific daily form for the contractor to fill out relating to the extra work being constructed. In these cases, the contract specifications will clearly define how much may be reimbursed per hour for a particular piece of equipment and per hour for each labor craft. This section of the specifications will also define what mark-up or profit will be added to each category (labor-equipment-materials-subcontractors) of the change order.

FIGURE 14-2   Typical Force Account Sheet

EXTRA WORK SHEET

Prepared By:		Issued To:	
Address:		Address:	
Date:		W.O. No.:	
Project:		Job No.:	

Material Supplier | | Subcontractor | | Owner Force Account | |

Description of Work:

Description of Labor, Equipment or Materials	Hours or Quantity	Hourly Rate Unit Price	Equipment Cost	Material Cost	Labor Cost	Subtotal Amount

		SUBTOTAL	
Summarize Taxes/Insurance	Labor Fringe Benefit		
	Labor Taxes/Insurance		
	Company Overhead		
	Company Profit		
		TOTAL	

BY: _____   BY: _____

　　Mater. Suppl./Subcontr./Owner　　　　　　Contractor

As in cost-plus contracting, timely and accurate tracking of the costs related to the change order will determine how well the contractor will be reimbursed monetarily in a force account claim settlement. The mark-up, or profit, is a set percentage of base cost, so the contractor must be very careful to include all of the costs in the change order. Force account settlement of a change order can be somewhat safe if the contractor accurately tracks the costs, but the profits will typically be small due to the cost-plus nature of the payment and the cost associated with tracking and reporting all aspects of the changed operation to the contracting agency.

FIXED PRICE SETTLEMENT

Another method with which to determine the amount to be paid on a contract change order is a fixed price settlement. A fixed price claim settlement is much like a fixed price contract proposal as discussed in previous chapters. The contractor would estimate the costs (labor-equipment- materials-subcontractors) that he believes it will take to complete the extra work contemplated. He would then add a mark-up or profit that he believes encompasses the risks associated with the work to be completed in the change order. In some cases, the contract specifications state the amount, or percentage, of mark-up that may be added to the direct cost of a change order.

The contractor would then submit this fixed price change order proposal to the owner for approval. If the owner agrees with the costs proposed, a change order will be written for the fixed price submitted, relating to the extra work. If the owner does not agree with the costs proposed, some negotiation may take place, the settlement may become force account in nature, or the situation may go into a claim as discussed later in this chapter. Figure 14-3 displays what a fixed price change order proposal may look like.

The contractor needs to give a great deal of consideration to the method of payment on contract change orders. For example, if the extra work to be done is very difficult and your company does not have any past productions with which to estimate the work, it would probably be a good idea to request a force account settlement of the change order in order to reduce the amount of associated risk. On the other hand, if your company has a substantial amount of past production information on the style of extra work to be constructed, then you might want to request a fixed price settlement of the change order so that you can estimate with your own company established equipment rates and add on a higher mark-up or profit percentage to the costs instead of the lower set reimbursements allowed under force account settlement in the contract specifications. The fixed price change order settlement method will typically require a great deal less paperwork and tracking time as compared to a force account settlement.

FIGURE 14-3 Fixed Price Change Order Proposal

ABC ASPHALT PAVING COMPANY
1234 E. Delta Ave.
Gresham, Oregon 97777
Phone (503) 663-6633 Fax (503) 663-6634

CONTRACTING AGENCY   February 29, 20--
4567 E. 129th Street   9901.2
Gresham, Oregon 97777

RE: N.E. 12th Street Reconstruction
  Contracting Agency
  Contract No.: 98345

Mr. Hal Johnson (Project Manager):

With respect to the above mentioned project, ABC Asphalt Paving Company has encountered a situation with the 12" Concrete Sewer Pipe at approximate Sta. 1+69.50 that we believe is outside the scope of this contract. ABC Asphalt
Paving Company has encountered some buried car bodies that appear to be part of an old dumpsite. Per our jobsite discussions, we agreed that the trench work for this mentioned sewer pipe is substantially different than that of the contract plans and specification.

ABC Asphalt Paving Company at this time is submitting our additional cost proposal for this extra work for your review. Our proposal to complete this additional trench work for the 12" Concrete Sewer Pipe is $7,500.00. We are also requesting an additional 5 working days be added to our current contract time.

We at ABC Asphalt Paving Company thank you in advance for your consideration in this matter and would appreciate a response as soon as possible so that this operation may resume. If you have any immediate suggestions or need further information, please contact me at our home office.

Sincerely,

Sam D. Maple
President
ABC ASPHALT PAVING COMPANY

CLAIMS

There are basically two ways in which an extra work situation can end up in a claim:

The contractor and the owner do not agree that a particular area of work is outside the original contract or,

The contractor and the owner do agree that a particular area of work is outside the original contract, but cannot agree upon a mutually equitable settlement.

If your company ends up in one of the disagreements listed above and if the operation of work in question needs to be completed in order for other operations of the project to begin, then the contractor should complete the extra work and track all costs associated with the work. At the end of the operation of the project, a formal claim that details the extra work and includes all of the extra costs spent by the contractor will have to be submitted to the next level of authority within the contracting agency. The contractor should quote sections of the contracts special provisions or standard specifications to substantiate his claim; avoid any kind of objective views or personal attacks, as being professional and proving your point using the contract will get you much farther in a claim situation. It is best to resolve these change orders or claims with those individuals at the jobsite level. Although, in some cases, the contractor may have to go beyond this jobsite level if he feels very strongly about a claim that cannot be resolved with his immediate contracting agency representative.

FIGURE 14-4   Typical Claim Letter

ABC ASPHALT PAVING COMPANY
1234 E. Delta Ave.
Gresham, Oregon 97777
Phone (503) 663-6633 Fax (503) 663-6634

CONTRACTING AGENCY   August 19, 20--
4567 E. 129th Street   9901.10
Gresham, Oregon 97777

RE: N.E. 12th Street Reconstruction
 Contracting Agency
 Contract No.: 98345

Mr. Hal Johnson (Project Manager):

   With respect to the above mentioned project, ABC Asphalt Paving Company previously submitted a Notice of Potential Claim (9701.2) regarding the 12" Concrete Sewer Pipe at approximate Sta. 1+69.50. ABC Asphalt Paving Company encountered some buried car bodies that were part of an old dumpsite.

   ABC Asphalt Paving Company tracked the extra work involved in trenching through this old dump site and stabilizing the foundation for the 12" Concrete Sewer Pipe. Listed below is the additional compensation that we believe is due for this additional work:

LABOR:	Operator	32 MH @ $22.50/HR =	$ 720.00
	Laborer	64 MH @ $18.00/HR =	$1,152.00
	Truck Driver	10 MH @ $21.00/HR =	$ 210.00

		Subtotal	$2,082.00
		Operator Fringe ($5.50)	$ 176.00
		Laborer Fringe ($4.75)	$ 304.00
		Tr. Dr. Fringe ($5.00)	$ 50.00
		Labor Accruals (29.51%)	$ 614.40

		Total Labor Expense	$3,226.40
EQUIPMENT:	Excavator	20 HR @ $65.00/HR =	$1,300.00
	Loader	12 HR @ $50.00/HR =	$ 600.00
	Dump Truck	10 HR @ $40.00/HR =	$ 400.00
	Compactor	12 HR @ $10.00/HR =	$ 120.00

		Total Equipment Expense	$2,420.00

N.E. 12th Street Reconstruction   **Page 2 of 2**

MATERIALS:	Baserock	91.56 TN @ $ 7.75/TN =	$ 709.59
	Rip Rap	32.90 TN @ $12.75/TN =	$ 419.48

		Total Material Expense	$1,129.07

MISC.:	Dump Fees	1 LS @ $956.78 =	$ 956.78

		Total Misc. Expense	$ 956.78

Total Additional Base Cost	$7,732.25
Total Additional Mark-up (20%)	$1,546.45
Total Additional Bond (2%)	$ 185.57
Total Additional Compensation	$9,464.27

ABC Asphalt Paving Company also believes that in addition to this monetary compensation that an additional 10 working days should be added to the contract days to make up for the lost time in this additional work.

If you need any additional information with respect to this claim, please contact me at our home office.

Sincerely,

Sam D. Maple
President
ABC ASPHALT PAVING COMPANY

SUBSTANTIAL COMPLETION/PUNCH LIST

When most of the construction contract work is near completion (approximately 95+%), the contracting agency can take possession of the major part of the project site. At this time, the contractor should ask for a letter certifying "substantial completion" from the owner. This is the point when the contracting agency will stop counting contract days, if a construction time is specified in your particular contract. This is also a good idea due to the fact that it will start the process of switching the maintenance of the project from the contractor to the contracting agency.

Most public works construction contracts have a warranty period associated with them in which the contractor must guarantee his workmanship and materials. Typically this period is for one calendar year, but can extend up to two years or more. This warranty period will generally start upon the certification of substantial completion by the owner. On large projects or projects that have substantially differing operations included within them, the contractor can request substantial completion on a portion of the project. Time spent getting the substantial completion on a portion of work can be very important because it can stop your company's maintenance of that completed work and begin the warranty period.

Also, in this request for substantial completion from the contractor, you should ask for a "punch list" from the contracting agency. This is a list of the remaining items that the contracting agency believes the contractor needs to complete for the project to reach final completion. Review this punch list in detail to make sure you are in agreement with all of the items on the list. Often punch lists can continue to grow from a contracting agency as time goes on and more of the owner's representatives review the substantially completed project. Request a final punch list from the owner and communicate with them on any areas with which you disagree so that you can provide them a quality project, while not dragging the final completion of the project beyond what is reasonable.

Many large construction companies will bring on one or more fresh jobsite supervisors near the time of substantial completion for a particular project. These supervisors specialize in closing out projects. Project supervisors tend to relax near the end of a project, after all of the daily stress and battles of constructing the major portion of a job. This last 5% - 10% of a project can often cost a construction company a substantial amount to complete, due to these human relaxation tendencies. Therefore, many firms bring on new, invigorated supervisors to knock out the last operations of a project and bring the construction to a profitable close.

Once all of the construction operations are complete and your company has mobilized off of the jobsite, it is time to focus on the contract closeout. Depending on the contracting agency, there will be various documents that will have to be completed by the contractor, and submitted to the contracting agency, verifying that certain contractual obligations were met.

Discussed in this chapter will be the following basic items to consider when it is time to closeout the contract:

Labor/DBE Utilization Reports,

As-Built Drawings,

Final Pay Estimate,

Lien Release,

Payment/Performance/Maintenance Bonds,

Project Profit vs. Loss.

LABOR/DBE UTILIZATION REPORTS

If the project your company completed was a publicly funded job, then there were probably some minority hiring goals or DBE subcontractor goals/requirements associated with the contract. Typically at the conclusion of one of these projects, there are some final utilization reports for the contractor to complete assuring the contracting agency that all of the contracts goals or requirements were met with respect to your labor and subcontractors. As a competent contractor you should have studied this situation and made any necessary adjustments well before this point to make sure that you were going to meet the established goals or requirements.

These reports will need to be received by the contracting agency before they will release final payment on the project. This is their way of making sure the contractor will complete all aspects of the job, including the finalization paperwork, in a timely manner.

AS-BUILT DRAWINGS

Most contracting agencies will require that the contractor submit to them, at the conclusion of the project, a set of as-built drawings. As-built drawings are simply a set of contract drawings that have marked in red all of the changes that were made during construction. These drawings are especially important for the contracting agency in determining the future locations of underground utilities, structures, etc.

As-built drawings are most efficiently developed as the project progresses. It can be a very difficult task to remember all of the slight changes and locations of buried or hidden objects after the contract is complete. Therefore, the contractor should keep on the jobsite a set of contract drawings that are marked up as operations are altered. The contractor should also make sure that any subcontractors associated with the project are aware of these as-built drawings so that they may also mark up the drawings as necessary. During the construction of a project, these on-site as-built drawing will most likely get very worn. At the end of the job the contractor can transfer all of the noted changes to a clean set of drawings for submission to the contracting agency.

Again, these as-built drawings, if required by your particular contract, will have to be submitted to the contracting agency before final payment on the project will be issued.

FINAL PAY ESTIMATE

When all contract construction items are complete, the contracting agency will generally submit to the contractor a proposed final pay estimate. Included in this final pay estimate will be all of the original bid items with the associated final quantities, any agreed upon contract change orders and any liquidated damages assessed against the contractor for finishing the contract in more time than was allowed by the contract.

If the contractor agrees with all of the final pay quantities, change orders and liquidated damages then he simply signs the pay estimate and the final payment will arrive within the time frame allowed by the contract, assuming all of the required contract finalization paperwork has been received by the owner.

If the contractor does not agree with any particular aspect of the proposed final pay estimate then he should return the final pay estimate to the owner with the corrections that he believes need to be made. If the contractor disagrees with some of the final quantities of the original bid items, then he needs to submit proof of the actual pay quantity that was constructed. This proof may come from many areas such as your superintendents/foremans daily logs, calculations or records of materials purchased, or possibly from your construction photographs. If the contractor disagrees with any part of the change orders on the project, then he needs to compile all of the information and documentation associated with extra work and submits this to the owner in the most discernable form available. If the contractor disagrees with the liquidated damages in the proposed final estimate, then he needs to submit proof to the owner of the actual completion time. This completion time section is often directly associated with the change orders written (or claimed) on the project, relating to the fact that the extra work caused the contract to be completed at a later date than originally contemplated.

If after the contractor submits his case on changes to the proposed final pay estimate to the highest authority within the contracting agency and an agreement cannot be negotiated, then the contractor would most likely have to seek legal advice on the matter. Once again, it is always best to try and resolve the situation by mutual negotiation with the contracting agency at the lowest level possible. On rare occasions the contractor would have to seek legal advise on recovering compensation that he believes is due his company from the contracting agency, on work that he believes to be outside of the original contract and that cannot be directly settled through negotiations with the contracting agency.

LIEN RELEASE

Once the final payment for the project has been agreed upon, the owner will typically ask your company to sign a lien release for the project. This release states that your company has been paid in full for the project and that you are giving up all legal right to further compensation from the contracting agency. It will also include language that states that your company has paid all of your subcontractors and material suppliers, in full, that were involved in the contract. This protects the owner from being sued for additional compensation by lower tier companies on the project, for which compensation had previously been paid to the prime contractor. If the prime contractor had not paid the subcontractors or material suppliers, then they would have to sue the prime contractor for any additional compensation and not the contracting agency.

If your company is the prime contractor on the project, you should have your subcontractors and material suppliers sign similar lien releases to the one your company signed for the owner, so that they may not come back to your company at a later date and say that they were not paid in full for the project. These lien releases can be associated with either a progress payment or final payment to that supplier or subcontractor. Lien laws vary from state to state, so the contractor should be well aware of the particular lien laws for the state you are working in, as these releases may prevent a supplier or subcontractor from fraudulently suing your company for damages they say occurred on a previous project.

Depending on which state your project was constructed within; there may be some additional lien releases you will need to obtain. For example, if there is a state run industrial insurance program that applied to your project, you will have to obtain a letter from that states Department of Labor and Industries that releases the contracting agency from the liability of any insurance premiums related to the project. Contractors may also have to request a certification from any applicable city, county, or state agency that any and all sales or excise taxes have been paid that relate to the project prior to final payment by the contracting agency.

FIGURE 15-1   Lien Release for Progress Payment

PARTIAL WAIVER OF LIEN

To All Whom It May Concern:

Whereas, the undersigned has been employed by (1) _____
_____ to furnish labor, services and materials for (2) _____
_____ work under a contract (3) _____
for the improvement of the premises described as (4) _____
in the (City – Village) _____ of _____ County of _____
State of _____ of which _____ is the Owner.

Now, therefore, the _____ day of _____, _____ for and in consideration of
the sum of (5) _____ dollars, paid simultaneously herewith, the receipt
whereof is hereby acknowledged by the undersigned, the undersigned does hereby waive and
release to the extent only of the aforesaid amount, any lien right to, or claim of lien with respect
to and on said above described premises, and the improvements thereon, and on the monies or
other considerations due or to become due from the owner, by virtue of said contract, on account
of labor, services, materials, fixtures, apparatus or machinery furnished by the undersigned to or
for the above referenced premises, releases and forever discharges the Owner and
_____ (General Contractor) from all claims and demands
arising out of labor performed or material or equipment supplied by the undersigned in
connection with the contract, but only to the extent of the payment aforesaid.

(6) _____
Name of Company

Affix Corporate
Seal Here

_____
Authorized Signature

(1) Person or Firm with whom you agreed to furnish either labor, or services, or materials.
(2) Fill in nature of work, strike the word labor, services or material if not in your contract.
(3) Describe contract by number if available, along with date and extent of work
(4) Describe improvements and location, distinguish from any other property.
(5) Amount shown should be the amount actually received on that date.
(6) If waiver is for a corporation, corporation name should be used and a corporate officer
      should sign; if waiver is for a partnership, partnership name should be used and a partner
      should sign.

State of _____
County of _____

The document was acknowledged before me this _____ day of _____, _____ by
_____ as _____ of _____

Notary Public _____

FIGURE 15-2 Final Lien Release

FINAL WAIVER OF LIEN

To All Whom It May Concern:

Whereas, the undersigned has been employed by (1) _____
_____ to furnish labor, services and materials for (2) _____
_____ work under a contract (3) _____
for the improvement of the premises described as (4) _____
in the (City – Village) _____ of _____ County of _____
State of _____ of which _____ is the Owner.

Now, therefore, the ____ day of _____, ____ for and in consideration of
the sum of (5) _____ dollars, paid simultaneously herewith, the receipt
whereof is hereby acknowledged by the undersigned, the undersigned does hereby waive and
release any lien rights to, or claim of lien with respect to and on said above described premises,
and the improvements thereon, and on the monies or other considerations due or to become due
from the owner and _____ (General Contractor) on account of
labor, services, material, fixtures, apparatus or machinery heretofore or which may hereafter be
furnished by the undersigned to or for the above described premises by virtue and said contract.

(6) _____
Name of Company

Affix Corporate
Seal Here

_____
Authorized Signature

(1) Person or Firm with whom you agreed to furnish either labor, or services, or materials.
(2) Fill in nature of work, strike the word labor, services or material if not in your contract.
(3) Describe contract by number if available, along with date and extent of work
(4) Describe improvements and location, distinguish from any other property.
(5) Amount shown should be the amount actually received on that date.
(6) If waiver is for a corporation, corporation name should be used and a corporate officer
   should sign; if waiver is for a partnership, partnership name should be used and a partner
   should sign.

State of _____
County of _____

The document was acknowledged before me this ____ day of _____, ____ by
_____ as _____ of _____

Notary Public _____

PAYMENT/PERFORMANCE/MAINTENANCE BONDS

Once the construction project has been fully accepted, paid for, and released by the contracting agency, the project payment and performance bonds become void unless they are necessary for the projects warranty period. Some projects will require a separate maintenance bond be submitted by the contractor, to the owner, covering the warranty period. Also the project certificates of insurance that your company provided the contracting agency become void at this time because your labor force and equipment are no longer working on the project. Your insurance policy is still valid, but the owner no longer needs proof of this since all work on the project is complete. Your insurance carrier will continue to send the contracting agencies updated certificates of insurance with each policy renewal period. Annually your company will receive a list of certificate holders from your insurance agent. The contractor then simply lets the agent know which projects are complete so that they do not have send those respective contracting agencies updated certificates of insurance.

PROJECT PROFIT VS. LOSS

The final item to analyze with respect to the project, now that final payment has been made, is did the job make the company money or did your expenditures exceed your compensation. You will already have a good handle on how the project did monetarily from analyzing the jobs cost control and schedule throughout the construction process, long before this contract close-out time. The contractor should analyze all operations of the completed project, including whether the individual operations made your original estimate or not. You should make notes on particular areas work that resulted in profits along with areas that produced job losses. These written notes on projects, along with all of the production rates established on the job, should go into a completed project book for future reference. When your company again is in a position to bid on a project that is similar to the one you just completed, the notes and the production rates you accurately tracked during construction of the job will assist you in successfully bidding the new project at prices with which your company can achieve reasonable profit margins and at reduced risks. The contractor should pay most of his attention to the production rates as compared to unit prices. Production rates will remain relatively constant depending on location, and the time of year for any particular operation. As discussed in Chapter 5, the contractor will simply have to adjust the material, F.O.G., labor and subcontractor prices. You should utilize unit prices as a double check of the estimate to make sure they are reasonable for the atmosphere in which you are bidding. There are too many items dependent upon unit prices to use them as your primary estimating source. Using past production rates will keep your estimates much more accurate.

CHAPTER SIXTEEN

WHAT IS ARBITRATION?

If the contractor and the contracting agency cannot come to an agreement on a contract change order or some extra work by the contractor, then a claim can be initiated by the contractor. Many contracts require these to be handled through arbitration.

Arbitration is a system, voluntarily adopted by parties to decide disputes, in which an impartial arbitrator, after hearings, issues a legally enforceable award. The arbitration system is composed of elements that have a regular relationship to each other and that operate to produce an end product: An enforceable award.

Arbitration produces an enforceable decision, which is the same thing produced by the court system. In a sense, therefore, arbitration and the courts are competitors, and parties to disputes have the option to adopt either system. If the parties fail to agree to utilize the arbitration system, then the court system is applied automatically.

The method most often utilized for the settlement of disputes is neither arbitration nor litigation, but negotiation. Negotiation, however, is not a system in the sense of arbitration or litigation. Negotiation is a technique or, some would say, an art. Successful negotiation prevents arbitration and lawsuits from arising. Once they have arisen, they can be terminated by negotiation. To most people, negotiation is by far the most attractive means of dispute settlement. It is faster and cheaper than arbitration, which, in turn, is usually faster and cheaper than litigation.

The negotiation of settlements is often facilitated by *mediation*. The parties to a dispute may select a mediator who has the experience and training that are appropriate to assist parties to resolve a dispute by negotiating an agreed settlement. Arbitration providers such as the American Arbitration Association administer mediation as well as arbitration. During the decade of the nineties, mediation has achieved strong recognition as an effective mechanism for dispute resolution. A detailed examination of the mediation process is, however, beyond the scope of this work.

The arbitration system is voluntary. Before a dispute can be settled by arbitration, all parties must agree to arbitrate. Agreement is usually reached before the dispute arises, by arbitration agreement included as a clause in a construction contract. Agreement also may be reached after the dispute arises, in which case it is referred to as a *submission*.

In the construction industry, arbitration often involves a technical aspect. The arbitrator may be called upon to interpret plans or specifications or to judge the quality of workmanship in the field. Judges and juries may not be very well qualified to resolve technical matters. A New Jersey court mused on the subject in 1951:

> It would appear that our law has evolved no entirely satisfactory method of determining disputes of this nature, when a contractor sues for a balance he alleges to be due for the construction of a building and the owner disputes the quantity and quality of the work. There are at least three methods for the determination of issues of this nature: (1) by the traditional trial by jury, (2) by the trial judge, sitting alone without a jury, (3) by arbitration. . . . Arbitration would seem to be a satisfactory method whereby designers, materialmen, building contractors and others trained in the construction field, could decide on any of the issues involved by an inspection of the premises. However, the parties in this case refused to send the matter to arbitration. One of them claimed that legal questions were involved concerning the interpretation of the contract, and that a board of three arbitrators would have one or more laymen thereon who would not be versed in legal principles.

It is a remarkable principle of arbitration that the arbitrator need *not* be "*versed in legal principle.*" Unless the parties direct otherwise, the arbitrator is not required to decide the controversy according to legal principles but may consult a sense of justice, fairness, and equity. The arbitrator need not, therefore, be learned in the law but must be impartial.

Arbitration is a system, voluntarily adopted for the resolution of disputes by an impartial arbitrator who affords hearings to parties. The dispute is resolved by a legally enforceable award. Awards that are produced by arbitration are routinely confirmed by judges and become judgments that are enforceable as any other court judgment. A judgment confirming an award is stronger than most because it is less subject to appeal.

MODERN ARBITRATION STATUTES

Common law was hostile to arbitration, since it ousted the courts of jurisdiction. Judges at old common law were paid by the parties according to the number of cases decided. Judges viewed competitive systems of dispute resolution as unwarranted competition. This began to change in 1947, when the United States Arbitration Act of 1925 was codified. Familiarly known as the FAA, the United States Arbitration Act served as a model for the Uniform Arbitration Act, which has been adopted by more than half the states.

While at common law arbitration agreements were revocable, the modern statutes make them enforceable. While at common law some courts treated arbitration awards as unenforceable, the modern statutes provide for judgments confirming awards. While at common law those courts that did enforce arbitration awards did so reluctantly and with difficulty and delay, the modern statutes provide for summary, expeditious confirmation.

The American Institute of Architects has included arbitration clauses in most of its recommended contract documents. Arbitration clauses are included in many construction contract forms published by construction industry trade associations and legal blank publishers. Many, probably most, arbitration clauses utilized in the construction industry refer to the Construction Industry Rules of the American Arbitration Association (AAA). The AAA rules provide a simplified, common sense structure for arbitration proceedings. Their length (vest-pocket pamphlet size to fit readily into an ordinary mailing envelope) compares favorably with the tome-sized Codes of Civil Procedure that flourish among the states.

Although statistics are not available, it is estimated that more than 50 per cent, of the construction industry cases that go to hearing do so under the arbitration system rather than under the court system. The proportion of *big cases* to *little cases* is about the same in both systems.

To summarize, the modern arbitration statutes have given arbitration respectability, enforceability, credibility, and popularity. The arbitration statutes have created a quiet revolution in the manner by which construction industry disputes are settled.

ARBITRATION AGREEMENTS

In all states an agreement to submit an existing dispute to binding arbitration is enforced. Under modern arbitration statutes, in almost all states, agreements to submit a dispute that may arise in the future to arbitration are also enforceable.

The arbitration clause favored for the construction industry by the American Arbitration Association is:

Any controversy or claim arising out of or relating to this contract, or the breach thereof, shall be settled by arbitration administered by the American Arbitration Association under its Construction Industry Arbitration Rules, and judgment on the award rendered by the arbitrator(s) may be entered in any court having jurisdiction thereof.

The FAA, (Federal Arbitration Act) reads:

> A written provision in any maritime transaction or a contract evidencing a transaction involving commerce to settle by arbitration a controversy thereafter arising out of such contract or transaction, or the refusal to perform the whole or any part thereof, or an agreement in writing to submit to arbitration an existing controversy arising out of such a contract, transaction, or refusal, shall be valid, irrevocable, and enforceable, save upon such grounds as exist at law or in equity for the revocation of any contract.

Although arbitration statutes require that the arbitration agreement be in writing, some cases have held that oral arbitration agreements also are enforceable, at least after the award has been made. In one case, a contractor suggested using a consultant as an impartial arbitrator, and the owner agreed. A consulting engineer determined the issue of how much material had been excavated from a construction site and the court enforced the engineer's decision.

A written arbitration clause also may apply to disputes arising under a subsequent oral agreement. In one case, the parties signed an AIA construction contract with a broad arbitration clause. Subsequently, they entered into an oral agreement under which the contractor agreed to serve as the carpentry manager. A dispute arose under the oral contract. The court held that the generic arbitration clause in the written contract covered the dispute arising under the oral agreement. The broad arbitration clause indicated that the parties intended to resolve all disputes arising out of the construction of the home by arbitration, both those arising under the original written contract and those arising out of the subsequent oral agreement.

An agreement that an designer will decide disputes may constitute an arbitration agreement. In a New York case, the contract contained a clause specifying that all questions would be resolved by the designer, whose decision would be conclusive, final, and binding. The court held that the designer's award was not to be disturbed unless totally irrational.

Most state statutes require that an arbitration agreement, to be enforceable, must be in writing. This does not mean the agreement also must be signed. The subscription of an instrument is only one way of manifesting assent to be bound by the terms of a written agreement. If the parties have manifested their assent to the contract, whether by subscription or otherwise, the arbitration agreement is enforceable.

A party, by participating in arbitration proceedings, may be deemed to have waived any objection to the arbitrability of the dispute. A party may not freely participate in selection of arbitrators, attend the hearing, submit evidence, hope for a favorable award, and in the event of an adverse finding, seek court review merely because the award is unfavorable. Although an oral contract to arbitrate is not specifically enforceable, if parties have received a written award after arbitration pursuant to an oral agreement, the award is enforceable.

Before the adoption of modern arbitration statutes, courts were reluctant to enforce arbitration agreements because they ousted courts of their jurisdiction. Alabama and Nebraska still refuse to enforce arbitration agreements on the ground that arbitration ousts the court from jurisdiction.

VALIDITY OF AGREEMENTS

Courts will enforce an agreement to arbitrate. But what if the making of the agreement itself is in dispute? Suppose, for example, that a subcontractor sends a prime contractor a bid form that contains an arbitration clause, and the prime contractor does not sign and return the form. Allegedly, however, there is an oral acceptance of the bid followed by a letter acknowledging the agreement to arbitrate. Since a bid is an offer, and an offer must be accepted before a contract is formed, there is an issue as to the formation of the contract that contains the arbitration clause. The important question is whether this issue shall be decided by arbitration or by the courts. A New York court, in such case, decided that the issue should be decided by the court. The modern trend, however, is to submit questions regarding the formation of an arbitration agreement to arbitration.

CLASS ACTIONS

It was held in a case decided in California in 1999 that the court has the power to order arbitration of potential class action claims. (Plaintiffs had filed a class action against Blue Cross, and Blue Cross filed a petition to compel arbitration.)

SUBMISSION AGREEMENTS

In the preceding section, arbitration agreements, i.e., agreements to arbitrate a dispute that may or may not arise in the future, were discussed. A *submission agreement*, on the other hand, is an agreement to submit an *existing* dispute to arbitration. The distinction is important.

One of the purposes of arbitration is to avoid litigation. Yet, arbitration has been the subject of considerable litigation. Does arbitration really succeed in avoiding litigation?

By far the largest single subject addressed by the courts in the field of construction arbitration is the enforceability of arbitration agreements. This question never seems to arise under a submission agreement. When the parties voluntarily agree to submit an *existing* dispute to arbitration, there is seldom any question of its enforceability.

A submission agreement need not be lengthy, complicated, or legalistic. For example:

I regret that we have not been able to reach agreement as to your claim for extras. I suggest that we ask the engineer to arbitrate this issue. You will present your extras along with their backup; we will present our side of the story, and the arbitrator's decision will be final.

The foregoing, when signed by the parties, is an enforceable submission agreement.

When a construction dispute arises, should the parties submit it to arbitration, or should they go to court? Each party to the dispute will analyze this question from its own point of view.

More often than not, the analysis by the prospective claimant produces a decision to submit the controversy to arbitration, while the analysis by the prospective respondent produces the opposite result. This occurs because the claimant almost always wants a speedy, efficient, and unappealable decision, while the respondent may be in no hurry to see the issue finally resolved. Therefore, since arbitration is expected to produce a speedy and final decision, the parties are likely to reach different decisions as to whether an existing dispute should be submitted to arbitration. It is for this reason that parties arrive at so few submission agreements.

APPOINTMENT OF THE ARBITRATOR

If the parties to a dispute fail to select an arbitrator, or if for some reason the agreed method for appointment of the arbitrator fails, the court will appoint the arbitrator. Court appointment of the arbitrator should be avoided, if possible. Although the appointment of an arbitrator by a court is a summary proceeding, the proceeding still takes time and costs attorneys' fees. Moreover, the performance of courts in selecting arbitrators is uneven. Some courts designate arbitrators from lists provided by the American Arbitration Association. Others proceed from lists of retired judges. Some judges appoint cronies, while others favor lawyers. Another may select a CPA, an architect, or an engineer. In short, judges have great discretion in the appointment of impartial arbitrators, but that discretion is best avoided if the parties can agree on the characteristics of the arbitrator they desire.

It is advantageous for parties to designate in advance the individual who will serve as arbitrator. Such a procedure saves time. The arbitrator starts with the trust and respect of both parties. In practice, however, it is difficult for parties to agree on an arbitrator.

Many construction contracts are handled on a mass production basis, and it is not practical to take time to negotiate the arbitration clause. Moreover, subcontractors usually want to appoint subcontractors as arbitrators; prime contractors, prime contractors; architects, architects; engineers, engineers, and so on. A subcontractor does not want a prime contractor to decide the dispute; a prime contractor does not want a subcontractor to do so; and neither wants an architect. Few contractors can be found who will readily place their affairs in the hands of a lawyer for adjudication.

A popular procedure, but one fraught with danger, is for each party to name an arbitrator, and the two arbitrators to select the third, neutral arbitrator. It sounds good, but in practice works poorly.

Party-appointed arbitrators may be, but usually are not, neutral. A party tries to select an arbitrator who will render a favorable award. Thus, the party-appointed arbitrators, rather than being impartial, may become advocates for the point of view of the party who appointed them as arbitrators.

The party-appointed arbitrators (sometimes called "partisan arbitrators") will have conferences with the neutral arbitrator. If a partisan arbitrator feels that the neutral arbitrator favors the other party, obstructive tactics may follow. Such an arbitrator may become unavailable, walk out of the hearings, abuse witnesses, or try to inject into the record error that will be grounds to vacate the anticipated award. After the award is rendered, a disgruntled partisan arbitrator may appear in court to testify against the award.

Party-appointed arbitrators do not easily agree on the neutral arbitrator. To exacerbate the problem, it frequently occurs that one party obstructs the proceedings. Oftentimes the respondent is in no hurry to have the dispute decided, thinking to have nothing to gain from the award and much to lose. As a result, the respondent may influence the partisan arbitrator to obstruct the selection of the neutral arbitrator. A party never acknowledges an intention to obstruct proceedings, and the fact that obstructive tactics are employed is not immediately obvious. Much time may be wasted in fruitless attempts to appoint a neutral arbitrator before a party, in desperation, applies to the court for appointment.

The proposition that each party should appoint an arbitrator, and the arbitrators thus selected should appoint the neutral arbitrator, though superficially attractive, can waste time and money.

A better formula is to call upon the president, the executive committee, or the board of directors of a respected trade association to appoint the neutral arbitrator. Such a selection can usually be accomplished quickly and certainly, and the arbitrator selected will likely be knowledgeable and impartial. A variation is to have the association nominate seven arbitrators, and for the parties alternately to strike names until but one remains.

The system employed by the American Arbitration Association is effective. The association provides a list of arbitrators deemed qualified and invites the parties to strike undesired names and rank the remainder in order of preference. In a single arbitrator case each party may strike three names. In a multi arbitrator case each party may strike five names. If the parties fail to agree the AAA may appoint the arbitrator from its Roster of Neutrals without the submission of additional lists.

The AAA usually appoints three arbitrators in large cases, i.e., those in which the amount in controversy exceeds $200,000, unless the parties agree otherwise.

SELECTION OF THE ARBITRATOR

It is the potential for exercising some degree of control over the selection of the arbitrator that gives the system of arbitration one of its most appealing qualities. This is not a point at which to save time, trouble, or expense. Consideration should be given to a potential arbitrator's age, education, experience, background, and profession. It is also essential to try to obtain specific information from someone who is personally acquainted with the prospective arbitrator. Such information includes: has the person had experience as an arbitrator and actually rendered awards; is the person likely to be biased against a subcontractor, contractor, architect, engineer, or owner; is the arbitrator likely to be cognizant of the professional reputations of the parties or their witnesses; is the arbitrator intelligent, patient, judicious, decisive, impulsive?

COMPELLING ARBITRATION

Suppose a party who has signed an arbitration agreement refuses to participate in arbitration proceedings. Is it necessary to get a court order to compel arbitration? Is it possible to do so?

In jurisdictions that have adopted modern arbitration statutes, courts summarily will issue orders compelling arbitration. But if the arbitration clause in the contract is so drafted, the parties will be bound by the award whether they participate in the proceedings or not and regardless of whether a court has ordered arbitration.

A self-executing arbitration agreement is one that does not need a prior court order to be enforceable. An example of such an agreement is: "If a party fails to appear and participate in a hearing after due notice, the arbitrator shall make an award based on evidence introduced by the party who does participate."

A party who spurns arbitration proceedings under a self-executing arbitration agreement does so at great risk. The arbitrator almost certainly will enter an ex parte award, and such an award will be enforced by the court just as if it were a judgment signed by a judge after a full trial.

When the arbitration is governed by the rules of the American Arbitration Association (AAA), the arbitration agreement is self-executing.

If the AAA rules apply or if the arbitration clause, as drafted, is self-executing, there is no need for a court proceeding to compel arbitration.

Modern arbitration statutes provide for summary enforcement of the obligation to arbitrate controversies. Courts order arbitration on summary proceedings. The appearance of witnesses usually is not necessary, rather the parties state their positions in the form of affidavits, declarations, and legal memoranda.

To begin arbitration, the petitioner files a petition or a motion to compel arbitration and gives the other side notice. A hearing is held within a few weeks. The court considers the papers submitted by the parties and the arguments of counsel and issues an order compelling arbitration. Thereafter, a party who fails to participate in the arbitration proceedings is nevertheless bound by the award.

MULTIPLE PARTIES

A distinctive characteristic of disputes in the construction industry is that they often involve more than one party. If something goes wrong with the roof on a project, at least four parties are likely to become involved in the ensuing controversy. When the roof leaks, the owner calls the general contractor for assistance. The general contractor then calls the roofing subcontractor. If the roofing subcontractor is unable to fix the leak, the owner makes a claim against the general contractor, who in turn looks to the roofing subcontractor for relief. The roofing subcontractor may contend that it installed the roof in accordance with the plans and specifications, which were negligently prepared by the architect. Thus, the classic four parties are involved: owner, architect, general contractor, and subcontractor. The sheet metal subcontractor, the roofing manufacturer, and the wholesaler and retailer of the roofing materials may also become parties to the dispute.

The court system is well adapted to joinder of parties and consolidation of proceedings. When a court issues a summons to a party, the party has no choice but to respond. This is not true in arbitration. If an arbitrator demands that a party, who has not signed an arbitration agreement, participate in an arbitration dispute, the party has every right to deny the demand. In short, a party has the right to have disputes determined by a court of law until that party signs an arbitration agreement.

Suppose that an owner demands arbitration against a contractor of a dispute about a leaking roof. The general contractor wants to bring the roofing subcontractor into the proceeding, but the subcontract does not contain an arbitration clause. The general contractor believes the fault for the leaking roof probably lies with the subcontractor. In fact, the general contractor's personnel do not know very much about the problem, since they were never on the roof when it was being installed. Therefore, the logical thing is to have the subcontractor participate in the arbitration proceedings and to defend its work. Since there is no arbitration clause in the subcontract, the general contractor is reduced to inviting the subcontractor to participate. The subcontractor may refuse to participate since it can see no advantage to arbitration. The subcontractor knows that if it refuses to arbitrate, it may or may not be sued. Therefore, it may decide to wait and bide its time.

The prime contractor's next move is to attempt to persuade the owner and then the arbitrator that the dispute is not suitable for arbitration because it is not possible to join the subcontractor as a party to the proceedings. The owner may not be persuaded since it feels it can get all the relief it needs, namely, an award against the prime contractor, in the arbitration proceedings. Similarly, the arbitrator may not be persuaded to abandon the proceedings, since from the point of view of the arbitrator, these parties agreed to arbitrate, and they are entitled to a hearing. The courts take a similar view. Therefore, the contractor will have to arbitrate the dispute with the owner and file suit against the subcontractor. This procedure is inefficient from the point of view of the prime contractor, although it is extremely efficient from the standpoint of the owner. Moreover, the prime contractor runs the risk of inconsistent decisions. The arbitrator might decide that the roof is defective, while the court, in the later proceeding between the prime contractor and the subcontractor, might decide that the roof complies with the requirements the contract documents. Such an outcome, though unlikely, would put the prime contractor at a disadvantage, since he or she would have lost both proceedings.

Multiple-party problems, such as the foregoing, are often susceptible to solution by careful drafting and coordination of the arbitration clauses in contract documents. The prime contractor could have employed a form of subcontract with language such as the following:

> Any dispute arising out of or related to the interpretation of this subcontract or the performance thereof shall be resolved by arbitration. If there is an arbitration provision in the prime contract documents, and if arbitration proceedings are commenced under the prime contract documents in a dispute that is related to work described in this subcontract, then at the demand of any party to the arbitration proceedings, or the arbitrator, the subcontractor shall become a party to the arbitration proceedings and be bound by the award, and a judgment may be entered on the award.
>
> If there is no arbitration clause in the prime contract documents, any dispute arising out of or relating to this subcontract or the work performed under this subcontract shall be subject to arbitration under the rules of the American Arbitration Association, and the arbitrator is authorized to resolve all such disputes between the parties to the arbitration.

Multiple-party problems also can be resolved by the doctrine of incorporation by reference. Typical construction contracts refer to plans, specifications, general conditions, special conditions, and other documents that, taken together, are described by the general term *contract documents*. The *contract documents* may be, and often are, incorporated by reference into subcontracts, sub-subcontracts, and purchase orders. If, then, the general conditions include an arbitration clause, that arbitration clause easily may be made to apply to subcontractors, sub-subcontractors, and material suppliers by the device of incorporation by reference. To make the intention to incorporate an arbitration clause into the contract documents clear, language such as the following may be employed:

> Subcontractor will perform its work in accordance with the plans, specifications, general conditions, special conditions, and other contract documents that are referred to, incorporated in, or otherwise made a part of the general contract between the contractor and the owner. If there is an arbitration clause in the contract documents, it is incorporated in this subcontract by this reference.

The arbitration clause that is included in the prime contract documents should be drafted so as to facilitate joinder of all parties who might become involved in a dispute:

> All disputes between all parties to the construction project arising out of or connected with the interpretation of the contract documents, or the work performed thereunder, shall be resolved by arbitration. This arbitration clause shall apply to, and be binding upon, subcontractors, sub-subcontractors, material suppliers, equipment renters, architects, engineers, and all other parties who are involved in or participate in the construction of the project. If the contract of such a party refers to or incorporates this arbitration clause or the contract documents that include this arbitration clause, then that party shall be bound by this arbitration clause.

The architect and the prime contractor are usually the parties who carry the primary responsibility for the preparation of the contract documents, and they should be alert to the problem of arbitration of disputes that involve multiple parties. The architect is a special case. Architects often take the view that they have little to gain by becoming involved as a party in disputes between the owner and the contractor. As a witness in the proceedings or as an advisor to the owner, an architect can participate in an arbitration proceeding without being subject to an award.

If the architect is involved in a dispute with the owner, why should the architect prefer arbitration with its potential for a prompt adverse award to litigation, where the architect may take advantage of depositions, discovery, demurrers, motions for summary judgment, the right to jury trial, appeal, and other protective procedures?

The contract forms recommended by the American Institute of Architects provide for arbitration of disputes between the owner and the architect and between the owner and the prime contractor. The provisions are drafted, however, so as to forbid joinder (without consent) of the architect in a dispute between the owner and the prime contractor.

The legislatures of some states have provided for consolidation of arbitration disputes. Other courts have ordered consolidation under common law doctrine. *Consolidation* presupposes the existence of separate disputes that are subject to arbitration, that are related and may therefore be consolidated. For example, the owner is involved in a dispute with the prime contractor about a roof leak, and the dispute is subject to arbitration. At the same time, the prime contractor is involved in arbitration proceedings with the subcontractor arising out of the same roof leak. Such disputes may be *consolidated* and thus resolved in a single proceeding.

The concept of joinder is related to consolidation. Suppose an arbitration is in progress between a contractor and an owner and the subject of the dispute is a roof leak. The contractor wishes to *join* the subcontractor in the proceeding. The difference between consolidation and joinder is that consolidation applies to two or more existing proceedings, while joinder brings a new party, not previously involved in a proceeding, into an existing arbitration proceeding.

Most problems of consolidation and joinder can be solved by appropriate language in the contract documents.

WAIVER

The question most frequently asked about arbitration agreements is whether they are binding. Arbitration agreements are, indeed, binding, and an arbitration agreement can be enforced by either party. But the parties, of course, can waive arbitration either by agreement or by conduct.

For example, suppose there is a dispute between an owner and a contractor about a leaking roof. The lawyers for the owner and the contractor agree that the dispute is more suitable for litigation than arbitration. An agreement is therefore prepared for signature by the owner and the contractor. Under the terms of the agreement, the parties waive arbitration.

Waiver by agreement, however, is the exception. Waiver often occurs by the conduct of the parties. For example, the contract documents between an owner and a prime contractor include an arbitration clause. A dispute arises because the roof leaks, and the owner files suit against the contractor. After a trial, judgment is entered in favor of the contractor. Dissatisfied with the judgment, the owner now files a demand for arbitration. By filing suit and prosecuting the case to judgment, the owner waived the right to demand arbitration. Nor could the contractor, if the judgment had been in favor of the owner, have demanded arbitration. Although the contractor did not begin the lawsuit, the contractor nevertheless waived the right to arbitrate by proceeding with the litigation.

HEARING

It is possible for parties to validly agree to an arbitration proceeding that does not include a hearing. They might, for example, agree to submit all relevant documents to an arbitrator for examination and to abide by the decision. Such, however, is seldom the case, and a hearing is usually an essential part, and the central feature, of any arbitration proceeding. The concept of hearings is an essential ingredient to the concept of due process of law. The fundamental premise that underlies the whole system for the administration of justice assumes that a just decision requires not only an impartial judge but also that the judge shall make the decision only after the parties have been given their day in court. In arbitration lingo, the *day in court* is known as a *hearing*. Many arbitrations are resolved after one hearing, while others are resolved after several days, or even months, of hearings.

The best analogue for arbitration hearings is court proceedings. The arbitrator generally sits at the head of the table and the parties along the sides. The parties usually, though not necessarily, are represented by counsel. Upon determining that the parties are ready to proceed, the arbitrator opens the hearings and swears in the witnesses.

The party who started the arbitration proceedings, or the claimant, usually presents its case first. In presenting its case, the claimant utilizes witnesses and documentary evidence. The documents are marked as exhibits and received in evidence by the arbitrator.

After a witness completes testimony, the other side may cross-examine. Cross examination is often followed by *re-direct* and *re-cross*. The arbitrator may question the witness at any time.

Most arbitrators like to hear an opening statement before testimony begins. The purpose of the opening statement is to introduce the subject of the dispute to the arbitrator. In the opening statement, first the claimant, then the respondent, explains to the arbitrator what will be proved by the evidence to be introduced by that party. Theoretically, the purpose of the opening statement is not to persuade the arbitrator which party is right and which party is wrong; that is the purpose of the argument, which follows the close of evidence.

The function of the opening statement is to describe the evidence, not to characterize the evidence and not to argue the conclusions that the arbitrator should draw from the evidence.

The arbitrator has the discretion to order the exclusion of witnesses. Suppose a number of witnesses are to present evidence on the same point. If a witness hears another witness on the same side present testimony on a point, the witness might consciously or unconsciously tailor his or her testimony to support the testimony of the prior witness. In order to avoid this possibility, the arbitrator has the discretion to exclude witnesses. One representative of each party, however, is entitled to be present al all times during the hearings, along with counsel.

The arbitrator may be, but need not be, a lawyer. Therefore, the arbitrator may not be acquainted with the rules of evidence. Arbitrators are free to admit any kind of evidence that would be considered relevant to the subject of the inquiry by a person conducting affairs of business. In short, arbitrators can hear virtually any kind of evidence that the parties want to produce.

Nevertheless, objections may be interposed in arbitration proceedings, and counsel do so with more or less frequency according to their views of advocacy and tactics. Arbitrators are usually reluctant to sustain objections because due process requires that each party be given an opportunity to present its case fully. By sustaining an objection, and thus refusing to hear evidence, an arbitrator may create grounds for vacation of the award. A study shows that courts, in fact, rarely vacate awards because of the failure to hear evidence. The rules of the American Arbitration Association provide that the arbitrator may reject evidence deemed to be cumulative, unreliable, unnecessary, or of slight value compared to the time and expense involved. (Rule R-31) Nevertheless, arbitrators are usually liberal in allowing the parties to present their cases.

Counsel is well advised to use objections sparingly in arbitration. Objectioning too frequently becomes obstructive to the proceedings and gives the impression that counsel is afraid to let the arbitrator learn the facts.

After the claimant has presented all its evidence, the claimant rests. Then the respondent presents its defense, and the respondent rests. The claimant thereafter usually presents additional evidence to rebut the defense. The respondent and the claimant may alternate in the presentation of evidence several times until finally both parties rest. Then the parties argue their cases. At this point, since all the evidence has been presented and is available for consideration by the arbitrator, each party will attempt to persuade the arbitrator that the award should be in its favor.

After arguments are concluded, the arbitrator usually will make sure that neither party has any additional evidence to present and than will announce the award or will take the matter under submission.

The parties may offer, or the arbitrator may order, written briefs on legal points. The arbitrator finally renders a written award that settles the dispute. The award is usually delivered in a few days, at most a month, after the closing of the hearings and the submissions of briefs.

PARTICIPATION OF LAWYERS

The arbitration system is designed so that a party can easily proceed without a lawyer. It is not necessary to know the rules of legal procedure or the rules of evidence in order to present a case to an arbitrator. Nevertheless, most construction industry arbitration proceedings are presented by lawyers and, more often than not, the arbitrator is a lawyer.

Arbitrators are not required to follow the law but may decide cases based on their own notions of justice and equity. Nevertheless, arbitrators usually do make their decisions based on their understanding of established legal principles, because arbitrators, by and large, are satisfied that by following the law they will produce a just decision. Rules of law tend to justify themselves, because contractors tend to rely on the law in the conduct of their affairs. For example, almost every contract contains a provision that the contractor will not be compensated for extra work without a written change order. Yet it is customary for contractors to perform extra work without written change orders and to be paid for it. Contractors who conduct their business in this manner are presumably familiar with the rule of law that says that a written contract can be modified by an oral contract. If an owner asks a contractor to perform extra work and agrees to pay for it, and the contractor performs the work, this rule of law says that the owner should pay the contractor for the work even though the contract contained a clause that provided that no additional payment would be made in the absence of a written change order.

An arbitrator, either being a lawyer or being advised as to the law by lawyers, would probably want to decide such a case according to law for two good reasons: the result seems equitable i.e., the owner should not get something for nothing, and the laborer is worthy of the hire; and the parties presumably conducted themselves within a frame of reference that included the assumption that the law would apply to their dealings.

Therefore, although arbitrators need not comply with the law and error in law is not a ground for reversing an award, arbitrators almost invariably attempt to decide cases according to the law.

At the conclusion of the submission of evidence, the parties argue their cases. It is rarely possible for a layperson to argue effectively the law against a lawyer. If, therefore, the decision is one that will be strongly influenced by principles of law and their precise application, the parties will need lawyers.

Some cases, of course, do not depend strongly for their resolution on the application of complicated principles of law. For example, suppose a contractor remodels an owner's garage into a spare bedroom, and the workmanship is obviously improper. The door does not close, the walls are not plumb, and the heater does not work. The statute of fraud, the statute of limitations, and the parole evidence rule are not in issue. It is just a plain case of a bad job. The arbitrator is an architect. Seemingly the facts speak for themselves; the arbitrator can walk the job, figure the cost of correction, and render an award. In such a case, the owner does not need a lawyer, and neither does the contractor.

In most cases of substance and complexity, the parties will want lawyers. A venerable aphorism, much appreciated in the legal profession, tells us that a person who acts as his or her own lawyer has a fool for a client. It is difficult to give effective testimony without a lawyer to ask friendly questions. It is difficult to parry unfriendly questions without the help of a lawyer to interpose suitable objections. It is difficult to cross-examine the other party effectively when one's own credibility is at stake. It is difficult to achieve a degree of professional detachment when one's own interests are at stake. In addition, most laypersons are not experienced or skilled in the organization and presentation of testimony and exhibits.

Yet the self-represented party does have some advantages. One may cast oneself in the role of David against Goliath, achieving an appearance of impecuniousness that could bias an arbitrator in one's favor. One may convey the impression that the justice of one's cause is so great as to be self-evident if not obscured by legalistic chicanery. One may disarm opposing counsel, who will be fearful of appearing to take advantage of an unrepresented party. And one may simply have the gift of advocacy; a person so endowed is an intimidating adversary.

EVIDENCE

Evidence is defined as information, presented to the senses, that is relevant to the establishment of a proposition material to the resolution of a dispute. For example, a dispute includes a question as to whether or not the owner agreed to pay for an extra light switch. The proposition advanced by the contractor is that the owner promised to pay. One form of evidence would be the contractor's testimony that the owner said "I will pay." Another form of evidence would be plans that call for one light switch, not two. While another form of evidence would be the physical existence of the light switch on the job. These three forms of evidence are described as testimony, documentary, and physical evidence.

The introduction of documentary evidence in arbitration proceedings is simplified as compared to the introduction of such evidence in court proceedings. Before they are introduced in court proceedings, documents are usually authenticated by testimony of a witness. The authenticity of documents is seldom disputed in arbitration proceedings, and, thus, authentication is seldom required. Photocopies of documents are customarily numbered in advance and introduced in mass, without formalities unless a party raises an objection to a specific item.

When job conditions or workmanship are at issue, the arbitrator usually visits the project.

Arbitrators are often liberal in receiving evidence that would be excluded by courts. This applies particularly to hearsay evidence, which for centuries has been excluded by courts on the ground that it is unreliable. The theory is that a witness should be permitted to testify only to events that the witness observed with his or her sight, hearing, smell, and the other senses. The badge of authenticity is the fact that the witness is then subject to cross-examination about exactly what he or she saw, heard, and smelled. On the other hand, if a witness testifies about something that somebody else said he or she saw, the percipient witness is unavailable for cross-examination, the testimony is branded *hearsay* and excluded (unless the testimony falls within one of the dozens of exceptions to the hearsay rule).

Arbitrators usually admit hearsay evidence *for what it is worth*, in other words, they listen to the evidence with a degree of skepticism.

Arbitrators may, and should, exclude cumulative evidence. Evidence is cumulative if it is repetitious, and the value of the evidence is overcome by tedium.

A favorite technique of lawyers versed in arbitration is to construct charts, models, and diagrams to illustrate testimony. Photographs are frequently introduced, including aerial photographs, which are particularly useful in presenting certain types of information. Under modern statutes, arbitrators have the power to issue subpoenas to compel the attendance of witnesses and to require the production of documents and other evidence. If a witness refuses to obey a subpoena issued by an arbitrator, it will be enforced by court proceedings.

The American Arbitration Association supplies forms of subpoenas. If the arbitration is not under the rules of the AAA, or another provider, the arbitrators or the parties prepare their own subpoenas.

An arbitration is initiated by notice to the other party. Any form of notice provided in the agreement should be sufficient if it meets the requirements of due process, that is, if the claimant is able to show that the respondent was notified of the arbitration proceedings with adequate time to prepare a defense. Some states, however, impose special service requirements. For example, New York requires that a demand for arbitration be served by registered or certified mail.

The AAA rules provide that any papers, notices, or process may be served on a party or its attorney by mail, overnight delivery, fax, telex, and telegram. Where all parties and the arbitrator agree notices may also be transmitted by e-mail or other methods. (Rule R-40)

Under the rules of the AAA, an arbitration is initiated by filing a demand which contains a statement setting forth the nature of the dispute, the amount involved, and the remedy sought. Two copies of the demand are filed with the AAA, along with a copy of the arbitration provisions of the contract under which the dispute arose. The AAA then gives notice to the other party, who may file an answering statement within seven days. If no answer is filed, the claim is treated as though it had been denied. (Rule R-6)

It has been said that the glory of the common law is its system of pleadings. Many categories of pleadings – complaints, answers, demurrers, and motions – exist; each category of pleading is governed by elaborate requirements, enforced by courts. Pleadings must be prepared in a manner that meets court approval before a case can proceed to trial.

The arbitration system, however, has evolved as an accepted method of resolving disputes without pleadings.

The demand is merely a simple notice to the other party that a dispute exists and a straightforward statement of the claimant's claim. The practical purpose of the demand is to initiate proceedings by letting the respondent know the nature of the claim so that the respondent can prepare a defense. The claimant is motivated to give a fair summary of the claim to the respondent, because otherwise the respondent may ask the arbitrator for a continuance on the ground that the notice was not adequate to allow for the preparation of a defense.

Under modern arbitration statutes, arbitrators have the power to order depositions *for use as evidence*. In court proceedings, depositions may be used as evidence. However, they mostly are used for discovery. If a party to a lawsuit wants to find out what a witness would say in court, the party can take a deposition of the witness. This means that the witness must appear with counsel, if the witness desires counsel, and answer questions. A court reporter transcribes the testimony, and the transcript may be introduced in court. Arbitration depositions, on the other hand, may not be used for discovery, they may only be used as evidence. For example, suppose a witness to disputed facts is located in Brazil. Since the arbitrator cannot reach the witness with a subpoena, a deposition may be ordered and the transcript used as evidence in the arbitration proceedings.

DISCOVERY

Discovery is an array of proceedings carried out by a party under the auspices of a court in order to obtain evidence to introduce at the trial. Discovery differs from investigation, because an investigation can be carried out without notice to the other party. Discovery is done in the presence of, and sometimes by, the other party.

The important tools of discovery are depositions, interrogatories, and demands for the production of records.

The procedure for a deposition is for the deponent to appear and testify in response to oral questions: the questions and answers are transcribed by a court reporter, and the transcript may be introduced into evidence at the trial.

Instead of asking questions orally, a party may direct written interrogatories to another party to a lawsuit, and the other party is required to answer the interrogatories in writing.

Likewise, a party may require another party to submit a list of documents relevant to a dispute and then permit inspection and copying of the documents.

The simple and straightforward premise for discovery is the assumption that the cause of justice is advanced when all parties to litigation are fully informed of the evidence that is available. Discovery procedures may therefore be applied to any line of inquiry that may lead to the discovery of relevant evidence that could be admissible at trial. The discovery process often harms, more than it helps, the cause of justice. This is because discovery becomes so protracted, intrusive, tedious, and expensive that a party may be intimidated to abandon a valid claim or defense simply in order to avoid the expense of making discovery to the other side.

An arbitration agreement, of course, can provide for discovery, and in such a case an arbitrator would have the undoubted authority to enforce discovery procedures. In the absence of such agreement, though, the modern arbitration statutes do not permit, and many of them expressly forbid, discovery. The absence of discovery in arbitration proceedings may be a virtue. However, in practice, a reasonable amount of discovery almost always is available in construction industry cases.

Construction industry disputes usually are oriented to paperwork. Drawings, correspondence, job logs, schedules, specifications, and other contract documents are important evidence. If overhead and lost profit are elements of damage, then the claimant's accounting books become relevant. In order to prepare a defense, the respondent may have a legitimate need to review the claimant's books and records. Likewise, the claimant may need to review the respondent's records in order to authenticate back charges, counterclaims, and other material.

Since the arbitrator has the power to subpoena documents at the request of a party, each party may require the other party to produce all relevant documents at the hearing. Since it is inefficient for the arbitrator to preside over an examination of documents, the usual procedure is for each party to make its documents available for inspection at a convenient place, such as the office of the party or its counsel.

If a party is not forthcoming, the arbitrator is likely to view that party's evidence with suspicion.

In a few cases, where circumstances are extraordinary and the applicable arbitration statute permits, courts order discovery in arbitration. As a general rule, however, state discovery statutes and the discovery rules contained in the Federal Rules of Civil Procedure are not applicable to arbitration proceedings.

AWARD

Arbitration proceedings are terminated by a written award signed by the arbitrator, or, if there is more than one arbitrator, the award is signed by a majority of the arbitrators.

The award usually requires one party to pay a sum of money to the other party or determines that the respondent is not indebted to the claimant. The award also may require some kind of action by the parties; for example it may require the contractor to pay off the mechanics' liens of subcontractors, or it may require the contractor to remedy construction defects to the satisfaction of the architect.

If a party does not comply with the requirements of an award, the other party may have the award enforced through summary court proceedings.

PROVISIONAL REMEDIES

Arbitrators are at a disadvantage because, unlike courts, they do not have the power to command the sheriff. As a result, certain provisional remedies that are available in court proceedings are not available in arbitration.

Attachment is a proceeding used by courts to sequester property of a defendant pending the outcome of litigation. Thus, if the plaintiff recovers a judgment against the defendant, property will be available to satisfy the judgment.

Attachment is accomplished by the sheriff, who physically seizes the property, usually money, and holds it subject to an order of the court. Attachment is not, therefore, directly available in arbitration proceedings, although, in some states, courts will issue writs of attachment in aid of arbitration.

A *mechanic's lien* is a lien provided by law in favor of a contractor or material supplier who improves real property. The lien is enforced by the sheriff, who sells the property at public auction and uses the proceeds of the sale to satisfy the lien. Arbitrators cannot foreclose mechanics' liens. An arbitrator can determine the amount of the debt owed by the owner to the lien claimant. The claimant then must obtain a court order foreclosing the mechanic's lien in the amount determined by the arbitrator.

Temporary restraining orders or *TROs* are used by courts to preserve the status quo pending litigation. Suppose, for example, that a contractor ejects a subcontractor from a construction project, claiming the subcontractor is guilty of a material breach of contract. The contractor has the right, under the contract documents, to take possession of the subcontractor's materials and equipment, but the subcontractor threatens to enter the job site and remove its property. An arbitrator can issue an order to prevent this, but the arbitrator has no way to enforce the order. If a court, on the other hand, issued a temporary restraining order, the sheriff would enforce it.

ATTACK ON AWARD

An arbitration award has greater finality than does a court judgment. The judgment of a trial court can be reversed by a court of appeal either if the trial court judge made the wrong decision or if the trial court judge used the wrong procedure in arriving at the decision. A court of appeal determines whether the trial court judge made the wrong decision by consulting rules of law. For example, a judgment will be reversed if it enforces a contract that is not supported by consideration on both sides.

If an arbitrator were to make an award enforcing a contract that was not supported by consideration on both sides, the award could not be reversed. This is because arbitrators are given the power to decide cases based on their own notions of justice and equity, and they need not necessarily observe rules of law in deciding cases.

The judgment of a court can also be reversed if the judge used the wrong procedure in arriving at the judgment. For example, a judgment may be reversed because the judge admitted hearsay evidence, or because the judge refused to admit relevant evidence, or because the judge permitted counsel to make improper arguments to the jury or refused a party the right to cross-examine, or because of other procedural mistakes. Arbitration awards, on the other hand, are not vacated because of procedural mistakes unless a mistake is so serious as to deny to a party due process of law. Due process means, in essence, that a party is entitled to a fair hearing by an impartial arbitrator who allows the party to cross-examine witnesses and to introduce important evidence. In practice, awards are seldom attacked and very seldom vacated on procedural or due process grounds. Most arbitrators are careful to give the parties a fair hearing.

An award, of course, is subject to attack, and a court will vacate it if the award was procured by corruption or fraud. An award also will be vacated if it was made by an arbitrator who was not impartial. An award may be vacated if there was no arbitration agreement or if the arbitrator decided a dispute that was beyond the terms of the arbitration agreement. For example, a construction contract may provide for arbitration of disputes concerning extra work. If the arbitrator, under such an arbitration agreement, decides a dispute about access to the job site, the award will be vacated.

When a court vacates an award, it may send the case back to the arbitrator for further hearings or may call for the parties to appoint a new arbitrator or may appoint a new arbitrator.

A losing party is given a relatively short period of time within which to attack an award: often thirty days. If not attacked within thirty days, the award becomes final.

The procedure for attacking an award is summary in nature. The losing party files a petition to vacate the award, supported by affidavits and legal arguments. The other party then files documents and makes arguments supporting the award, and the court decides the issue. The hearing on a petition to vacate an arbitration award usually occurs within one month after it is filed, and the hearing usually takes less than an hour. A court judgment generally follows within 30 days, and it is subject to appeal.

ENFORCEMENT OF AWARD

If the court does not grant the petition to vacate the award, it must confirm the award, and if the losing party makes no attack on the award, the prevailing party may petition the court to confirm the award. Confirmation is a summary proceeding based on affidavits, briefs, and legal argument. A party who files a petition to confirm an award may expect judgment within 45 days after the petition is filed. After the award is confirmed, the court enters judgment on the award. The judgment is then enforceable by the sheriff in the same manner as any other judgment.

APPEAL

As with any other judgment, a judgment vacating or confirming an award is subject to appeal. Grounds for appeal are limited, however, just as grounds for attack on an award are limited. If the losing party attacked the award on the ground that the arbitrator was biased, the only ground for appeal would be that the trial court wrongly decided the question of whether the arbitrator was biased. In essence, the grounds for appeal are the same as the grounds for attack on the award: corruption, fraud, bias, misconduct, or the fact that the arbitrator decided a dispute that was outside the terms of the arbitration agreement.

JUDICIAL ATTITUDES

For centuries courts were hostile to arbitration, since arbitration represents a competing system of dispute resolution. This attitude has changed in recent decades, and most courts now view arbitration as an aid to the efficient administration of justice. As late as 1951, a prominent judge held it was unquestioned that a claim could be so unconscionable or a defense so frivolous as to justify a court in refusing to order the parties to proceed to arbitration.

Early courts were reluctant to enforce arbitration agreements because arbitration represented a competing system for the adjudication of disputes. Courts gave a highly restrictive interpretation to arbitration clauses so as to retain jurisdiction of disputes in the court system whenever possible.

In modern times, at least for the past several decades, this attitude has given way, and courts have recognized arbitration as a legitimate part of the dispute resolution system. Contributing to this attitude is court congestion and resulting delay in the resolution of cases. The arbitration system is now seen by the courts as a practical way to deal, in part, with the problem of court congestion. On a more fundamental level, the courts now recognize that parties should be free to contract with respect to a system of dispute resolution, and to select, if they choose, arbitration as preferable to litigation. Although a court may not always agree with such a selection, the court will defend the parties' freedom to make that selection.

THE FEDERAL ARBITRATION ACT

The side-by-side existence and operation of the Federal Arbitration Act along with the arbitration statute adopted by a state is sometimes a source of confusion. Under the Constitution of the United States, when Congress adopts a statute that it has the power to adopt, the federal law is supreme and therefore deviations in state law from the requirements of the Federal Arbitration Act are forbidden.

It is clear from modern decisions that in instances where there is a conflict between federal and state law, the Federal Arbitration Act (FAA) prevails. In such circumstances it is not just the federal courts, but also the state courts, that must enforce the Federal Arbitration Act and the federal policy that favors the arbitration of disputes. Thus, even in states that do not enforce arbitration agreements, such agreements are enforceable under the Federal Arbitration Act if the prerequisites of federal jurisdiction are present. In most cases, federal jurisdiction is based on the commerce clause of the United States Constitution.

The commerce clause gives Congress power over interstate commerce, and federal courts thus derive jurisdiction over disputes involving interstate commerce. The Federal Arbitration Act is applicable, specifically, to " . . . a contract evidencing a transaction involving commerce. . . ."

Diversity occurs when the parties to a controversy are citizens of different states.

In diversity cases and in cases involving interstate commerce, the federal rule that arbitration agreements are enforceable takes precedence over contrary state enactments. After a court has determined that an arbitration agreement has been entered into, under the FAA the court must issue an order directing the parties to proceed to arbitration in accordance with the terms of the agreement. State policy must yield to the federal policy of enforcing arbitration agreements.

It often happens that the use of arbitration depends on whether the Federal Arbitration Act applies to the controversy, since, under state law, the controversy is not arbitrable. Moreover, if the FAA does apply, there may be important procedural and substantive differences between results obtained under the FAA and results obtained under state law.

In a case decided by the United States Supreme Court, an Alabama contractor agreed to expand a North Carolina hospital. The contractor submitted a number of claims for delay and impact costs to the architect, many of which were settled. After several months of meetings on the remaining claims, the hospital informed the contractor that the claims were denied, and that it intended to file suit in North Carolina state court, alleging that the claims were barred by the statute of limitations. The hospital obtained an injunction in state court, which was subsequently dissolved. The contractor then filed suit in federal court to compel arbitration under the FAA. The hospital moved, in federal court, to stay the federal action pending the resolution of the state court action. The federal court granted the stay, and the contractor appealed. The United States Supreme Court reversed the decision of the lower court, holding that the stay was a final appealable order, since it effectively barred the contractor access to the federal court, and that the district court abused its discretion by staying the federal suit. The Supreme Court stressed that when parties agree to arbitration, public policy requires that the arbitration proceed as quickly as possible, and it entered an order requiring the parties to proceed to arbitration.

In a case in which a contractor made a claim against a city in Iowa, the dispute involved the construction of a water pollution control facility. The city contended that enforcement of the FAA would violate the doctrine of separation of powers. The court held that the FAA does not violate the doctrine of separation of powers since its application is limited to consensual agreements. Therefore, the act did not violate the city's prerogative in such a way as to hamper its ability to fulfill its role in the construction project and endanger its separate and independent existence. Under the FAA, the city retained authority to structure its own affairs, including the authority to make the decision whether to enter into a contract that included an arbitration agreement. Once the city agrees to arbitrate a controversy that may arise out of a transaction involving interstate commerce, then the FAA applies, and the city is bound to arbitrate. The federal government has a strong interest in regulating interstate commerce and a strong policy supporting the prompt resolution of disputes concerning federally funded construction projects. As to the alleged, perhaps nonexistent, Iowa policy disfavoring the enforcement of executory common law arbitration agreements, such a policy has negligible weight when compared to the importance of the federal interest in the prompt resolution of disputes.

A Kansas contractor filed suit against an owner for the balance claimed for repair of a building, and the owner filed a third-party claim against an architect. The architect filed a motion to compel arbitration which was denied. The lower court decision was reversed since the contract evidenced a transaction involving commerce, and consequently the Federal Arbitration Act was applicable. Although the Kansas Uniform Arbitration Act forbids the arbitration of a "claim in tort," and although the malpractice claim against the architect is a "claim in tort," the Federal Arbitration Act has no such prohibition. The principle of federal supremacy requires state courts to enforce the arbitration clause even though enforcement may be contrary to state policy.

In a Nebraska case, homebuyers sued developer. When owners purchased their house they also purchased a one-year workmanship/two-year systems warranty and a ten-year structural coverage warranty, as well as a separate five-year extended structural coverage warranty.

The United States Supreme Court has held that an arbitration award will not be vacated for errors of law unless the award demonstrates a "manifest disregard" of the law. Manifest disregard is beyond mere error in law or a failure on the part of the arbitrator to understand or apply the law.

Alabama law explicitly prohibits the enforcement of predispute arbitration agreements on the ground that parties may not oust the courts of jurisdiction by advance agreement. However, where the arbitration agreement at issue is governed by the FAA, the principle of federal supremacy requires the trial court to enforce its terms.

Questions of unconscionability in an arbitration clause, and a contention that an arbitration clause is unenforceable as a contract of adhesion, must be decided under the Federal Arbitration Act rather than under state law. Even if the Federal Arbitration Act applies, a contractor may waive the right to insist on arbitration.

The Federal Arbitration Act applies to arbitration agreements "evidencing a transaction involving commerce." In 1995 the United States Supreme Court interpreted this language in two different ways. In one case it was held that the words "involving commerce" should be broadly construed to include any contract "affecting" interstate commerce. Four months later the same court held that when a law does not concern regulating the channels of interstate commerce, the instrumentalities of interstate commerce, or persons or things in interstate commerce, it does not "substantially affect" interstate commerce. Relying on the second case, the Texas Supreme Court held that the Federal Arbitration Act did not apply to a dispute between a Texas contractor and a Georgia corporation in a case where the owner was a resident of Georgia and the contractor was a resident of

Texas, the repairs were done on property located in Texas and the labor, the compensation for the labor, and the effects of the labor all remained in Texas. Even though the financing of the project which was done by the owner in Georgia might have had some incidental effect on commerce in Georgia, the effect was minor and indirect and therefore the Federal Arbitration Act did not apply.

In a case that arose in South Carolina, the court considered notice requirements in arbitration agreements. The first page of the contract contained a notice, in capital bold-faced letters, that the contract was subject to arbitration. The trial court denied a motion to compel arbitration because the "heading [was] not underlined" as required by the South Carolina arbitration statute. The court held that even though the arbitration clause was not enforceable under state law, it was enforceable under federal law. The object of the contract in this case was the removal and disposal of water and sludge materials located on property situated in South Carolina. The subcontractor hired to dispose of the materials was required to transport them to a facility located in North Carolina. Therefore, the contract involves interstate commerce and the FAA is applicable. The FAA preempts the South Carolina notice statute; under the FAA, courts may not invalidate arbitration agreements under state laws applicable only to arbitration provisions. Arbitration agreements must be placed on "the same footing as other contracts." By requiring arbitration clauses to meet special notice requirements that are not applicable to other contracts, South Carolina impermissibly singled out arbitration agreements.

A case decided by the Supreme Court of Montana held that the Federal Arbitration Act does not apply to a "local" contract. In that case, a contractor served a notice to arbitrate a construction dispute with the City of Cut Bank. The city objected that the arbitration agreement failed to comply with the Montana statute that requires that arbitration clauses be typed in underlined capital letters on the front page of a contract. The court held that, examining only the face of the complaint and the documents attached to the complaint, it appeared that the transaction was local and therefore the Montana statute was enforceable.

IMMUNITY OF ARBITRATOR

Federal policy establishes that arbitrators are immune from claims arising out of their exercise of adjudicatory functions. Many states incorporate immunity provisions in their arbitration statutes, and the decisions of many state courts have held that under common law, arbitrators enjoy the same immunities as judicial managers.

SUMMARY

The construction industry has typically been a very competitive market with which to work in. In general, the risks are reasonably high and any person new to the industry should realize these associated risks before they begin operations. A construction contractor will inevitably encounter some major economic cycles; often it seems you are either experiencing a feast or a famine. The competent, professional, and experienced public works construction contractor will learn how to level out these up and down cycles to produce a conservative and profitable economic pattern.

Some contractors in the past and present have fallen prey to the competitiveness of the business by trying to gain an edge by illegal, immoral or unethical means. This author believes that managing any business in a legal, moral, ethical, and in an old-fashioned, hard working manner will return your company the greatest rewards. Protecting your company's reputation by taking care of your employees, subcontractors, material suppliers and contracting agencies will open up more opportunities and make your company as successful as possible over a long period of time. Many contractors, especially in the public market, do not give enough attention to their companies reputation because they believe it is not necessary due to the fact that all of their work is awarded on the low bid wins system. This author further believes the contractor's reputation is always important, even in the public market. Even a public contacting agency can make a particular project very difficult to complete on-time and within your estimated budget if you do not oversee the construction process with a mutually equitable and ethical manner. The contract specifications that are used in public contracting are always slanted towards the benefit of the contracting agency. Therefore, the individuals involved in any particular project can manipulate the contract specifications in a partnering fashion or the owner's representatives can simply apply the stringent specifications, word for word, to the sole benefit of the contracting agency. Because of the contractual language imbalance, the competent contractor will strive to keep that partnering relationship with the contracting agencies that he works with, and not become their enemy.

Also, to reduce the stress that is generally associated with managing any business, not only a construction company, we recommend that you keep the company as organized as possible. Stress is typically created when an individual encounters a situation that they have no control over. Keeping an organized business will help you stay in control when obstacles are placed in your way. This book was designed to give that person considering starting, or who actually has in operation, a small construction business a basis for organizing their company and getting it started on a successful track.

APPENDIX

GUIDE TO EFFICIENT USE
OF EQUIPMENT IN EXCAVATION
AND EARTHMOVING OPERATIONS

INTRODUCTION

This is a guide for engineering or construction personnel responsible for planning, designing, and constructing sitework, along with other earthmoving operations. The material in this appendix applies to all construction equipment regardless of make or model. The equipment used are examples only. Use the operator and maintenance manuals for the make and model of the equipment you are using to obtain information for your use.

APPENDIX ONE

MANAGEMENT OF EARTHMOVING OPERATIONS

Earthmoving may include site preparation, excavation and backfill, dredging, preparing base course, subbase, and subgrade, compaction, and road surfacing. The type of equipment used can have a great effect on the man-hours and machine-hours required to complete a given amount of work Before estimates can be prepared, a decision must be reached on the best method of operation and the type of equipment to be used. Equipment selection should be based on efficient operation and availability of equipment. It is best to use available equipment that can reduce the amount of manual labor required

1.) Management Phases. Project managers must follow basic management phases to ensure construction projects successfully meet deadlines set forth in project directives. They must also ensure that safety standards are followed. Basic management phases are-

    a.) Planning. Planning consists of selecting the most appropriate item of equipment for each task in the project. Production records kept on previous jobs are usually a good basis for estimating production for each successive job.

    b.) Scheduling. The project manager should schedule dates for material delivery, for completion of individual tasks, and for final completion of the project. Use previous production records updated to account for weather, status of troop training, condition of equipment, and other factors peculiar to the present project.

    c.) Controlling. By analyzing production and progress reports and equipment availability lists, the project manager can keep the project as close as possible to the original schedule. He must keep a watchful eye over the entire project including work in progress, work accomplished, and work to be done, in order to complete the mission on or before the deadline date.

2.) Equipment Selection. Proper selection of equipment is vital to all phases of earthmoving and construction operations. Equipment availability and efficiency must be considered when selecting equipment. Soil considerations, zone of operation, and specific job requirements also determine the best equipment for a particular job.

Equipment production estimates in this manual help determine efficiency of equipment.

3.) Production Estimates. Production estimates, production control, and production records are the basis for management decisions. Therefore, it is helpful to have a common method of recording, directing, and reporting production progress.

a.) Production Formula.

Rate of production

- Work Done. This denotes, in a general way, the unit of progress accomplished. For example, it can be the volume or weight of material moved, the number of pieces of material cut, the distance traveled, or any similar measurement of production or progress.

- Unit of Time. This denotes an arbitrary time unit such as a minute, an hour, a 10-hour shift, a day, or any other convenient period during which the work done (work measured) is accomplished.

- Rate of Production. The entire expression is a time-related rate of production. It could be cubic yards (cu yd) per hour, tons per shift (the hours of the shift must also be indicated), or feet of ditch per hour.

b.) Time Formula. The inverse of the production formula is sometimes useful when scheduling a project because it defines the time required to accomplish an arbitrary amount of work. Time required could be expressed as hours per 1,000 cubic yards, days per acre, or minutes per foot of ditch.

c.) Units of Work and Time. Specific tasks determine the most convenient and useful units of time and work done to use for a particular piece of equipment or a particular job. So that planners can make accurate and meaningful progress comparisons and conclusions, it is important to standardize the use of these production data terms.

4.) Soil Considerations. Soil volume changes as soil is moved. The prime question an earthmover has is not about the nature of the material, but rather its physical properties; for example, how easy is it to move? For earthmoving operations, material is placed in three categories: rock, soil, and rock-soil mixtures. These materials must be thoroughly analyzed and incorporated into the construction plan.

- Rock. Rock is hard and firm like ledge rock, masonry and concrete structures, large boulders, and similar material that may require drilling and blasting to remove. All other material is classified as soil material.

- Soil. Soil is further classified by particle size and type. For instance, gravel has large, coarse, rocky-type particles, while clay has small, fine, platy-type particles. Sand and silt have particle sizes between these two extremes.

- Rock-Soil. The rock-soil combination is the most common material found throughout the world. It is a mixture of various rock and soil materials. The name given to a mixture identifies its composition. For example, glacial till consists of all sizes ranging from rock flour to huge boulders.

a.) Loadability. Loadability is a general property or characteristic. If soil digs and loads easily, it has high loadability. Conversely, if it is difficult to dig and load, it has low loadability. Certain types of clay and loam are considered very loadable. They can be dozed, or loaded into a scraper from their natural state. Other types of material, such as rock or hardpan, must be loosened with a ripper or even blasted before they can be moved. Choosing the appropriate earthmoving equipment depends to a great degree on the loadability of the material to be moved.

b.) Moisture. Moisture content is very important to the earthmover, since moisture affects a soil's weight and handling properties. All soil in its natural state contains some moisture. The amount of moisture-retained depends on weather conditions, drainage, and the retention quality of the soil itself. The moisture retention quality of soil can sometimes be changed, but the procedure must be closely studied from an earthmoving standpoint.

Table 1-1. Soil conversion factors for earth-volume changes

Soil	Converted From:	Converted To:		
		Bank (in place)	Loose	Compacted
Sand or gravel	Bank (in place)	-	1.11	0.95
	Loose	0.90	-	0.86
	Compacted	1.05	1.17	-
Loam	Bank (in place)	-	1.25	0.90
	Loose	0.80	-	0.72
	Compacted	1.11	1.39	-
Clay	Bank (in place)	-	1.43	0.90
	Loose	0.70	-	0.63
	Compacted	1.11	1.59	-
Rock (blasted)	Bank (in place)	-	1.50	1.30
	Loose	0.67	-	0.87
	Compacted	0.77	1.15	-
Coral comparable to limestone	Bank (in place)	-	1.50	1.30
	Loose	0.67	-	0.87
	Compacted	0.77	1.15	-

Table 1-2. Soil weights, swell percentages, and load factors

Soil	Loose (pounds/cubic yards)	Swell (percent)	Load Factor	Bank (pounds/cubic yards)
Cinders	800-1,200	40-55	0.65-0.72	1,100-1,860
Clay, dry	1,700-2,000	40	0.72	2,360-2,780
Clay, wet	2,400-3,000	40	0.72	3,360-4,200
Earth (loam or silt), dry	1,900-2,200	15-35	0.74-0.87	2,180-2,980
Earth (loam or silt), wet	2,800-3,200	25	0.80	3,500-4,000
Gravel, dry	2,700-3,000	10-15	0.87-0.91	2,980-3,450
Gravel, wet	2,800-3,100	10-15	0.87-0.91	3,080-3,560
Sand, dry	2,600-2,900	10-15	0.87-0.91	2,860-3,340
Sand, wet	2,800-3,100	10-15	0.87-0.91	3,080-3,560
Shale (soft rock)	2,400-2,700	65	0.60	4,000-4,500
Trap rock	2,700-3,500	50	0.66	4,100-5,300

Note: The above numbers are averages for common materials. Weights and load factors vary with such factors as grain size, moisture content, and degree of compaction. If an exact weight for a specific material must be determined, run a test on a sample of that particular material.

c.) Weight. Every earthmoving supervisor is interested in soil weight, because it affects the performance of the equipment moving it. To accurately estimate the equipment requirements of a job, you must know the weight of each cubic yard of soil being moved. Soil weight affects how bulldozers push, motor graders cast, and how scrapers load the material. Assume the volume capacity of a scraper is 25 cubic yards and the recommended weight limit is 50,000 pounds per load.

If the load is relatively light, such as cinder, the volume capacity of the scraper will be exceeded before the weight capacity is reached. Conversely, if the load is gravel, which may weigh more than 3,000 pounds per cubic yard, the weight capacity will be exceeded before the volume capacity is reached.

d.) Swell. When earth is removed from its natural resting place, it swells. Its volume expands because of voids created during the loading process. Swell can be predetermined if the general classification of the soil has been established. Swell is expressed as a percentage increase in volume. For example, the swell of dry clay is 40 percent, which means that a cubic yard of clay in the bank (undisturbed/natural) state will fill a space of 1.40 cubic yards in a loosened state. By referring to a table of soil properties, supervisors can determine the swell of a soil. For example, excavating 100 cubic yards of clay from a through-hill cut will take up approximately 140 cubic yards because the soil will swell approximately 40 percent during excavation. As all material that is moved is in a loosened state, it is obvious that earthmovers must move more loose yards than bank yards.

e.) Compaction Requirements. In earthmoving work, it is common to compact soil tighter than it was in its natural state to prevent future settling, provide a firmer road base, or for other purposes. Therefore, a typical earthmoving material cycle might be- 1 cu yd (bank) = 1.3 cu yd (loose)- 0.75 cu yd (compacted).

5.) Zones of Operation. The relationship of specific zones of operation to various types of earthmoving equipment is significant when selecting earthmoving equipment, so they should not be overlooked. The mass diagram is a good working tool in determining the zones of operations.

Zones of operation to consider on construction projects are:

a.) Power Zone. In the power zone, maximum power is required to overcome adverse site or job conditions. Such conditions include rough terrain, steep slopes, pioneer operations, or extremely heavy loads. The work in these areas requires dozers (crawler-tractors) that can develop high drawbar pull at slow speeds. In these adverse conditions, the more traction a dozer develops, the more likely it will reach its full potential. Dozers are usually used in this zone.

b.) Slow-Speed Hauling Zone. The slow-speed hauling zone is similar to the power zone, since power, more than speed, is the essential factor. In this zone, site conditions are slightly better than in the power zone, and the haul distance is short. Since improved conditions give the dozer more power, and distances are too short for most scrapers to build up sufficient momentum to shift into higher speeds, both machines achieve the same speed. Considerations that determine a slow-speed hauling zone are-

- Ground conditions do not permit rapid travel, and material must be moved distances beyond economical bulldozing operations.

- Haul distances are not long enough to permit scrapers to travel at high speeds.

- Number of scrapers on the job are too few to permit the economical use of a push-tractor.

c.) High-Speed Hauling Zone. In the high-speed hauling zone, construction has progressed to where ground conditions are excellent, or where long, well-maintained haul roads have been established. This condition should be achieved as soon as possible in earthmoving operations. At this point the full potential of the scraper is used. Production increases as soon as the scraper can be used at its maximum speed. Considerations that determine a high-speed hauling zone are-

- Haul distances are long enough to permit fast average travel speeds.

- Push-tractors are available to assist in loading.

APPENDIX TWO

BULLDOZERS

Bulldozers (dozers/crawler-tractors) are perhaps the most basic and versatile item of equipment in the construction industry. They are usually the first equipment to arrive on a construction site and the last to leave. They are used primarily where it is advantageous to sacrifice high travel speed to obtain high drawbar pull and extra traction. They are the standard equipment for land clearing; prime movers for pulling or pushing loads (often used to assist in scraper loading); power units for winches and hoists; and moving mounts for dozer blades, side booms, and scoop loaders. They get much of their all-terrain type versatility from their low ground bearing pressure at the track The dozer, considering both production and efficiency, has demonstrated its ability to do a wider variety of jobs than any other earthmoving tool. Dozer characteristics are given in the appropriate operators' manuals.

1.) Description. Dozers have three major assemblies: a center section containing the power source and the operator's controls, and two side sections. The side sections consist of track frames which mount tracks extending almost the full length of the tractor. There are three major classifications of dozers, based on weight and minimum and maximum pounds of drawbar pull: light, medium and heavy.

2.) Blades. A dozer blade consists of a moldboard, cutting edge, and side bits. The push arms and tilt cylinders connect the blade to the tractor. Blades vary in size and design based on their different earthmoving functions. The hardened steel side bits and cutting edge are replaceable, which allows them to absorb the wear so the moldboard itself will last the life of the machine. Dozer blade designs allow either edge to be raised or lowered horizontally. The top of the blade can be pitched forward or backward, and the blade can be angled from the direction of travel. These features are not applicable to all blades, but any two of these features may be incorporated in a single blade type.

(a.) Straight Blade. Straight blades are used for cutting ditches and breaking through crusted material. This blade is mounted in a fixed position, perpendicular to the line of travel. It can be tilted laterally, and the blade top can be pitched either forward or backward within a 10° arc. Tilting the blade allows concentration of tractor power on a small segment of the blade. Pitching, on the other hand, allows a variance in ground pressure of the blade, increasing or decreasing penetration. Pitching the blade provides cutting or dragging, whichever is desired.

(b.) Angle Blade. Angle blades are used most effectively to side cast material when backfilling or when making sidehill cuts. They can also be used for rough grading, spreading piles, or windrowing material. This blade can be set at angles facing the direction of travel. It can also be set at right angles to the tractor and used as a straight blade. When angled, the blade can be tilted, but it cannot be pitched.

(c.) Special Purpose Blades. Special purpose blades such as those designed for clearing brush and trees and should not be used for earthmoving.

3.) Selection Factors. Dozers have a distinct flotation advantage over scrapers. They can operate in muck or water as deep as the height of their tracks and in deeper water for short periods of time (if they are properly waterproofed). They have more traction than the scraper.

4.) Clearing, Grubbing, and Stripping Operations. Clearing consists of removing all trees, fallen timber, brush, and other vegetation from a designated area. Clearing also includes removing surface boulders and other materials embedded in the ground, and disposing of all cleared material. Clearing techniques vary with the type of vegetation being cleared and soil and moisture conditions.

Table 2-1. Quick estimates for clearing

Equipment	Man or Equipment (hours per acre)		
	Small Trees (6 inches or less)	Medium Trees (7 to 12 inches)	Large Trees (12 to 30 inches)
Bulldozer: Medium tractor (D7)	2.50	5.00	10.00
Heavy tractor (D8)	1.50	3.00	8.00
Spade: Medium tractor (D7)	1.33	2.20	3.90
Shear blade: Medium tractor (D7)	0.40	0.80	1.30
Heavy tractor (D8)	0.30	0.50	0.80
Note: These clearing rates are averages for tree counts of 50 trees per acre. Adverse conditions can reduce these rates significantly.			

(a.) Brush and Small Trees. Moving with the blade slightly below ground level will usually remove small trees and brush. The blade cuts, breaks off, or uproots most of the tree and bends the rest, so it can be removed on the return trip. Dozers operating in first or second gear can clear brush and small trees at the rate of 900 to 1,000 square yards per hour.

(b.) Medium Trees. To remove medium trees (7 to 12 inches in diameter), raise the blade as high as possible to gain added leverage and push the tree over slowly. As the tree starts to fall, back up the dozer quickly to avoid the rising roots. Then, lower the blade and drive the dozer forward, lifting the roots out. Average clearing time for medium-sized trees is 2 to 9 minutes per tree.

(c.) Large Trees. Removing large trees (over 12 inches in diameter) is much slower and more difficult than clearing brush and smaller trees. First, gently and cautiously probe the tree for dead limbs that could fall and injure you. Then, position the blade high and center it for maximum leverage. Determine the direction of fall before pushing the tree over; the direction of lean, if any, is usually the direction of fall. If possible, push the tree over the same as you would a medium tree. Large trees may need to be dug out.

(d.) Safety Precautions. Take necessary safety precautions to avoid personnel injury and equipment damage. Working units should never operate too close together. Do not follow a tree too closely when pushing it, because when it begins to fall, its stump and roots may catch under the front of the tractor. If this happens, the tractor may require assistance to back off the roots and may sustain extensive bottom damage.

5.) Sidehill Excavations.

(a.) Creating a Slope. One of a dozer's more important uses is making sidehill cuts, which includes preparing roads along the sides of hills. An angle blade is preferred for this cut because of its side-casting ability.

(1) Starting the Cut. It is best to start the cut at the top of the hill, creating a bench several dozer lengths long. This should be done by working up and down the slope. If more than one bench is required to complete the sidehill cut and maintain a safe stable slope, design the benches to ensure water runs off without damaging the slope. If possible, start the bench on the uphill extreme of the cut and then widen and deepen the cut until you achieve the desired profile.

Be sure you start the bench far enough up the slope to allow room for both the inner slope and roadway. (NOTE: When working on extremely steep slopes, you may need to use a winch line to stabilize the dozer.)

(2) Cutting the Slope. Once the initial bench is constructed, turn the tractor onto the bench and work your way down the hill. After the bench is constructed all the way along the hill, use either a dozer or scraper to complete the cut. Keep the inside (slope side) of the roadway cut lower than the outside. This allows

the dozer to work more effectively on the edge and decreases the erosion of the outer slope. Make sure the proper slope is maintained on the inner slope. It is very difficult to change the slope after construction. When cutting the slope, work from the toe of the bench to the outside edge of the road. Maintain the proper slope on the bench by moving a specified distance from the inside slope on each successive lift. Determine the slope ratio from the distance moved away from the slope for each successive layer and the depth of each layer.

(b.)Finishing Side Slopes. The two methods for finishing side slopes are-

(1) Parallel to the Right-of-Way. A dozer in can finish a side slope by working parallel to the right-of-way. Start the dozer at the top and, on each pass, earth will fall to the lower side of the blade, forming a windrow. (On succeeding passes, pick up this windrow and use it to fill holes and other irregularities in the terrain.) Be careful to prevent the blade corner from digging in; this would steepen the slope beyond job specifications.

(2) Diagonally Up the Slope. The dozer finishes the side slope by starting at the bottom and working diagonally up the slope. The windrow formed is continually pushed to one side, which tends to fill low spots, holes, and irregularities. This is one of the few instances where a dozer effectively cuts uphill.

6.) Operation Techniques. Dozers work best when the ground is firm and without potholes, sharp ridges, or rocks. Uneven surfaces make it difficult to keep the blade in contact with the ground. This tends to bury vegetation in hollows rather than removing it. For long moves, dozers should be transported on heavy trailers. They may be moved under their own power at slow speed, but this shortens their operational life. Use the following techniques, when conditions permit, to save time and increase output:

7.) Dozing. When straight dozing, if the blade digs in and the rear of the machine rises, raise the blade to continue an even cut. If moving a heavy load causes travel speed to drop, shift to a lower speed and/or raise the blade slightly. When finishing or leveling, a full blade handles easier than a partially loaded blade.

a.) Blade-to-Blade (Buddy) Dozing. When moving material 50 to 300 feet, blade-to-blade dozing will increase production considerably. When the distance is less than 50 feet, the extra time needed to maneuver and position the dozers offsets the increased production. However, one disadvantage of buddy dozing is operators have a tendency to stop and talk or wait for each other when their machines are undergoing repairs or adjustments.

b.)Slot Dozing. Slot dozing uses spillage from the first few dozer passes to build a windrow on each side of a dozer's path. This forms a trench, preventing spillage on subsequent passes. To increase production, align cuts parallel, leaving a narrow uncut section between slots. Then remove the uncut section by normal dozing. When grade and soil conditions are favorable, slot dozing can increase output by as much as 50 percent.

c.) Downhill Dozing. When dozing downhill, you must travel to the bottom of the hill with each load. Pile several loads at the brink of the hill and push them to the bottom in one pass. Since downhill dozing increases production, use it whenever possible.

d.) Dozing in Hard Materials. The dozer blade may be used to loosen hard material when rippers are not available. Tilt the blade so that one corner is forced into the material. Tilting can be done through adjustments or by driving one track onto a ridge of material bladed up for this purpose, by placing a rock or log under the track. A thin layer may be broken by turning a dozer on it. Turning causes the grousers to break through the top layer. In a thin layer of frozen material, it is best to break through at one point. By lifting and pushing, the blade breaks through the top frozen layer. Dozer work in rocky areas increases track wear. To cut down on this wear, install rock shoes/rock pads, if possible.

8.) Ripping. Use first gear for most ripping operations. When one-shank ripping, always use the center shank. Use additional shanks, where practical, to increase speed. If material breaks up satisfactorily, more shanks may be used. When ripping for scraper loading, rip in the same direction that the scrapers are loading, whenever possible. It is usually desirable to rip as deep as possible. However, it is sometimes better to rip at partial depth and remove the material in its natural layers. Cross rip only when necessary.

a.) Packed Soil, Hardpan, Shale, and Cemented Gravel. Three-shank ripping (Figure 2-13) works well in these materials. Use as many shanks as possible to break material to the desired size. Do not stall the engine or hang the machine up.

b.) Rock with Fractures, Faults, and Planes of Weakness. Use two shanks for ripping where rocks break out in small pieces and the machine can handle the job easily. Use only the center shank if the machine begins to stall or the tracks spin.

c.) Solid Rock, Granite, and Hard-to-Rip Material. Use one shank (center shank) in hard-to-rip material or material that tends to break out in large slabs or pieces.

- Asphalt. Raise the ripper shank to lift out and break the material.

- Concrete. One-shank ripping is especially effective in severing reinforcing rods or wire mesh

- Back Ripping. Back ripping is loosening the ground with the teeth on the dozer blade while backing the tractor for the next pass.

- Winching. When winching, make sure personnel are clear of the cable. Cables can break and cause severe injury. Exercise care with suspended loads. If the engine speed is too low, the weight of the load may cause it to drop, even though the winch is in the reel-in position. Always keep the winch cable in a straight line behind the machine. When moving away from a load, operate the machine in low gear to prevent overspeeding winch components.

APPENDIX THREE

SCRAPERS

Scrapers (tractor-scrapers) are designed for cutting, self-loading, hauling, dumping, and spreading for long-haul earthmoving operations. The scraper is used for essentially the same purpose as the dozer, but it is faster and more economical than a dozer. Haul distance (zone of operation), volume to be hauled, and the type of surface moved over are the primary factors in determining whether or not a scraper should be used on a particular job. Other factors such as type of tires and drive (two-wheel, four-wheel) also affect the selection.

1.) Description.

The Scraper's three basic operating parts are:

- Bowl. The bowl, equipped with a cutting edge on the front bottom, is the loading and carrying component. It is lowered for cutting and loading, raised for carrying, and lowered for dumping and spreading.

- Apron. The front wall of the bowl is the apron. It can be raised or lowered independently of the bowl.

- Ejector. The rear wall of the bowl is the ejector. It is moved back to load materials and forward to discharge materials.

    (a.) Capacity. The capacity of the scraper is measured either struck or heaped. Struck means the bowl has a full load of material that is level with its top. Heaped means the bowl is heaped with all the material it can hold without the material falling off. Overloading the bowl lowers efficiency and puts stress on the machine.

    (b.) Operating Range. The optimum operating range for scrapers is from 300 to 5,000 feet.

    (c.) Traction. Scrapers are rated in pounds of pulling force. Attainable traction is influenced by the coefficient of friction between the drive wheels and the travel surface, and by the weight on the drive wheels.

2.) Selection Factors.

(a.) Advantages. The scraper has several advantages over the dozer. It's faster; requires less transportation equipment because it can travel considerable distances under its own power; has good traction on existing hard surfaces such as concrete and asphalt (and will not damage those surfaces as the dozer will); and, because of its high ground-bearing pressure, compacts materials.

b. Disadvantages. The main disadvantage of the scraper is that it has less traction than the dozer. Therefore, in rough work where extra traction is essential, a dozer is more suitable than a scraper.

3.) Cutting, Loading, and Spreading Operations. Whenever possible, use a figure-8 cycle for scraper operations. By using this pattern, as much as one-half of the turning can be eliminated. If possible, load alternately at either end of the fill, or at a central point and spread in opposite directions. Output is increased in either case. For maximum production, push-tractor assistance is normally used in scraper operations. To cover scraper operations and how to achieve maximum production, the earthmoving cycle is broken down into specific operations- cutting and loading, hauling, and spreading.

(a.) Cutting and Loading. All loading methods are usually done with pusher (tractor) assistance.

Scrapers can load without assistance, but push-loading usually results in maximum production.

With few exceptions, pusher assistance is necessary to reduce loading time and tire spinning.

Reducing scraper tire spinning increases tire life. The scraper should not depend on the pusher to do all the work. Conversely, the scraper's wheels should not be spun in an effort to pull away from the pusher. Pusher assistance can be used for either straight loading, downhill loading, or straddle loading.

- Downhill Loading. Downhill loading enables a scraper to obtain larger loads in less time.

The additional weight of larger loads, in turn, adds further to the gravitational force. The added force of gravity is 20 pounds per gross ton of weight per 1 percent of downhill grade.

- Straddle Loading. Straddle loading (Figure 3-5, page 3-6) requires three cuts with a scraper. The first two cuts should be parallel, leaving an island between the two cuts. The scraper straddles this island of earth to make the final cut. The island should be no wider than the distance between a scraper's wheels. With straddle loading you gain time on every third trip because the center strip loads with less resistance than a full cut.

Listed below are a few tips on scraper operation.

- The cut should be as deep as possible, allowing the machine to move at a constant speed, without lugging the engine. Decrease the cut depth if the scraper or pusher engine lugs, or the drive wheels slip. Gauge the most efficient depth of cut by the depth of the router bits. Use this depth on successive passes.

- When more torque is required at the wheels, use a lower gear. Keep the engine revolutions per minute (RPM) within the operating range. Maintain a steady speed while the pusher makes contact.

- Regulate the apron opening to prevent material from piling in front of the lip or falling out of the bowl.

- Keep the machine moving in a straight line while maintaining pusher and scraper alignment.

- Do not overload the bowl. Overloading lowers efficiency and makes the machine do more than it was designed to do.

- Raise and lower the bowl rapidly when loading loose material such as sand.

- Allow the pusher to help the machine out of the cut area, if necessary.

- Accelerate to travel speed as quickly as possible. Travel a few feet before lifting the bowl to the carrying position to spread any loose material piled up. This allows the following scraper to maintain speed.

Listed below are a few materials with their associated loading techniques.

- Loam and clay. Loam and most clay soils cut easily and rapidly with minimum effort. However, loosen very hard clay with scarifier teeth before loading.

- Sand. Since sand has little cohesion between its particles, it has a tendency to run ahead of the scraper blade and apron. The finer and drier the particles, the worse this condition is.

- Rock and shale. Loading rock and shale with a scraper is at best a difficult task, causing maximum wear and tear on the equipment. Ripping will ease this problem. However, you may be able to load some soft rock and shale by using a scraper with pusher assistance, without ripping.

When loading stratified rock, start the scraper cutting edge in dirt, where possible, moving in to catch the blade in planes of lamination. This forces material into the bowl. Pick up loose rock or shale on the level or on a slight upgrade, with the blade following the lamination planes.

Loading time is critical for obtaining maximum scraper output. Push-loading should normally take less than one minute within a distance of 100 feet (time and distance change with each project). You may use more time and distance to load extra material, if the haul distance is long enough and the added weight is enough to allow fewer loads.

b.) Hauling. Hauling, or travel time, includes haul time and return time. Here the power and characteristics of the scraper become very important. The following considerations should be made to minimize travel time:

- Haul Route Planning. Haul routes should be laid out to eliminate unnecessary maneuvering. The job should be planned to avoid adverse grades that could drastically reduce production. Remember, where grades permit, the shortest distance between two points is always a straight line.

- Road Maintenance. Haul roads should be kept in good condition. A well-maintained haul road permits traveling at higher speeds, increases safety, and reduces operator fatigue and equipment wear.

Ruts and rough surfaces. Use a grader or dozer to eliminate ruts and rough (washboard) surfaces. Dust. Water distributors are used to reduce dust. Dust tends to get into all parts of the scraper causing additional wear and requiring frequent lubrication. Reducing the amount of dust helps alleviate this, provides better visibility, and lessens the chance of accidents.

Also, roads kept moist (but not wet) will pack into hard, smooth surfaces permitting higher travel speeds.

- Travel Conditions. Once on the haul road, the unit should travel in the highest safe gear appropriate for road conditions. When possible, the scraper bowl should be carried fairly close to the ground (approximately 18 inches). This lowers the center of gravity of the scraper and reduces the chance of overturning.

Avoid unnecessary lugging. The operator should shift down when momentum is lost. Lugging the motor usually results in a slower speed than the top range of the next lower gear. Even though the machine can make it, it is best to shift down and accelerate faster.

Never coast on a downgrade. When approaching a downgrade, slow down and downshift the transmission. To prevent unwanted upshifting, use the transmission hold on a downgrade if it is available. Also use it when approaching an upgrade or in rough underfooting. To control speed on a downgrade, use the retarder and service brakes. Engine speed should not exceed the manufacturer's recommended RPM.

- Avoid long, slow turns. Turns should be made on the shortest radius possible and at the highest safe speed. Valuable time can be saved on short-radius turns. Long turns mean longer, slower cycles.

c.) Spreading. When spreading, lower the bowl to the desired spread height and open the apron at the beginning of the dump area. Spread in the highest gear permitted by haul road conditions and fill-material characteristics. Slowly dribbling the load at low speed slows down the cycle.

- Spread Sequence. The scraper spread sequence is-

  Step 1. Spread the first load at the front of the fill.

  Step 2. Travel with subsequent loads over the previous fill, provided lifts are small.

  Step 3. Make each following spread start at the end of the previous fill.

  Step 4. Finish spreading one full lane before starting a new one, so that rollers can start compaction.

  Step 5. Repeat method in next lane.

- Maintain Desired Fill Slope. To maintain desired fill slope, make the fill high on the outside edges. This will prevent the scraper from sliding over the slope and damaging the slope, and will eliminate the necessity for handwork. In the event of rain, build up the low center for drainage, or use a grader to cut the outside edge down, creating a crown.

  Time should not be wasted on the fill. As soon as the load is dumped, the scraper should get back on the haul road and return to the cut area as fast as possible. Plan the egress from the fill area to avoid soft ground or detours around trees or other obstacles.

- Materials. Different materials require different dumping and spreading procedures. Sand should be spread as thin as possible, to allow better compaction and make traveling over the fill easier. Wet or sticky material may be difficult to unload or spread. When operating in this type of material. Keep the bowl high enough to allow the material to pass under the scraper. Material not having enough room to pass under the scraper will roll up inside the bowl into a solid mass that will be difficult to eject.

APPENDIX FOUR

GRADERS

Graders are multipurpose machines used for grading, shaping, bank sloping, and ditching.

They are also used for mixing, spreading, side casting, leveling and crowning, light stripping operations, general construction, and road and runway maintenance. They are restricted to making shallow cuts in medium-hard materials; they should not be used for heavy excavation. A grader cannot perform dozer-type work because of the structural strength and location of its moldboard However, they can move a tremendous amount of material They are capable of working on slopes as steep as 3:1. However, it is not advisable to use a grader to construct ditches running parallel on such a slope because graders have a comparatively high center of gravity, and the right pressure at a critical point of the blade could cause the machine to roll over. Graders are capable of progressively cutting ditches to a depth of 3 feet. However, it is more economical to use other types of equipment to cut ditches deeper than 3 feet. The components of the grader that actually do the work are the moldboard or blade, and the scarifier.

(a.) Moldboard/Blade. The moldboard (blade) is used to side cast the material it encounters. The blade is controlled through a hydraulic power system. The ends of the blade can be raised or lowered together or independently of one another.

- Blade Angle. The moldboard (blade) can be angled (positioned) as follows: perpendicular to the line of travel, parallel to the direction of travel, shifted to the side, or in a vertical position.

- Blade Pitch. The blade is ordinarily kept near the center of the pitch adjustment, which keeps the top of the blade directly over the cutting edge. However, the top of the blade can be pitched (leaned) forward or backward. When leaned forward, the cutting ability of the blade is decreased, and it has more dragging action. It will tend to ride over material rather than push it, and it has less chance of catching on solid obstructions. This pitch is used to make light, rapid cuts and to blend materials. When leaned to the rear, the blade cuts readily but tends to let material spill back over the blade.

(b.) Scarifier. Scarifiers are used to break up material too hard for the blade to cut. A scarifier is composed of a scarifier log with 11 removable teeth. The teeth can be adjusted to cut to a depth of 12 inches. When operating in hard material, it may be necessary to remove teeth from the scarifier log. A maximum of five teeth may be removed from the log. If more than five teeth are removed, the force against the remaining teeth could shear them off. When removing teeth, take the center one out first and then alternately remove the other four teeth. This balances the scarifier and distributes the load evenly. With the top of the scarifier pitched to the rear, the teeth lift and tear the material being loosened. This position is also used for breaking up asphalt pavement. (The pitch of the scarifier log can be adjusted for the material being used.)

1.) Construction of Roads and Ditches. Road and ditch construction are basic grader operations, normally performed as follows:

(a.) Marking Ditch Cuts. For better grader control and straighter ditches, make a 3- to 4-inch deep marking cut at the outer edge of the bank slope (usually identified by slope stakes) on the first pass. The toe of the blade should be in line with the outside edge of the lead tire. This marking cut provides a guide for subsequent operations.

(b.) Making Ditch Cuts. Make each ditch cut as deep as possible without stalling or losing control of the grader. Normally, ditching cuts are done in second gear at full throttle. Start with the blade positioned so the toe is in line with the center of the lead tire. Each successive cut is brought in from the edge of the bank slope so the Blade pitch toe of the blade will be in line with the ditch bottom on the final cut.

(c.) Creating a Bank Slope. Sloping the bank on a road cut prevents slope failure. It also prevents excessive erosion of the bank, which could fill the road-side ditch. Make cuts as close to the design slope as possible to prevent having to make additional passes. The material cut from the outer slope is initially cast into the bottom of the ditch and must be removed later.

(d.) Cleaning the Ditch. To remove unwanted material that was pushed into the ditch during the bank slope operation, place the blade in the same position as used for the ditching cuts. This casts the material onto the shoulder.

(e.) Finishing the Shoulder. The windrow formed by cleaning the ditch is moved onto the road at the same time the shoulder is finished to the desired slope.

(f.) Moving Windrows. When making ditch cuts, windrows form between the heel of the blade and the left rear wheel. These windrows must be moved and leveled off either when the ditch is at the planned depth or when the windrow becomes higher than the road clearance of the grader. The shoulder of the road is formed during this operation. Sometimes ditch cuts produce more material than needed for the roadbed and shoulders. This excess material can be used as fill at other locations throughout your construction project. In this case, the excess material is bladed into a windrow and hauled to the appropriate location.

(g.) Leveling Windrows. Windrows are leveled to form the crown of the road. The moldboard is set so that the material is leveled across the length of the blade and no windrow is formed at the heel.

(h.) Finishing the Crown. The final operation is spreading all material brought from the ditch onto the roadway. This material is used to bring the roadway to the desired crown.

2.) Maintenance of Earth and Gravel Roads.

(a.) Leveling and Maintaining Surfaces. Ordinarily, leveling and maintaining a surface is done by working the material across the road or runway from one side to the other. However, to maintain a satisfactory surface in dry weather, work traffic-eroded material from the edges and shoulders of the road toward the center. Traffic and wind can cause loss of binder, so be cautious when disturbing dry road surfaces.

(b.) Smoothing Pitted Surfaces. When binder is present and moisture content is appropriate, rough or badly pitted surfaces may be cut smooth. The cut surface material is then respread over the smooth base. The best time to reshape earth and gravel roads is after a rain. Dry roads may be watered down using a truck-mounted water distributor. This ensures that the material will have sufficient moisture content to recompact readily.

(c.) Correcting Corrugated Roads. When correcting corrugated roads, be careful not to make the situation worse. Deep cuts on a washboard surface will set up blade chatter, which emphasizes rather than corrects, corrugations. Scarifying may be required if the surface is too badly corrugated. With proper moisture content, the surface can be leveled by cutting across the corrugations. Alternate the blade so the cutting edge will not follow the rough surface. Cut the surface to the bottom of the corrugations. Then, reshape the road surface by spreading the windrows in an even layer across the road. After the road has been reshaped, traffic will compact it. However, rolling after shaping gives better, longer-lasting results.

3.) Snow Removal. Graders remove snow in much the same way as snowplows do. Be sure to raise blade 1/2 to 1 inch when removing snow from uneven pavements or portable runway surfaces. Improper adjustment may not only damage the grader, but may also scuff or gouge the road.

c. Mixtures Spread over Large Areas. Set stakes to mark the edges of the spread from each windrow. Then, spread the mixed windrow using a grader. When mixtures are spread over large areas, drive blue top stakes to indicate final pavement elevation. The stakes are usually placed 20 feet apart in both directions, in a grid. Remove the stakes before rolling the pavement. Usually the windrow is flattened with one pass of a grader, after which it is spread to each side in increments.

4.) Operation Techniques.

(a.) Spreading and Leveling. Graders may be used to spread piles of loose material.

If there is space to work around the sides of the piles, extend the blade well to the side and reduce the pile, using a series of side cuts. Spread the piles as much as possible. The load to be pushed is limited by the power and traction of the grader. Graders have less power and traction than dozers, but graders move the load faster. Although the grader blade is quite low, it is more concave than the dozer blade. This gives more rolling action to the load so that a large quantity can be pushed without spilling over the top. Leveling large piles or large windrows may require two or more passes with a grader.

(b.) Side Casting. When the blade is set at an angle, the load being pushed tends to drift off to the trailing end of the blade. Rolling action caused by the moldboard curve assists this side movement. As the blade is angled more sharply, the speed of the side drift increases, so that material is not carried forward as far, and deeper cuts can be made. To shape and maintain most roads, set the blade at a 25° to 30° angle. Decrease the angle for spreading windrows; increase the angle for hard cuts and ditching.

(c.) Planing Surfaces. Set the blade at an angle to plane off irregular surfaces. (Use that material to fill the hollows.) Cut enough material to keep some in front of the blade. Move the loosened material forward and sideward to distribute it evenly. On the next pass, pick up the windrow left at the trailing edge of the blade. On the final pass, make a lighter cut and lift the trailing edge of the blade enough to allow the surplus material to go under rather than around it. This will avoid leaving a ridge. Do not pile windrows in front of the ridge. Also, do not pile them in front of the rear wheels because they will adversely affect traction and grader accuracy.

5.) Tips for Efficient Grader Operations.

(a.) Proper Working Speeds. Always operate as fast as the skill of the operator and the condition of the road permit. Operate at full throttle in each gear. If less speed is required, use a lower gear, rather than run at less than full throttle.

(b.) Blade-Setting Speed. Each job requires a specific blade-setting speed for optimum production. Deviations from these settings (and speeds) may be misinterpreted as machine inefficiency rather than operator error.

(c.) Turns. When making a number of passes over a short distance (less than 1,000 feet), backing the grader to the starting point is normally more efficient than turning it around and continuing the work from the far end. Never make turns on newly laid bituminous road or runway surfaces.

(d.) Number of Passes. Grader efficiency is in direct proportion to the number of passes made. Operator skill, coupled with planning, is most important in eliminating unnecessary passes. For example, if four passes will complete a job, every additional pass increases the time and cost of the job.

(e.) Tire Inflation. Keep tires properly inflated to get the best results. Overinflated tires cause less contact between the tires and the road surface, resulting in a loss of traction. Air pressure differences in the rear tires cause wheel slippage and grader bucking. (Correct tire inflation figures are given in the appropriate operator's manual.)

(f.) Wet and Muddy Conditions. Wet and muddy conditions cause poor traction, which may decrease grader efficiency. However, in spite of reduced efficiency, the grader is the best machine to use under these conditions. One example of this would be casting surface mud to the side of a haul road.

(g.) Haul Road Maintenance. Keep haul roads in good condition. This will increase the efficiency of scrapers, dozers, or dump trucks on large earthmoving operations. Graders are the best machines for maintaining haul roads. The most efficient method of road maintenance is to use enough graders to complete one side of a road with one pass of each grader (tandem operation). In this method, one side of the road is complete while the other side is left open to traffic.

(h.) Tandem Operations. Using graders in tandem expedites such operations as leveling, mixing, spreading, and maintaining haul roads.

APPENDIX FIVE

FRONT END (SCOOP) LOADERS

Handling construction supplies and excavating are major tasks in any construction operation. Lifting and loading equipment performs the major portion of this work. Included in this group is the scoop loader, sometimes referred to as a front-end loader. The scoop loader is a versatile item of equipment. It can be used in all zones of operation and can dig at ground level, above ground level, and below ground level. The large rubber tires provide good traction on unstable surfaces, and they provide low ground-bearing pressure. Also, because of its rubber tires, it has the added advantage of minimum tearing effect when working on hard surfaces such as asphalt. It can attain high speeds permitting it to travel from one job site to another under its own power and does not require other equipment to level smooth, or clean up the area in which it has been working.

1.) Description. The scoop loader is a diesel-driven, four-wheel drive machine mounted on rubber tires. It is available in varied sizes and capacities. Models that are 2 1/2 cubic yards and smaller have rigid front axles and a rear axle that provides the steering capability. Models larger than 2 1/2 cubic yards have a hinged frame, with the front axle providing the steering. Both types are referred to as articulated. The power-shift transmission with torque converter gives the loader fast movement in both forward and reverse, with a minimum of shock. This lets the machine maintain a high rate of production. The hydraulic system gives the operator positive control of mounted attachments and assists in steering. Most loaders have pintles or towing hooks for towing small trailers or light loads.

2.) Attachments. The most common scoop loader attachments are a shovel-type bucket and a forklift. The scoop loader's hydraulic system provides the control necessary for operating these attachments.

   (a.) Buckets. Buckets may be the one-piece conventional type (general-purpose) or a hinged-jaw, multipurpose type (multisegmented). The selected bucket is attached to the tractor by a push frame and lift arms. Bucket capacity is 2 1/2 to 5 cubic yards.

   • General Purpose. The general-purpose (one-piece) bucket is made of heavy-duty, all-welded steel. The teeth are bolted or welded onto replaceable cutting edges. Bolt-on type replaceable teeth are provided for excavation of medium-type materials.

- Multisegmented. The multisegmented (two-piece) hinged-jawed bucket is made of heavy-duty, all-welded steel. It has bolted or welded replaceable cutting edges. Bolt-on type replaceable teeth are provided for excavation of medium-type materials. The two-piece bucket provides capabilities not available with a single-piece bucket. It can be used as a clamshell, dozer, and scraper, as well as a scoop shovel.

(b.) Forklift. The forklift is attached to the loader in place of a bucket or other attachment. Designed for material handling, it is made of steel with two movable tines.

(c.) Other. Other attachments available, but of limited use, are the crane hook (designed for lifting and moving sling loads), the snow plow, the backhoe, and rippers.

(3.) Use. Scoop loaders are typically used for stockpiling materials; digging basements or gun emplacements; backfilling ditches; loading trucks; lifting and moving construction materials; and, when equipped with rock-type tread tires, operating in and around rock quarries. Scoop loaders are also used for many miscellaneous construction tasks including stripping overburden; charging hoppers and skips; carrying concrete to forms; lifting and moving forms, large concrete and steel pipes; and assisting in plant erection and maintenance. They are also used to tow small trailers and light loads.

(4.) Selection Factors. Certain factors enter into the selection of a scoop loader. One of these is the volume of material to be excavated; another is material characteristics. Scoop loaders are excellent machines for soft-to-medium material. Scoop loader production rates decrease rapidly when used in medium to hard material. Another factor to consider is the height that the material must be lifted to or dumped from, since other equipment may construct stockpiles faster or load hoppers or grizzlies more efficiently. A scoop loader attains its highest production rate when working on flat, smooth-surfaced areas with enough space to maneuver. In poor underfoot conditions or when there is a lack of space to operate efficiently, other equipment may be more effective.

(5.) Operation Techniques.

(a.) Loading Method. When loading trucks from a bank or stockpile with a single scoop loader, use and the following V-loading method:

Step 1. Head the scoop loader toward the bank or pile in low gear with the bucket lowered and almost touching the ground. Moving forward, raise the bucket as it enters the material.

Step 2. Place the upright, loaded bucket in a hold position, and back away from the stockpile or bank.

Step 3. Then, approach the haul unit at a 90° angle.

(b.) Positioning of Haul Units. Proper positioning of equipment to receive material from the loader is necessary for maximum production. This cuts down on maneuver time and avoids damaging the travel surface.

(c.) Loading the Bucket. When loading the bucket, the bucket should be parallel with the ground so its cutting edge can skim the travel surface and remove ruts, obstacles, and loose material on the forward pass. As the cutting edge contacts the bank or stockpile, move the scoop loader forward at slow speed and increase power. Upon penetration of the material, raise the bucket. Crowd material into the bucket and roll the bucket back to prevent spilling. Maintain the upward position while backing away, to prevent spilling.

(d.) Working in Difficult Material. The multisegmented bucket works best when handling sticky material, which has a tendency to cling to the bucket. Its clam-type opening digs best in this type material. When handling (digging) medium-to-hard material, greater efficiency can be achieved by first breaking or loosening the material.

(e.) Excavating Basements. A scoop loader can dig excavations such as basements if the material to be excavated is not too hard. The operator should first construct a ramp into the excavation to bring the material out.

(f.) Backfilling. When backfilling ditches or trenches, lower the bucket to grade level and use the forward movement of the machine to push the stockpiled earth into the trench. This type of work is ideal for the scoop loader as long as the bucket is as wide, or wider than, the loader's wheels, to ensure cleanup in the least number of passes. Narrow buckets cause the wheels to ride up the stockpile. This raises one corner of the bucket and reduces the cleanup area.

(g.) Job factors. Job factors are the physical conditions that affect the production rate of specific jobs, other than the type of material to be handled. They include-

- Topography and work dimensions, including depth of cut and amount of movement required.

- Surface and weather conditions, including the season of the year and drainage conditions.

- Specifications that control handling of work or indicate the operational sequence.

- Equipment maintenance and repair.

Management factors. Management factors are-

- Planning, organizing, and laying out the job; supervising and controlling the operation.

- Selecting, training, and directing personnel.

6.) Safety. Operate scoop loaders carefully, since they are easy to turn over. Do not extend any portion of the body between the cab and the raised bucket arms. Scoop loader operators should be aware of the following unsafe practices:

- Bucket not blocked when being worked on.

- Lack of caution when removing and replacing the lock rings.

- Working between wheels and frame while the engine is running.

- Operating too close to the edge of a trench when backfilling.

- Digging into banks or stockpiles creating overhangs, then working under the overhangs.

- Riding with the bucket above axle height when traveling.

- Carrying or lifting passengers in the bucket.

- Forgetting to ground the bucket or set the parking brake before leaving the machine.

APPENDIX SIX

CRANES

Cranes are used primarily for lifting objects or loads, transferring them to new locations by swinging or traveling, and then placing them in the new location. The crane with its variety of attachments is the most common piece of lifting and loading equipment. The crane's variety of attachments enable it to also be used for ditching, excavating, and pile driving.

1.) Basic Crane Unit. The basic crane unit consists of a mounting (or carrier) and a revolving superstructure. The superstructure, or upper revolving frame, is substantially the same on all makes and models. The mounting may be one of three types: crawler, truck, or wheel. To make the basic unit complete and operable, any one of six attachments may be installed. Once an attachment is installed, the entire unit acquires the name of the attachment: crane (hook), clamshell, dragline, backhoe, shovel (crane-shovel), or pile driver.

(1) Crawler Mounting. As a normal rule, crawler-mounted cranes are best suited for jobs lasting a long time. The crawler mounting provides excellent on-the-job maneuverability and low ground-bearing pressure. Crawler cranes must be transported by low-bed trailers for distances greater than 1 mile because of their relatively low travel speeds and because of excessive wear and strain on the tracks.

(a) Stability. The crawler mounting provides a stable base for operating the revolving superstructure. The weight of the machine is spread over a large area because of the size of the crawler tracks. Normally, crawler-mounted cranes are not equipped with outriggers.

(b) Terrain capabilities. Crawler-mounted cranes can climb grades on firm, dry ground. Also, their low ground-bearing pressure enables them to travel over softer ground than cranes on other mountings. In locations where the ground is extremely soft or unstable, timber construction mats can be used to provide firm footings.

(c) Operating hints. Exercise extreme caution when moving crawler cranes down steep slopes. Steer crawler-mounted cranes by alternately engaging and disengaging power to the tracks. When using a crane that does not have brakes (some are not equipped with brakes), there is the danger of disengaging power from both tracks, allowing the crane to freewheel down the slope.

(2) Truck Mounting. Truck-type mountings are specially designed heavy-duty motor trucks. This mounting provides higher ground-bearing pressure and more on-the-job mobility than the crawler mounting. It also has excellent interjob mobility.

The 20- and 25-ton truck-mounted cranes can operate a hook block; a 3/4-cubic yard clamshell, dragline, backhoe, or shovel front; a concrete bucket; a wrecking ball; or a 7,000 foot-pound blow diesel pile driver.

(a) Stability. This mounting provides a less stable base (70 to 100 pounds per square inch [psi]) than the crawler mounting. It has two outriggers on each side to improve stability when operating the crane or its attachments. Increase stability by extending the outriggers and screwing the jacks down securely on the base plates.

(b) Terrain capability. The truck mounting restricts the efficient use of these cranes to firm, level terrain.

(c) Towing. A pintle hook (towing connection) is located on the rear of the mounting, and towing eyes are located on the front. The pintle hook enables the machine to tow the attachment trailer, allowing it to transport the associated attachments to the job site. The towing eyes are used to attach the truck to an assisting vehicle to tow the truck if its engine is inoperable, or if it is stuck or mired down. The towing eyes will withstand twice the dead weight pull of the vehicle.

(3) Wheel Mounting. The 20-ton wheel-mounted crane can operate a hook block, a 3/4-cubic yard clamshell or dragline, a concrete bucket, a wrecking ball, or a 7,000 foot-pound blow diesel pile driver.

(a.) Revolving Superstructure. The revolving superstructure rests on the mounting (or carrier) and includes the counterweight, engine, operating mechanism, boom hinge brackets, and cab.

- Counterweight. The counterweight is normally a cast-steel member bolted to the rear of the superstructure to produce a force that offsets the weight of the load and prevents the crane from tipping. If too much weight is lifted, the machine will tip forward. Conversely, too much counterweight in the rear could cause it to tip over backwards when heavy loads are dropped or released suddenly, or when the boom is removed. The danger of tipping backwards is the main reason for not adding more counterweight at the rear to increase lifting capacity.

- Operating Mechanisms. Two independent cable drums control the operation of the various attachments. The drums are mounted parallel to each other or one behind the other. They are referred to by their relative mounting, such as right or left, front or rear, or by their function, such as drag cable drum during dragline operations and closing line cable drum during clamshell operations. When the drums are used in conjunction with the crane hook block, they are usually referred to as the rear or main hoist drum. Crane-shovel units are equipped for hoisting the boom as well as operating the attachments.

A third drum, the boom hoist drum, controls the raising and lowering of the boom. Some models have a two-piece grooved drum lagging for quick attachment to the front drum shaft. This lagging replaces the split-type sprocket used on the front drum shaft for shovel operations. The diameters of the grooved lagging differ depending on the make and model of the machine. Differences in diameter provide different line speeds. For example, the lagging used for dragline operations may be smaller to provide a slower line, giving greater power.

Clutches and brakes may be powered mechanically, hydraulically, or pneumatically. On some makes and models, the clutch is the internal-expanding type. They are mounted internally and, when actuated, expand to drive the drum. Other makes and models have external-contracting type clutches.

(b.) Booms.

- Lattice-Boom. The lattice boom is usually made in two sections. The sections are fastened approximately in the center by one of two methods: bolted butt plate (flange) connections or pin and clevis connections. The upper section with the boom head and a system of sheaves is usually, but does not have to be, the same length as the lower section. Lattice boom standard lengths can be increased two ways: (1) insert an intermediate center section between the upper and lower sections (the most common way); or (2) add a boom tip extension called a jib. (A jib is constructed similarly to the standard boom, but is much shorter. Use this extension as a straight continuation of the main boom or, for greater horizontal reach, offset it from the boom centerline. Some crane booms are not equipped with jib boom anchor plates and can, therefore, only use the center section for extending the boom.) When lengthening lattice booms, you must extend the gantry or "A" frame to provide an efficient lifting angle for the boom lines.

- Telescopic Boom. Cranes with hydraulic telescopic booms operate with hydraulic controls and are limited to certain attachments. This boom is used on some wheel- and truck-mounted cranes.

(4.) Rough Terrain Crane. Both the outriggers and blade are operated hydraulically from the driver's cab on the mounting. The four large earthmoving-type tires can all be driven. The travel speed of the machine is 30 mph on the highway. The main difference between this crane and the truck-mounted crane is its high mobility, which enables it to work in areas inaccessible to the truck-mounted crane. The carrier is steered hydraulically with a choice of three methods: conventional two-wheeled steering, four-wheeled steering, and crab steering. The carrier transmission has four forward and four reverse speeds, depending on the position of the directional control lever. The machine has a torque converter drive.

a. Stability. Since the carrier is not suspended on springs, the front axle must oscillate. This oscillation prevents the carrier from tipping or rolling over if one of the front wheels drops into a hole. Whenever possible, raise the machine off the ground on the outriggers to increase stability.

If the machine is stabilized using outriggers, position safety wedges on the front axle to prevent oscillation. The outriggers operate individually allowing the complete machine top to be leveled.

b. Terrain Capabilities. The machine's low ground-bearing pressure lets it travel over relatively soft terrain. It can traverse slopes up to 48 percent if the ground is firm and dry. (Check operator's manual.)

c. Towing. The pintle hook on the rear of the carrier lets it tow the 10-ton tandem-wheel attachment trailer. However, towing the trailer cross-country is not recommended.

5.) Crane (Hook). The crane (hook) uses a lattice-type boom. Basic crane equipment includes hoist drums, hook blocks to provide the required parts of line (reeving), and boom suspension and hoist-wire ropes. The accessories available for use on the crane hook determine the types of loads it can handle. Accessories include, but are not limited to, slings, concrete buckets, and electromagnets.

(a.) Factors Affecting Lifting Capacity. All cranes, regardless of size, are rated on their maximum safe lifting capacity, which is based on the following: boom length, operating radius or boom angle, type of mounting, stability (use of outriggers), amount of counterweight, size of hook block, construction and size of wire rope, and position of lift.

- Boom Length. Booms exceeding the normal or standard length reduce crane lifting capacities. Capacity is further reduced when a boom jib attachment is used on a lattice boom. This is caused by the increased movement, or operating radius, and the added weight of the additional boom sections.

- Operating Radius. Operating radius is defined as the horizontal distance from the axis of rotation of the cab to a vertical line extending down from the outside edge of the crane boom-head sheave. Cranes are rated according to lifting capacities at various radii. Boom swing, centrifugal force, and wind speed, can increase the operating radius. As the working radius increases, the lifting capacity decreases. Therefore, it is essential to know the weight of a load prior to lifting, especially if boom swing is necessary. A crane is rated on the maximum load it is capable of lifting.

Crawler-mounted Crawler cranes are classified and identified by their rated lifting capacity at a 12-foot radius with a specified boom length. This means that a 12 1/2-ton crawler crane will lift 12 1/2 tons at a 12-foot radius, but not at any greater working radius.

Truck-mounted cranes are classified and identified by their rated lifting capacity at a 10-foot radius, with a specified boom length, and with outriggers set. This means that a 25-ton truck crane will lift 25 tons at a 10-foot radius, but not at any greater working radius.

- Footing. It is extremely important that operators position their cranes on a firm, level footing to prevent accidental tipping and to reduce excessive stress on the machine. If necessary, the crane's operating site should be prepared in advance.

- Stability. Cranes can be made stable by two methods: use of heavy counterweights (all cranes) and proper use of outriggers (wheel- and truck-mounted cranes).

- Hook Block. The hook block on the crane should be the size prescribed for the crane. Lack of proper rigging and hook capacity may damage the block and sheave system. Tables of lifting capacities usually do not allow for the weight of the hook block used; therefore, you must add it into the overall weight lifted.

- Lift Position. The crane can lift its heaviest load when the boom and load are in line with the longitudinal axis of the mounting. If the boom is positioned at a right angle to the mounting, the maximum weight lifted must be reduced, or the crane will tip. (Check operators' manuals for maximum lifting capacity when lifting at right angles to the crane mounting.)

- Maintenance. Cranes should be properly maintained to achieve maximum safe lifting capacity. With use, the crane's condition can deteriorate so that it can no longer perform certain jobs safely. Check the following items to ensure the crane can attain its maximum lifting capacity:

* Type, size, and condition of the wire ropes.

* Type, size, and condition of the hook block.

* Structural condition of the boom.

* Mechanical condition of the engine.

* Adjustments and functional qualities of the clutches and brakes.

(b.) Safe Lifting Capacity Calculation.

- Where repetitive lifting is involved, position the crane for the shortest possible boom swing and swing the load slowly.

- Maintain level footing to avoid swinging uphill or downhill.

- Use tag lines on loads to prevent excessive swaying of the load.

- Use adequate hoist line lengths to assure full travel of the block to the lowest point required.

- Organize the work for minimum travel time. When possible, lifts needed in one area should be completed before moving to a new position.

- Where precise load handling is required, use the power-down device on equipment if it is available.

- Do not use excessive counterweight or tie-down devices to increase stability.

- Check weather reports to determine approximate wind speed. Wind can affect crane operations and even cause overturning when working close to or at maximum lifting capacity.

6.) Pile Drivers.

(a.) Description. The pile driver attachment consists of a lattice-type boom, adapter plates, leads, a catwalk, a hammer, a pile cap, and wire ropes. The adapter plates are bolted to the top section of the leads and fastened to the boom tip. The leads are fastened below the base of the boom. The hammer may be diesel- or gravity-operated.

- Diesel Hammers. Diesel-driven, self-powered, pile-driving hammers are made in two types: open-top and closed-end. Both types have self-contained, free-piston engines, operating on a two-cycle compression-ignition principle. Diesel-driven hammers eliminate the need for air compressors or steam boilers. These hammers, however, cannot be used to pull piles. They are suitable for use with truck- and crawler-mounted pile drivers. When driving piling in very soft soil, the hammer may fire once and stop. This happens because all the energy from the fuel ignition is absorbed in the pile and the ground. Not enough power remains to recycle the hammer upward.

When this happens, it is more effective to revert to the drop hammer to drive the pile until you reach sufficient driving resistance. When using the diesel pile driver, a transmitter attachment is required.

- Drop Hammers. Drop hammers are best for driving vertical piling. When piling is angled, part of the driving force is lost in friction with the pile leads. Drop hammers are relatively slow compared to other types of hammers. Use a hammer that is at least equal to the weight of the pile being driven; however, for best results, it should be twice as heavy. Raise and drop the hammer at a steady rate of speed. Raising the hammer to the top of the leads each time will considerably slow down the driving action. For best results, only raise the drop hammer about 10 feet when driving.

(b.) Use. The pile driver is used to drive various types of wood, steel, and concrete piling for foundations, bridge bents, piers, and wharves.

(c.) Driving Rate. Driving time varies greatly depending on the terrain, weather, soil conditions, type of pile, and type of hammer used. The only way to determine the driving rate is to actually drive a pile under your particular set of conditions. However, a rule of thumb to use for planning is 30 minutes to drive a 12-inch diameter pile, 20 feet. This includes setting the pile in the leads.

(d.) Tips for Efficient Pile-Driving Operations.

- Position the pile driver so that only minimum time is lost in moving. Placement is generally parallel to the long axis of the pile group.

- Place the piles close enough to the site that the operator only needs to swing to pick up the next pile.

- Position the pile driver on stable, level soil while driving.

- Make light continuous blows with the hammer, rather than heavy infrequent blows. The latter causes more pile failures.

7.) Clamshell. Like the crane (hook), the clamshell is a vertically operated attachment capable of working-at, above, and below ground level. However, it is equipped with a bucket instead of a hook block. It can dig in loose- to medium-type soils. The length of the boom determines the height a clamshell can reach. The length of wire rope the cable drums can accommodate limits the depth a clamshell can reach. A clamshell's lifting capacity varies greatly. Lifting capacity factors, such as boom length and operating radius, determine a clamshell's safe lifting capacity.

The holding, closing, and tag lines control the bucket. At the start of the digging cycle, the bucket rests on the material to be dug, with the shells open. As the closing line rope is wound up on the drum lagging, the shells are drawn together causing them to dig into the material. The weight of the bucket, which is the only crowding action available, helps the bucket penetrate the material.

The holding and closing lines then raise the bucket and swing it to the dumping point. Open the bucket by releasing tension on the closing line.

(a.) Description. The clamshell consists of a clamshell bucket, a lattice-type boom, hoist drum laggings, a tag line, and wire ropes for the boom. The clamshell bucket consists of two scoops hinged together. Counterweights are bolted around the hinge. The boom can only be lengthened by using boom center sections fitted together by one of the two methods used with the crane (hook).

Clamshell drum laggings may be the same as those used for the crane, or they may be changed to meet speed and pull requirements of the clamshell. This requirement changes with the design of the equipment (check the operator's manual). The same wire ropes used for the crane (hook) operation can, for the most part, be used for clamshell operations. However, two additional lines must be added: a secondary hoist line and the tag line. The tag line winder is used to control the tension on the tag line to prevent the clamshell bucket from twisting during operation. The tag line winder, like the clamshell bucket, is interchangeable with any make or model in the same size range.

(b.) Use. The clamshell can be used on numerous jobs. However, it is best for such jobs as excavating vertical shafts or footings, and filling aggregate bins or hoppers.

  (1) Loading Aggregate Bins or Hoppers. Position the clamshell to avoid excessive raising and lowering of the boom or movement of the machine between the aggregate stockpiles and the hoppers.

  (2) Excavating Vertical Shafts or Footings. Since the dimensions of this type of excavation may vary, it would be difficult to give the most efficient digging position for the clamshell. Two important facts to consider are: the amount of wire rope on the machine; and the need to keep the outside edge of the cut lower than the center, which keeps the bucket from drifting toward the center and causing a V-shaped excavation. In deep excavations, a bucket spotter or signalman usually guides the operator, especially when the bucket is out of the

operator's sight. Spotters may also need to use hand tag lines to guide the bucket.

8.) Dragline. The dragline is a versatile machine capable of a wide range of operations at and below ground level. It can handle material that ranges from soft to medium hard. The greatest advantage of a dragline over other machines is its long reach for digging and dumping. Another advantage is its high cycle speed, second only to the shovel. The dragline, however, does not have the positive digging force of the shovel or backhoe, since the bucket is not weighted or held in alignment by rigid structures. Therefore, it can bounce, tip over, or drift sideward when it encounters hard material. These weaknesses increase with digging depth and are particularly noticeable with smaller machines.

(a.) Description. Dragline components consist of a lattice-type boom, a dragline bucket, and a fairlead assembly. Wire ropes are used for the boom suspension, drag, bucket hoist, and dump.

The fairlead accomplishes what its name implies-it guides the cable onto the drum when loading the bucket. The hoist rope, which operates over the boom-point sheave, is used to raise and lower the bucket. In the digging operation, the drag rope is used to pull the bucket through the material.

When the bucket is raised and moved to the dump point, empty the bucket by releasing the tension on the drag rope. Dragline buckets are rated by type and class for all bucket sizes, as follows:

- Bucket Types. Type I (light duty), Type II (medium duty), and Type III (heavy duty).

- Bucket Classes. Class P (perforated plate) and Class S (solid plate).

(b.) Use. The dragline can effectively accomplish many tasks. It can be used on dredging operations where the material is wet and sticky. It can dig trenches, strip overburden, clean and dig roadside ditches, and slope embankments. The dragline is the most practical attachment to use when handling mud. Its reach lets it handle a wide area from one position, and the sliding action of the bucket decreases suction problems.

(c.) Factors Affecting Dragline Capacity. The dragline boom may be angled relatively low when operating. However, boom angles of less than 35° from the horizontal are seldom advisable for dragline work because of the possibility of tipping the machine. When excavating wet, sticky material and casting it onto a spoil bank, the chance of tipping the machine increases because the material tends to stick in the bucket, increasing the momentum at the end of the boom. When using a crawler dragline, heavy counterweight must be used. Without it the capacity of the machine should be reduced to prevent tipping. When using the dragline, track digging locks should be positioned to the front.

Draglines may also be used for the following:

- Excavating a Trench. When excavating a trench with the dragline, center the machine on the excavation. The crane or mounting should be in line with it, called in-line approach. The dragline cuts or digs to the front and dumps on either side of the excavation. The machine moves away from the face as the work progresses.

- Sloping an Embankment. The dragline can slope an embankment more effectively by working from the bottom to the top. The machine is positioned on the top with the crawler parallel to the working face, which enables it to move the full length of the job without excessive turning. This is the parallel approach.

- Digging Underwater or in Wet Materials (Dredging). A dragline is ideal for removing earth from water-filled trenches, canals, gravel pits, ditches, or the like. However, drainage ditches may be required to carry water away from the construction site. Or, you can begin at the lowest grade point on the job so that drainage will be provided as the dragline progresses toward higher levels. Crawler-mounted draglines can and do work with their tracks underwater, but this requires slow moves and involves treacherous footing. Digging underwater or in wet materials increases the weight of materials and frequently prevents carrying heaped bucket loads. Plan your operations to work as dry as possible and to provide drainage to eliminate existing water as well as rain water, which might otherwise slow down or delay production. Drainage projects involving ditch excavation through swamps or soft terrain are common. Under these conditions the excavated material is normally cast onto a levee or spoil bank, which eliminates the problem of constructing roads for haul equipment. However, it may be necessary to construct rudimentary supply and service roads to support the construction effort.

- Loading Haul Units. Where job conditions require loading excavated material into trucks, plan the excavation so the loaded trucks can travel on high, dry ground and/or on advantageous grades. Spot trucks and dragline for minimum boom swing. Position the truck bed under the boom tip with the long axis of the bed parallel to the long axis of the boom or right angle to the boom.

  More spillage is to be expected from draglines than from shovel loading. Dragline operators have no rigid control of the bucket. Even skilled operators have a hard time positioning every pass over the center of the target. The dragline, not being a rigid attachment, will not dump its material as accurately as other excavators. Therefore, you will need more time to position the dragline than other equipment before dumping.

9.) Crane-Shovel. Cable and hydraulic shovels operate under the same principle. The shovel is designed to operate against a bank or cliff face, which it displaces as it moves forward. It forms its own roadway as it works since the bottom, or floor, of the excavation becomes the level upon which the machine travels. It has the ability to form the roadway to grade and to shape the sides of the cut. The shovel can dig above and below track or wheel level. Because its bucket can be forced into the bank, the shovel is capable of digging in some medium-hard to hard materials. (Some materials must be blasted prior to excavation.)

(a.) Description. The shovel attachment includes the shovel boom; dipper stick; shovel-type bucket; mechanism for crowding, retracting, and dumping the bucket; and a dipper trip.

When positioning a crawler shovel at the face of an embankment, the drive end, or the rear of the crawler assembly, should be away from the embankment. This prevents damage to drive chains and drive sprockets caused by the operator's failure to stop the bucket soon enough when lowering it to the embankment toe. During operation, fill the bucket by pulling it up through the material in a combination hoisting and crowding action

(b.) Use. The shovel is the best attachment for excavating hard compact material at the face of an embankment and loading the material into haul units. It can also be used to strip overburden and dump it onto spoil banks. Although the shovel is capable of digging below ground level, the underground depth is limited by the length of the dipper stick.

(c.) Tips for Maximum Shovel Efficiency.

- Keep the crane-shovel's working area as level as possible. When used in areas where material is soft, support the shovel on construction mats. Avoid digging in areas with oversized material.

- At the start of digging, set the dipper at crawler (or wheel) level and 2 or 3 feet in front of the crawler (or wheels).

- Do not excavate too far beyond the boom point because power diminishes rapidly as the dipper stick is extended beyond an imaginary line plumbed from the boom tip to the ground.

- Spot hauling units on each side of the shovel (for frontal approach). Do not spot haul units so close to the shovel that the shovel's counterweight can strike them when the superstructure swings in the opposite direction. On the other hand, they should not be spotted out so far that the shovel operator will have to overcrowd the dipper stick to reach them. Proper spotting of trucks plays an important part in increasing shovel production.

- Keep the shovel close to the working face. After each move, start the digging sequence at the center of the face. Make passes on each side of the center until all of the face within reach is cut back evenly. Fill the dipper using straight, forward passes against the working face.

10.) Safety.

(a.) Crane Operators. Crane operators must be thoroughly trained in the fundamental rules of crane safety before they can operate a crane. They are also responsible for knowing the limitations and capabilities of cranes.

Operators must not operate an unsafe crane. They must be able to identify and promptly report any equipment malfunction or defect. Operators must have the authority to stop and refuse to handle loads until safety has been assured. They should be aware of the following:

(b.) Crane General Safety Hazards. The common hazards associated with operating cranes (crane-hook), clamshells, draglines, backhoes and shovels are-

- Boom contacting high-tension wires. (This is considered the most hazardous aspect of crane operation.)

- Dropping or slipping the load.

- Breaking cables.

- Not using outriggers.

- Clutch or brake slipping, allowing boom radius to increase.

- Leaking hydraulic lines and fittings.

- Not using mousing or safety-type hooks.

- Obstructing the free passage of the boom or the load of the crane.

- Backing and turning machines without looking.

- Not being familiar with equipment.

- Operating on uneven ground.

- Using an estimated weight rather than determining actual weight of load.

- Making test lift of an unknown weight load in an unstable lifting position.

- Not referring to capacity charts on cranes when different boom lengths are used.

- Using booms with bent or dented chord members.

- Using crane hoist cable for towing.

(c.) Power-Shovel Safety Standards. Shovel operators must never leave the cab while the master clutch is engaged. Leave shovel buckets on the ground when they are not in use. Drivers must leave cabs of dump trucks being loaded by power shovels, clamshells, draglines, or other overhead means even if the truck top is equipped with a steel protective shield. They should set the brakes, leave the cab, and remain outside the swing range of the bucket until the truck has been loaded. Shovel operators should be aware of the following safety precautions:

- Repairs. In case of a breakdown, move the shovel well away from the foot of a slope before making repairs.

- Wire Cable Replacement. Regularly inspect wire rope cables used on power shovels and change them when 10 percent of the wires in any 3-foot length are broken.

- Overhead Wires. When moving the shovel under electric wires, ensure that there is at least a 10-foot clearance. Do everything possible to prevent contact with overhead wires.

- Range of swing. All persons (to include dump truck drivers) must keep out of the shovel's swing range, to avoid being struck by the cab as it rotates. When workmen must work at the rear of the cab of a shovel, they should notify the operator that they are working there.

- Ramps. Do not make ramps for power shovels too steep. Check the brakes and travel mechanisms before traversing ramps.

- Slides. When excavating in a bank of sand, gravel, or rocks that have been blasted, shovel operators must be careful not to dig the shovel too far into the bank since this could cause slides.

(d.) Pile-Driver Safety Standards. In setting up to drive piles, use suitable guys, outriggers, counterbalances, or rail clamps (as necessary) to maintain stability of the rig. When in operation, use safety lashings for all hose connections to pile drivers, pile ejectors, or jet pipes. Use tag lines to control unguided piles and flying hammers. When hoisting steel piling, use a closed shackle or other positive means of attachment. Only pile-driving crew members and other authorized persons should be permitted in the actual work areas when driving piles. Pile-driver operators should be aware of the following safety precautions:

- Repairs. Never repair any diesel or air equipment while it is in operation or under pressure.

- Defective Air Hose. Promptly replace defective air hoses on compressed air equipment. Make frequent inspections of air hoses to locate defects.

- Overhead Wires. When moving this equipment, make sure that it does not come in contact with overhead electric power lines.

- Hammer and Driving Heads. When a pile driver is not in use, use a cleat or timber to hold the hammer in place at the bottom of the leads. Tie-in driving heads whenever the rig is used to shift cribbing or other material. Workers should never place their heads or other parts of their bodies under suspended hammers that are not dogged or blocked in the leads.

- Drums, Brakes, and Leads. Keep hoisting drums and brakes in the best condition possible and shelter them from the weather. Keep leads well greased to provide smooth hammer travel.

APPENDIX SEVEN

BACKHOES

This chapter covers the different types of backhoes and backhoe attachments. Backhoes can dig in soft to hard material using the weight of the boom plus the positive pull on the dipper to force the dipper into the material. Hydraulic and cable-operated backhoes use the same operating principles. Cable-operated backhoes are used less frequently than hydraulic ones. The major components of the hydraulic backhoe are the dipper stick (arm) and the dipper.

1.) Hydraulic Excavator's Versatility on Tracks or Wheels. Fast-acting, variable-flow hydraulic systems and easy-to-operate controls give hydraulic excavators the high implement speed and breakout force to excavate trenches, the precision to set pipes, and the capacity to backfill.

2.) Use. The backhoe is best suited for trench excavating, since it can dig well below the unit's tracks. A large variety of booms, sticks, buckets, and attachments gives excavators the versatility to excavate trenches, load trucks, clean ditches, breakout old concrete, install outlet pipes, handle toxic waste, and so forth.

3.) Types of Excavations. The backhoe is normally associated with two types of excavations-trenching and basement-type.

   (a.) Single Lift. In the single-lift method, center the backhoe on the trench and keep the crawler (or carrier) in line with it. As the digging progresses, move the machine away from the end of the excavation and load the material in haul units or stockpile it along the side of the trench for backfill.

   (b.) Multiple Lifts. When using the multiple-lift method, dig the trenches in two or more cuts or lifts. To excavate the top 35 to 45 percent of the trench depth, make the first cut with the boom carried high. To finish the cut and remove the remainder of the material, move forward about one-half the length of the machine with the boom carried low. Even though this method involves more and shorter moves, it has its advantages in better digging angles of the dipper, better filling of the dipper, and shorter hoisting distance on top lifts. It also affords the operator better visibility because of close-in dipper action.

4.) Operation Techniques

(a.) Underground Utilities. Since the backhoe is used primarily for below-track-level excavations, survey for underground hazards as well as surface obstacles before starting operation. This applies particularly to populated areas with underground utilities.

(b.) Confined Quarters. Working in confined quarters is not efficient from a production standpoint. If considerable close quarter work is expected, use small machines that can operate efficiently within a minimum work radius.

(c.) Positioning of Backhoe. It is important that the machine be positioned properly on the job to gain its greatest effectiveness. Efficient positioning of the backhoe is contingent on the type of work to be done.

(d.) Drainage. If the job is to continue during wet seasons or in wet areas, give prime consideration to drainage.

(e.) Hard Materials. A backhoe will dig into fairly hard materials. However, blasting or ripping may be more efficient than breaking through hardpan and rock strata with the dipper. Once the trench is open, ledge rock can be broken by pulling the dipper up under the layers. Remove the top layers first, lifting only one or two layers at a time.

(f.) Length and Depth of Cut. The length and depth of cut should be judged to produce a full dipper at every pass.

5.) Wheel-Mounted Backhoes with Front Bucket. These backhoes have the precision to work in tight places and can move quickly from job to job.

Backhoe/Loader. The backhoe/loader tractor is used for excavation of pipeline trenches, drainage ditches, building footings, and hasty fortifications. It is also used for backfilling, loading small quantities of earth into trucks, and moving earth and construction materials within confined areas of a construction site. In addition to the backhoe and front-mounted bucket, it can be equipped with a variety of attachments, including a hydraulically operated concrete breaker and a tamper. However, it is primarily used in the backhoe/loader configuration. The backhoe/loader tractor is roadable for short distances at speeds of 15-20 mph. For long distances, it must be transported.

(a.) Excavating.

- Backhoe Bucket. Before operating the backhoe, level the machine. Lower the front bucket to the ground (flat), then move the gearshift and the range shift levers to their neutral positions. Turn the operator's seat to face the backhoe and lower the stabilizers (outriggers) onto firm ground. Plan and lay out the work area. Always operate with the least amount of dipper-arm swing.

- Front Bucket. Use bucket cylinders to help break the ground loose instead of depending on the forward movement of the machine. When starting an excavation, use as flat a ramp as possible. Plan the job so that most of the work can be done when the unit is driving forward out of the excavation. A steeper ramp can be used when driving in forward than when driving in reverse.

(b.) Loading. When loading dump trucks with a backhoe, plan and lay out the area of operation.

- Park the truck so the dipper arm does not have to turn (revolve) more than 90°.

- Dipper arm should be rotated over rear of dump bed, rather than over the passenger compartment.

- Keep the working area smooth.

- Raise the bucket while moving toward the truck.

- Lower the bucket while moving away from the truck.

- Shake the bucket only when necessary to loosen dirt stuck in the bottom.

Loading trucks with a backhoe is somewhat inefficient which results from the maneuvering necessary to complete a dump within the length of a truck body. It also has a slow digging cycle. The slow cycle is caused by the necessity of pulling the bucket toward the boom before hoisting, to retain the load during the hoist. The most convenient position for a truck to be loaded by a backhoe is parallel to the backhoe. This allows loading along the full length of the truck body and reduces spillage.

(c.) Leveling and Grading. Use the front bucket for leveling and grading, as follows:

- Plan and lay out the work area.

- Before attempting to finish the grade, fill holes and hollows and loosen up any high spots by rough grading.

- To spread dirt evenly, hold the bucket close to the grade (tipped slightly forward) and let the dirt spill.

- Level and pack with the loader bucket in a lowered position. Operate the loader in reverse with the bucket dragging (back blading) on the ground.

APPENDIX EIGHT

TRENCHING EQUIPMENT

Trenchers can dig in a variety of materials ranging from relatively soft material to tough or sticky clay, shale, and in some cases, semihard coral and thin pavements. Their continuous chain of steel buckets dig the material carry it upward, and deposit it on a conveyor belt that transports it to one side and dumps it. They can traverse slopes of up to 11 degrees, or 20 percent, while ditching. The trenchers large pneumatic tires give it good cross-country mobility. However, use only minimum speeds when traversing rough country, since the trencher has no springs over the rear wheels and could sustain severe damage.

On a construction site, it may be more economical to use equipment (such as dozers, graders, or backhoes) that is already on-site for ditching. The type of machine selected for a particular job depends on the size and depth of trench needed, the soil, and other conditions such as specific job requirements.

SECTION I. TRENCHERS

1.) Use. Trenchers are used to dig ditches for-

- Water, gas, and oil lines.

- Telephone cable lines.

- Sewer lines.

- Irrigation lines.

- Drainage.

2.) Grade and Depth of Cut.

(a.) Mark the Depth of Cut. For the desired grade and depth of the ditch, use a dozer or grader to level the entire line of the ditch. This could entail considerable site preparation on relatively rough terrain. Grade stakes can be offset from the center line of the ditch to indicate the depth of cut. Place grade stakes wherever there is a significant change in surface elevation. Accuracy of this method depends on the interval between the stakes, the condition of the terrain, and the skill of the operator.

(b.) Establish a Grade Line. Establish a grade line (on the surface), which conforms to the desired grade and depth of the ditch. The grade (guide) line can consist of a string line or wire stretched on stakes. Establish the line about 24 inches off the ground. After the operator is told the initial depth of cut, he will drive the bucket line into the ground. When he reaches the required depth of chain, he should adjust the depth gauge to ride on top of the string line and proceed to dig.

On level ground, the operator simply watches the chain on his depth indicator and keeps it just touching the grade line. On rough ground, however, the operator is required to keep a closer watch on the depth indicator.

(3). Side Slopes. Trenchers can excavate ditches working parallel on slopes up to 15°. However, when working parallel on slopes, the ditch will be undercut. This undercutting can cause the trench wall to cave in because of soil instability or from erosion during inclement weather. If a trench is being constructed for troop emplacements, use hand labor to remove the overhang. For best results on side slopes, use a dozer to bench-off the slope ahead of the trencher. This lets the trencher excavate a vertical-walled ditch. In addition to using benches on side slopes, it is best to clear brush off ahead of the trencher. Trenchers can be used to walk down small brush and tear through it, but it is best to clear the area first.

(4.) Curved Trenches. Trenchers can excavate curved trenches if the curve is not too acute. Cutting on a very acute curve can cause the bucket line of the unit to jam in the trench. The sides of the bucket wear rapidly when excavating curved trenches.

(5.) Below-Ground Obstructions. Be cautious when digging in areas where pipes, conduits, or other objects have been installed. Although trenchers have overload-release sprockets to protect the machine, the machine could fracture an underground utility line before breaking power to the bucket line. Before digging, locate any obstructions and mark them plainly on the grade line. Dig until the buckets are about 1 foot from the obstruction, then raise the bucket line while still digging. Do not work closer than 1 foot to pipes or telephone cables.

SECTION II. OTHER TRENCHING EQUIPMENT

1.) Graders. Graders are restricted to shallow cuts in medium-hard materials; they should not be used for heavy excavation. They are capable of working on slopes as steep as 3:1. However, it is not advisable to use a grader to construct ditches running parallel on such a slope because graders have a comparatively high center of gravity and the right pressure at a critical point of the blade could cause the machine to roll over. Graders are capable of progressively cutting ditches to a depth of 3 feet. It is more economical to use other types of equipment to cut ditches deeper than 3 feet.

2.) Backhoes. Backhoes can be used for ditching, as well as excavating. A backhoe can dig well below the tracks or wheels of its carrier in most soils. Various dipper sizes are available to match the desired trench width. Listed below are some operating techniques when ditching with a backhoe:

- Select the dipper to match the trench width, rather than make additional passes with a smaller dipper.

- Keep the backhoe centered over the trench.

- Dig hard materials, such as rock, after the surface is broken with more suitable equipment.

- Remove very hard materials in relatively shallow layers.

- Cut so that each pass results in a full dipper.

For stabilization, operate with the least amount of swing, and lower the outriggers.

3.) Dozers. Dozers can be used to make shallow V-ditch cuts. They can also be used for deeper V-shaped ditches when other equipment is not available. To cut shallow "V" ditches, tilt the blade. For larger ditches, doze at right angles to the center line of the ditch. When the desired depth is reached, doze the length of the ditch to smooth the sides and bottom.

APPENDIX NINE

HAULING EQUIPMENT

DUMP TRUCKS

1.) Use. The 2 1/2-ton, 5-ton, and 20-ton dump trucks are used for a variety of purposes. This manual, however, discusses dump trucks used primarily for hauling, dumping, and spreading base course and surfacing materials; for hauling other materials for construction operations; and for miscellaneous hauling.

2.) Capacity. The capacity of hauling equipment, including dump trucks, can be expressed in three ways: by the load it will carry (in tons), by its truck volume (in cubic yards), or by its heaped capacity (in cubic yards). The capacity of dump trucks is normally expressed in tons: 2 1/2-ton, 5-ton, and 20-ton. (Conversely, the capacity of loading equipment is normally expressed in cubic yards.) The unit weight of the various materials to be transported may vary from as little as 1,700 pounds per cubic yard for dry clay, to 3,888 pounds per cubic yard for concrete.

3.) Operation Techniques.

    (a.) Loading. For maximum efficiency, fill trucks as close to rated capacity as practical.

However, if haul roads are in poor condition or if trucks must traverse extremely steep grades, you should underload the truck. Trucks rendered out of service because springs, axles, or transmissions were broken due to overloading, more than offsets the production gained from maximum loads.

- Positioning of Load. Pay attention to the location of the load within the truck body. Place heavier materials near the rear of the body to minimize the work of the dumping mechanism.

- Spotting Trucks. Use spotting logs or blocks when trucks are hauling from a hopper, grizzly ramp, or stockpile. They are also beneficial when hauling from a crane (dragline, backhoe, clamshell, or front shovel) where they facilitate prompt and accurate vehicle spotting. This improves the shovel operator's efficiency and speeds loading.

When loaded by a backhoe, spot trucks as close to the bank as possible. Ensure that they are within the working radius of the dipper arm as it leaves the bank. This saves time in racking the dipper in and out. Make sure that the shovel does not unload over the truck cab.

(b.) Maintaining Proper Speed. Haul at the highest safe speed without speeding, and in the proper gear. Speeding is unsafe and hard on equipment. When several trucks are hauling, it is essential to maintain the proper speed to prevent hauling delays or bottlenecks at the loading or dumping site. Slow trucks, as well as speeding ones, disrupt normal traffic patterns. Until the maintenance crew can repair a sluggish truck, replace it with a standby truck.

(c.) Planning Haul Roads.

- Use separate haul roads to and from the dump site, if possible.

- Ensure haul roads are well maintained with minimum grade.

- To increase efficiency of haul roads, use one-way traffic.

(d.) Planning Traffic Patterns (Load and Dump Sites). Lay out traffic patterns in loading and dumping sites to minimize backing, passing, and cross traffic.

(e.) Dumping.

- When unloading (dumping) material that requires spreading, move the truck forward slowly while dumping the load. This makes spreading easier.

- Establish alternate dumping locations to maintain truck spacing when poor footing or difficult spotting causes slow dumping.

(f.) Recording Loads. Station tallymen at unloading points to keep accurate records of the number of loads hauled by each truck. These records help in preparing production records and in locating irregularities that may warrant further investigation.

(g.) Preventive Maintenance. Keep truck bodies clean and in good condition. Accumulations of rust, dirt, dried concrete, or bituminous materials hamper dumping operations. The time spent cleaning and oiling truck bodies must be considered in computing transportation requirements (cycle time). (See operators' manuals for additional preventive maintenance procedures.)

- Clean truck bodies thoroughly at the end of the day. When used to haul concrete, spray the dump beds with water before loading and clean them thoroughly as soon as practical after dumping.

- Coat the walls and sides of truck bodies with diesel fuel or crankcase oil to prevent bituminous materials from sticking.

EQUIPMENT TRAILERS

1.) Use. Equipment trailers are used to transport heavy construction equipment such as cranes, dozers, and any other tracked, wheeled, or skid-mounted equipment not designed for movement by other means. They are also used to haul long items such as pipes or lumber, or packaged items such as bagged cement.

2.) Operation Techniques.

   (a.) Loading. For maximum efficiency, load trailers as close as possible to their rated capacity.

When loading, always station a man on the trailer to direct the equipment operator and to keep the machine centered on the ramps and trailer.

- Rear-Loading. With rear-loading trailers, use low banks or built-up earth ramps where possible. However, these trailers also carry loading ramps for loading from level ground. When using the loading ramps to load a crawler-tractor (dozer), the dozer operator should run it slowly up the ramps and, as the balance point is reached, reduce speed to allow it to settle gently onto the trailer bed. The dozer should then be moved slowly ahead until it rocks forward onto the trailer. (Some low-beds are front-loaded.)

- Side-Loading. Trailers may be loaded from the side in areas that restrict rear-loading, but be careful not to damage the trailer bed.

   (b.) Positioning and Securing Equipment on Trailer. After positioning the equipment on the trailer bed, block and chock it and chain it to the bed. Ensure that the weight of large equipment is properly distributed on the trailer. (Trailers are normally marked with the centering position.) c. Unloading. Unload heavy equipment slowly to prevent damage to the trailer or the equipment. Always use ramps to unload, as well as to load.

APPENDIX TEN

SOIL-PREPARATION AND COMPACTION EQUIPMENT

Horizontal construction projects such as roads, airfields, and heliports can be constructed using a wide variety of soil types. The suitability of these soils for various construction applications depends on the gradation, the physical characteristics (Atterberg Limits), and the load-bearing capacity of the soil. While some soil types are suitable for construction in their natural state, others may require preparation, such as adjusting the moisture content or mixing/blending the soil.

Additionally, most soils require compaction to eliminate voids in the soil mass, which increases the density of the soil as well as its load-bearing capacity.

SOIL PREPARATION

1.) Moisture Content. The amount of water present in a soil mass (moisture content) is directly related to the achievable density of that soil. Each type of soil has an optimum moisture content (OMC) at which the maximum density can be obtained. The OMC is the moisture content at which the soil has become sufficiently workable under a given compactive effort to cause the particles to become so closely packed that most of the air has been expelled. Most soils (except cohesionless sands) are more difficult to compact when the moisture content is less than optimum. When a soil's moisture content is above optimum, water interferes with the close packing of the soil particles, which decreases the attainable density. The OMC varies from about 12 to 25 percent for fine-grained soils and from 7 to 12 percent for well-graded granular soils. Since it is difficult to attain and maintain the exact OMC, there is an acceptable moisture range for each type of soil. This range (4 percent) is based on attaining the maximum density with the minimum compactive effort.

  (a.) Adding Water to Soil. If the moisture content of a soil is below the optimum moisture range, water must be added to the soil prior to compaction. When it is necessary to add water, the project manager must consider the following:

   • Amount of water required.

   • Water application rate.

   • Method of application.

Water can be added to the soil at the borrow pit or in-place (at the construction site). When processing granular materials, best results are usually obtained by adding water in place. After water is added, it must be thoroughly and uniformly mixed. It is essential to determine the amount of water required to increase the moisture content to within the acceptable moisture range. The amount of water that must be added is normally computed in gallons per station (100 feet of length). The computation is based on the dry weight of a soil contained in a compacted layer. Once the total amount of water has been calculated, the water application rate must be calculated. The water application rate is normally calculated in gallons per square yard. Once the application rate has been calculated the method of application must be determined. Regardless of which method of application is used, it is important to ensure that the proper application rate is achieved and that the water is uniformly distributed.

(b.) Reducing the Moisture Content. As previously stated, soil which contains more water than desired (above the optimum moisture range) is also difficult to compact. Excess water interferes with the close packing of the particles, and makes the desired density unachievable. In these cases, steps must be taken to reduce the moisture content to within the desired moisture range. These actions may be as simple as aerating soil or as complicated as adding a soil stabilization agent that promotes drying, such as lime, Portland cement, or flyash. If the excess moisture is being caused by a high-water table, some form of subsurface drainage structure may be required before the soil's moisture content can be reduced. The most common method of reducing the moisture is to scarify the soil prior to compaction. This can be accomplished with the scarifying teeth on a motor grader, or by the use of a rotary cultivator.

(c.) Effects of Weather. Weather conditions substantially affect soil moisture content. Cold, rainy, cloudy, and calm weather allows soil to retain water; hot, dry, sunny, and windy weather quickly dries it out.

SOIL COMPACTION

1.) Compaction is the process of mechanically densifying a soil, normally by the application of a moving (or dynamic) load. This is in contrast to consolidation, which is the gradual densification of a soil under a static load. No other construction process, applied to natural soils, so drastically affects a soil's properties. When controlled properly, compaction increases a soil's load-bearing capacity (shear resistance), minimizes settlement (consolidation), changes the soil's volume, and reduces its permeability (ability to transmit water). Although compaction does not affect all soils alike, the advantages gained by compaction have made it a standard and essential part of the horizontal construction process.

(a.) Compactive Effort. Compactive effort is the method used to compact the soil mass. The appropriate compactive effort is based on the physical properties of the soil, including gradation (well-graded or poorly graded), the Atterberg limits (cohesive or cohesionless), and the moisture content. Compaction equipment uses one or more of the following methods of compactive effort:

- Static weight (or pressure).

- Kneading (or manipulation).

- Impact (or sharp blow).

- Vibration (or shaking).

(b.) Compaction Equipment. Soil-compaction equipment ranges from handheld vibratory tampers, suitable for small or confined areas, to large self-propelled rollers and high-speed compactors, ideally suited for large horizontal construction projects. Consider the following factors when selecting the appropriate compaction equipment:

- Type and properties of the soil.

- Desired density.

- Desired lift thickness.

- Size of the job.

- Stage of construction (subgrade, subbase, or base course).

2.) Compaction equipment available.

    (a.) Sheepsfoot Roller. The sheepsfoot roller is suitable for compacting all fined-grained materials, but is generally not suitable for use on cohesionless granular materials. The sheepsfoot roller compacts from the bottom up and is especially appropriate for plastic materials.

    The lift thickness for a sheepsfoot roller is limited to 6 inches in compacted depth. A uniform density can usually be obtained throughout the full depth of the lift if the material is loose and workable enough to permit the roller's feet to penetrate into the layer on the initial pass.

    Since the sheepsfoot roller tends to aerate the soil as it compacts, it is ideally suited for working soils that have moisture contents above the acceptable moisture range. The sheepsfoot roller does not adequately compact the upper 2 to 3 inches of a lift and should, therefore, be followed by a lighter pneumatic-tired or steel-wheeled roller.

    (b.) Tamping-foot Roller. The self-propelled, high-speed tamping-foot roller is similar in design and operation to the sheepsfoot roller except that the feet are square or angular and taper down to the end. This design allows the roller to achieve better penetration on the initial pass and consequently allows the roller to attain thorough, uniform compaction throughout a lift. This compactor operates on the same principle as the sheepsfoot roller in that it compacts from the bottom of the lift to the top, "walking out" as it goes. The high-speed, self-propelled design makes the tamping-foot roller faster (12-15 mph) and more versatile than the towed sheepsfoot, and it is replacing the latter in most construction units.

(c.) Pneumatic-Tired Roller. Pneumatic-tired (rubber-tired), rollers are suitable for compacting most granular materials but are not recommended for use with fine-grained clays which tend to adhere to the rubber tires. Pneumatic-tired rollers are extremely versatile in that they compact using two types of compactive effort: static load and kneading.

(d.) Steel-Wheeled Roller. Although not as versatile as the pneumatic-tired roller, the smooth steel-wheeled roller can compact a wide variety of materials from sand to cobbles. It is generally used for compacting cohesionless subgrade, base course, and wearing surfaces. It is most effective when used immediately behind a blade to create a smooth, dense, and watertight subgrade finish. Because of its relatively low unit pressure and the fact that it compacts from the top down, it is normally used for relatively shallow lifts (<4 inches) and is often used in conjunction with a kneading action (bottom up) roller such as the sheepsfoot or tamping-foot roller. The smooth steel drum is ideal for base- or wearing-course finish work, but care should be exercised to prevent excess. crushing of base course material.

(e.) Smooth-Drum Vibratory Roller. The smooth-drum vibratory roller uses a vibratory action in conjunction with the ballast weight of the drum to rearrange the soil particles into a dense soil mass. Vibratory compaction, when properly controlled, can be one of the most effective and economical means of attaining the desired density. This roller is very effective in compacting the noncohesive/nonplastic sands and gravel often used in subbase and base course applications. It is also being used more frequently on cohesive materials. Cohesive soil particles are bound by many forces, including electrical charges, intermolecular attraction, and surface tension. These forces do not permit particles to flow over and around each other freely. The vibrating action of this roller has been proven effective in overcoming these forces. Because this roller is relatively light, the recommended maximum loose lift depth is 9 inches. However, satisfactory results have been achieved in sand with thicker lifts.

(f.) Plate Compactor and Backfill Tamper. The Plate Compactor and backfill tamper are small, self-contained hand-operated vibratory compactors. The Plate Compactor resembles a power mower, while the backfill tamper is similar in appearance to a: jackhammer.   Both the Plate Compactor and backfill tamper are ideally suited for compacting materials in small confined spaces such as in a trench or inside of an existing structure. Both units are powered with a gasoline engine so there are no supply lines or other auxiliary items to hinder their operations.

3.) Management Techniques. To fully understand the management techniques associated with the different types of compactors, it is essential to understand the various phases in a horizontal construction project. These include preparing the subgrade; placing, spreading, and compacting fill material for the subgrade and base courses; and finishing and surfacing operations.

(a.) Preparing Subgrade. The term subgrade (subbase) describes the in-place soil on which a road, airfield, or heliport is built. The subgrade is considered to include soil to the depth that may affect the structural design of the project or the depth at which climate affects the soil. This area may be at depths up to 10 feet for pavements carrying heavy loads. The quality and natural density of this material dictates what action(s) must be taken to prepare the subgrade. For example, a highly organic subgrade material may have to be totally removed and replaced with a higher quality, select material. In most situations this is not the case. Often the in-place material is suitable, but requires some level of compactive effort to achieve the necessary density. Heavy pneumatic-tired rollers are preferred for subgrade compaction because of their capability to compact the soil to depths up to 18 inches. Often it is necessary to scarify the top six inches of the material, so that the moisture content can be adjusted. Then with the adjusted moisture content the material is recompacted to a higher density than could be achieved in its natural state. This process is known as scarify and compact in place (SCIP). The sheepsfoot roller is well suited for this operation since it loosens the material, yet compacts it as it walks out of the material.   The sheepsfoot roller also helps to break up oversized rock in the area. If a sheepsfoot roller is used to prepare the subgrade, you should use a rubber-tired or smooth-drum roller in conjunction with the sheepsfoot to compact the top 1 to 2 inches of material, dress the surface, and seal the subgrade to protect it from the damaging effects of water infiltration. If the subgrade material is sand, then a vibratory compactor would provide the most effective compactive effort.

(b.) Spreading the Fill Material. Once the subgrade has been prepared, fill material must be brought in to form the subbase and base courses of the project. When placing fill, it is important to spread the material in uniform layers and to maintain a reasonably even surface. The thickness of the layers is dependent on the desired compacted lift thickness. The thickness of the uncompacted lift is normally 1 1/2 to 2 times greater than the final compacted lift. For example, you would need to place fill in 9- to 12-inch lifts to achieve a compacted lift thickness of 6 inches. Fill material can be placed by a scraper or dump truck and spread by a dozer or motor grader. When spreading material on a prepared subgrade, you can spread the material from the farthest point from the source to the nearest or vice-versa. The advantages of spreading fill from the farthest point to the closest are: hauling equipment will further compact the subgrade, will reveal previously undetected weaknesses in the subgrade, and will not hinder spreading or compacting operations. On the other hand, spreading from the closest point to the farthest point has the advantage of haul equipment traveling over the newly spread material, which compacts the material and greatly reduces the overall compactive effort required.

(c.) Compacting the Fill. Before beginning compaction operations, the project manager must determine the moisture content of the fill and compare it to the acceptable moisture range for that material. If the moisture content is below the acceptable range, add water to the fill. If the moisture content is higher than the acceptable range, use one of the previously discussed methods to dry the soil. Compaction operations can commence once you have achieved the appropriate moisture range.

(d.) Compacting Base Course. The base course is the most important element in a road structure. It functions as the primary load-bearing component of the road, ultimately providing the pavement (or surface) strength. Because of this, the base course must be densely compacted and capable of supporting heavy loads. Base course material is, therefore, higher quality material than subgrade or subbase fill. Base courses normally consist of well-graded, granular materials that have a liquid limit less than 25 percent and a plastic limit less than 6 percent. The thickness of the base course is dependent on the strength of the subgrade. If the strength of the subgrade is low, a thick base course is necessary. Conversely, where the subgrade strength is high, a thinner base course is suitable. Heavy pneumatic-tired rollers are ideally suited for base course compaction. The heavy static weight combined with the kneading action of the rubber tires provides effective compaction in this area. Although a smooth-drum steel-wheel roller can be used for base course compaction, it is slower and, therefore, less efficient than the pneumatic-tired roller. The smooth drum, steel-wheeled roller is, however, very effective at leveling the wheel ruts created by the pneumatic-tired roller and dressing the final surface.

4.) Operating Techniques.

(a.) Compacting Against Structures. Jay tampers and backfill tampers are specifically designed for use in confined areas such as against existing structures. However, if these are not available, a roller can be used to achieve satisfactory results. If space permits, run the roller parallel to the structure. If it is necessary to work perpendicular to the structure, place fill material (sloped to the height of the roller axle) against the structure. Apply compactive effort against this excess fill material. Care must be exercised to avoid damaging the existing structure.

(b.) Aerating Materials. When the roller is used to aerate soils, it should travel at the highest practical speed. This high speed tends to kick up the material so that particles are exposed to the air for a longer time.

(c.) Overlapping Passes. To eliminate noncompacted strips, each pass with the roller should overlap the preceding pass by 1 foot.

(d.) Turning. When turning at the end of each pass, make gradual turns on the construction site. This prevents tearing up the surface, eliminates the possibility of damaging the roller tongue with the tracks of the tractor, and prevents the roller drums from binding.

APPENDIX ELEVEN

SAFETY

Supervisors should make sure personnel use lowered safety standards only in cases of extreme urgency, and not as a general practice. Time is usually the controlling factor in construction operations in the theater of operations. The necessity for economy of time coupled with the temporary nature of much of the work, sometimes results in safety precautions that are substantially lower than those used in civilian practice. Relaxed general safety rules may result in temporary increases in production, but advantages so gained are often negated by damage to equipment or by injury or death to personnel.

1.) General Safety Rules. General safety rules are as follows:

- Equipment should be inspected before it is used, and periodically thereafter.

- Mechanized equipment must be operated by qualified and authorized personnel only, in a manner that will not endanger persons or property.

- Safe operating speeds will not be exceeded.

- Operators should not leave running equipment unattended.

- Mounting or dismounting equipment while it is in motion is prohibited.

- An operator should not operate any machinery or equipment for more than 10 consecutive hours without an 8-hour rest interval.

- All equipment not rigged to prevent overloading or excessive speed should have safe load capacities and/or operating speeds posted at the operator's position.

- Seatbelts will be used on all equipment whenever they are available.

- Rollover protection (ROPS) will be mounted on equipment whenever it is available.

- Only the operator will be on equipment while it is running except in emergency situations, some training situations, and when required for maintenance. (Construction equipment, except for dump trucks, is designed to accommodate only the operator.) NOTE: Look for safety rules for specific equipment in appropriate chapters in this manual. Also, check appropriate technical manual prior to operating all equipment.

2.) Safety Program. Do not construe the lack of documentation of hazards as an indication of their nonexistence or unimportance. Where safety precautions are necessary but have not been provided, or where existing precautions are judged to be inadequate, the commanding manager must issue new or supplementary precautions.

Each job has its own particular safety hazards. These hazards must be identified and a safety program prepared to reduce or eliminate them. Once the program has been prepared, supervisors must ensure that it is carried out. Additionally, operators' manuals give do's and don'ts, cautions, and safety warnings. Check them!

3.) Personnel Indoctrination. Operators should be taught safety practices during their operator training. Each person should be given an initial indoctrination advising him of job hazards and ways to reduce or avoid them. Personnel should also be instructed to watch out for fellow workmen and to warn them when they get into dangerous situations.   Horseplay, wrestling, scuffling, practical jokes, or unnecessary conversation must be avoided during working hours. Personnel should also receive continuing instructions during the job to make sure they meet the objectives of the safety program.

4.) Operator Qualifications and Requirements.   Construction and material-handling equipment operators must be tested and licensed according to appropriate regulations. An apprentice or applicant can only operate equipment under the direct supervision of a licensed operator. An operator who is not physically or mentally capable must not be permitted to operate any item of equipment. Operators are responsible for the safe operation of equipment while it is assigned to them, and for the safety of their passengers and cargo.

5.) Equipment Inspection. Before using any mechanized equipment, make sure it has been inspected and tested by a qualified, licensed operator and determined to be in safe operating condition. Ensure that equipment is periodically inspected (use equipment, operator maintenance checks, service charts, and common sense) to assure safe operation and proper maintenance. Tag any unsafe machinery or equipment "Out Of Service, Do Not Use" at the operator's position, to prevent its use until repaired.

6.) Repairs and Maintenance.

- Shut down or control all equipment while it is being repaired, adjusted, or serviced.

- Make repairs in a place that is safe for the repairman.

- Crib or block heavy machinery, equipment, and parts that are suspended or held apart by slings, hoists, or jacks before allowing personnel to work underneath or between them.

- Lower devices such as hooks, drag buckets, and fork tines to the ground or onto suitable blocking material when equipment is undergoing maintenance or repairs.

7.) Equipment Refueling. Shut down and turn off the ignition when refueling all motor vehicles and mechanized equipment.

8.) Guards and Safety Devices. Guards, appliances, and similar devices are placed on equipment to protect personnel. Do not remove these devices or render them ineffective except during repairs, lubrication, or adjustment, and then only after the power has been shut off. Replace all guards and devices immediately after completion of repairs and adjustments. Provide guards to prevent personnel from walking under loading equipment that has a hoist or lifting capability.

9.) Signals. Provide a warning device or a signalman wherever there is possible danger to persons from moving equipment.

(a.) Hand Signals. When using manual (hand) signals, designate one person (only) to give the signals to the operator. Only dependable, fully qualified personnel should be used as signalmen. A signalman should be used whenever the point of operation is not in full and direct view of the equipment operator. The signalman must use a uniform system of signals, and he must be clearly visible to the operator at all times. Ensure that operators and signalmen have a full understanding of the meaning of all signals.

(b.) Warning Devices. All equipment should have properly working warning devices, such as backup alarms and properly working turn signals.

10.) Ropes, Cables, and Chains. Ropes, cables, and chains used in construction operations present some of the largest potential safety hazards. To eliminate these hazards - ensure strict observance of load limitations.

- Ensure periodic inspections of their physical condition.

- Use according to manufacturers' recommended procedures or within the safe limits recommended by the manufacturers of the equipment with which they are used.

- Always wear leather gloves when handling wire rope.

(a.) Wire Ropes and Cables. Inspect wire rope or cable at the time of installation and as part of daily operator maintenance. Remove the rope or cable from hoisting or lifting service when it is kinked, or when it has a number of broken wires in a strand.

(b.) Slings. Slings should be used by qualified personnel only. When using slings and their fittings and fastenings-

- Make sure they are inspected daily for evidence of overloading, excessive wear, or damage. Replace defective slings or accessories.

- All eye splices must be made properly. The correct size of wire rope thimbles must be fitted in the eye.

- U-bolt clips used to attach wire rope clips must be on the dead (or short end) of the rope. Tighten U-bolts immediately after initial load application and at frequent intervals thereafter.

(c.) Hooks and Shackles.

- Replace hooks, shackles, rings, pad eyes, and other fittings that show excessive wear or that have been bent, twisted, or otherwise damaged.

- Close all hooks used to support human loads or loads that pass over workmen. Open hooks are prohibited in rigging any load where there is danger of relieving the tension on the cable due to the load or hook catching or fouling.

- When using a wedge-socket type fastening to fasten the load to a line, clip the dead (or short end) of the line with a U-bolt or otherwise secure it against loosening.

11.) Equipment Loading. When equipment is being loaded onto its hauling trailer using suspended or overhead loading equipment or methods, the equipment operator (or driver) must leave the cab. When loading equipment for transporting, ensure that-

- Load is properly distributed, chocked, and tied down (or otherwise secured).

- Load does not obscure the driver's forward or side vision, or interfere in any way with the safe operation of the vehicle.

- Load does not extend beyond the sides of the vehicle except under emergency circumstances. When working with such oversized loads, give adequate warnings and precautions to prevent endangering passing or opposing traffic, or damaging the vehicle.

12.) Equipment Transporting. Equipment that is wider than standard vehicles or protrudes past the ends or sides of its hauling trailer must be given special considerations. When traveling or being transported from one job site to another, this oversized equipment should be-

- Transported during daylight hours.

- Equipped with warning flags and/or lights.

- Accompanied by a lead and/or follow vehicle(s) with warning signs whenever possible, especially in congested or heavy traffic areas.

13.) Night Operations. When operating at night-

- Equip all mobile equipment with adequate headlights and taillights.

- Keep construction roads and working areas adequately lighted until work is completed and all workmen have left the area.

- Ensure that signalmen, spotters, inspectors, servicemen, and others who work in dark areas, exposed to vehicular traffic, wear reflectorized vests or other such apparel.

14.) Excavations. When excavating-

- Shore, brace, or slope excavations that are more than 4 feet deep, unless you are working in solid rock, hard shale, hardpan, cemented sand and gravel, or other similar materials.

- Design shoring and bracing that is effective to the bottom of the excavation.

- Use sheet piling, bracing, shoring, trench boxes, and other methods of protection, including sloping, based upon calculation of the pressures exerted by and the condition and nature of the materials to be retained.

- Provide additional shoring and bracing to prevent slides or cave-ins when excavating or trenching in locations adjacent to backfilled excavations or when subjected to vibrations from traffic, vehicles, or machinery.

ABBREVIATIONS

AA - As Authorized

AASHTO - American Association of State Highway and Transportation Officials

AB - Anchor Bolt

AC - Asphaltic Concrete

ACI - American Concrete Institute

ACP - Asphalt Concrete Pavement

ADA - American Disabilities Act

ADD - Addendum

AGG - Aggregate (As in crushed rock)

A/E - Architect or Engineer

AGC - Associated General Contractors of America

AI - As Installed

AIA - American Institute of Architects

ALT - Alternate

ALUM - Aluminum

ANSI - American National Standards Institute

APWA - American Public Works Association

ASCE - American Society of Civil Engineers

ASTM - American Society for Testing Materials

AWPA - American Wood Preservers Association

AWWA - American Water Works Association

BAT - Bureau of Apprenticeship and Training

BC - Base of Curb

BLDG - Building

BOLI - Bureau of Labor and Industries

BR - Base of Ramp

BW - Both Ways

CB - Catch Basin

CD - Calendar Days (As in contract time)

CE - Civil Engineer

CEM - Cement

CF - Cubic Feet

CI - Construction Inspector

CIP - Cast in Place (As in concrete structures) or Cast Iron Pipe

CJ - Construction Joint

CL - Class (As in class of pipe or class of asphalt) or Centerline

CLR - Clear

CM - Centimeter (2.538 CM = 1 IN)

CMP - Corrugated Metal Pipe

CMU - Concrete Masonry Unit

CO - Contracting Officer or Clean Out

COD - Cash on Delivery

CONC - Concrete

COR - Contracting Officers Representative

CPM - Critical Path Method (As in scheduling)

CSBC - Crushed Surfacing Base Course

CSP - Corrugated Steel Pipe / Concrete Sewer Pipe

CSTC - Crushed Surfacing Top Course

CTB - Cement Treated Base

CULV - Culvert

Cu. M - Cubic Meter (0.764 Cu. M = 1 CY)

CY - Cubic Yard

DBE - Disadvantaged Business Enterprise

DET - Detail

DEQ - Department of Environmental Quality

DI - Ductile Iron / Ditch Inlet

DIA - Diameter

DIM - Dimension

DIP - Ductile Iron Pipe

DOT - Department of Transportation

EA - Each

EB - Expansion Bolt

ECP - Erosion Control Plan

EEO - Equal Employment Opportunity

EF - Each Face

EL - Elevation

ELEC - Electric

EP - Edge of Pavement

EPA - Environmental Protection Agency

EQUIP - Equipment

ESB - Emerging Small Business

EW - Each Way

EX - Existing

EXT - Exterior

FG - Finish or Final Grade

FHWA - Federal Highway Administration

FICA - Federal Insurance Compensation Act

FLR - Floor

FO - Field Order

FOB - Freight On Board (As in material supplied location)

FOC - Face of Concrete

FS - Finish surface

FTG - Footing (As in concrete footing of a wall)

FUTA - Federal Unemployment Tax Act

GA - Gauge

GALV - Galvanized

GC - General Contractor

GL - General Ledger (As in accounting)

GPM - Gallons per Minute

GR - Grade (As in surveying elevations)

GV - Gate Valve

HA - Hectare (0.405 HA = 1 ACRE)

HDPE - High Density Polyethylene (As in drainage pipe)

HGT - Height

HORIZ - Horizontal

HP - High Point or Horsepower

HR - Hour

ID - Inside Diameter

IE - Invert Elevation (As in pipe flow line elevation)

IN - Inch

INCL - Including (Inclusive)

IP - Inside Face

JMF - Job Mix Formula (As in asphalt or concrete mix design)

JT - Joint

Kg - Kilograms (0.454 Kg = 1 LB)

Km - Kilometer (1.610 KM = 1 Mile)

L - Liters (3.785 L = 1 Gallon) or Length of Curve

LB - Lag Bolt

LBS - Pounds

LD - Liquidated Damages (As in late contract completion)

LF - Lineal Feet

LOC - Location

LONG - Longitudinal

LP - Low point

LS - Lump Sum

M - Meter (0.305 M = 1 FT)

MATL - Material

MAX - Maximum

MB - Machine Bolt

MBE - Minority Owner Business Enterprise

MBGR - Metal Beam Guard Rail

MECH - Mechanical

Mg - Mega grams (0.907 Mg = 1 Ton)

MH - Manhole

MIN - Minimum

MISC - Miscellaneous

MM - Millimeter (25.38 MM = 1 IN)

MP - Milepost

MPa - Mega Pascal (0.0069 MPa - 1 PSI)

MSDS - Material Safety Data Sheet

MSL - Mean Sea Level

MTL - Metal

NA - Not Applicable

NBS - National Bureau of Standards

NCR - Non-Conformance Report

NIC - Not in Contract

NLMA - National Lumber Manufacturer's Association

NTP - Notice to Proceed

NTS - Not to Scale

OC - On Center

OD - Outside Diameter

OF - Outside Face

OFCI - Owner Furnished – Contractor Installed

OFOI - Owner Furnished – Owner Installed

O & M - Operation And Maintenance Manuals

OPP - Opposite

OSHA - Occupational Safety & Health Administration

OT - Overtime

PC - Point of Curvature

PCC - Portland Cement Concrete

PCF - Pounds per Cubic Foot

PCS - Petroleum Contaminated Soils

PCP - Pollution Control Plan

PE - Professional Engineer

PERF - Perforated (As in drain pipe)

PI - Point of Intersection

PIP - Poured in Place (As in concrete)

PLYWD - Plywood

PM - Project Manager

PO - Purchase Order

PSI - Pounds per Square Inch

PT - Point of Tangency

PVC - Polyvinyl chloride (As in drainage or water piping)

PVMT - Pavement

PWR - Prevailing Wage Rate

QA - Quality Assurance

QC - Quality Control

QCT - Quality Control Technician

QPL - Qualified Products List

QTY - Quantity

R - Radius

RAP - Recycled Asphalt Pavement

RCP - Reinforced Concrete Pipe / Reinforced Concrete Pavement

RE - Resident Engineer

REINF - Reinforced (As in rebar)

RFC - Request For Clarification

RFI - Request For Information

RFQ - Request For Quotation

RP - Radius Point

R/W - Right of Way

SCHED - Schedule

SF - Square Foot

SIM - Similar

SM - Square Meters (.093 SM = 1 SF)

SPEC - Specifications

SS - Stainless Steel

ST - Straight Time or Street

STA - Station

STD - Standard

STL - Steel

SUB - Subcontractor

SY - Square Yard

TB - Thrust Block

TBM - Temporary Benchmark

TC - Top of Curb

TCD - Traffic Control Device

TCP - Traffic Control Plan

TCS - Traffic Control Supervisor

TEMP - Temporary

TF - Top Face

TIN - Tax Identification Number

T & M - Time and Materials (As in cost plus contracting)

TN - Ton

TR - Top of Ramp

TW - Top of Wall

TYP - Typical (As in several places on a plan sheet)

UBC - Uniform Building Code

UNS - Utility Notification System

UON - Unless Otherwise Noted

VERT - Vertical

VF - Vertical Feet

WBE - Woman Owned Business Enterprise

WD - Working Days (As in contract time)

WM - Water Meter

WS - Water Surface

WWF - Welded Wire Fabric

WWPA - Western Wood Products Association

YD - Yard

CONSTRUCTION DICTIONARY

A

Abandonment. The failure of both parties to a contract to abide by its terms.

Above Ground Tank. A large above ground vessel used for the storage of liquids.

Abrasion. Wearing away by friction.

Abrasive. A substance used for wearing, grinding, or rubbing away by friction.

Abrasive Surface. A surface that has been roughened for safety or for warning.

Abrasive Surface Tile. Floor tile that has been roughened to be slip-resistant.

ABS. Acrylonitrile-butadiene-styrene, hard plastic used because of its resistance to impact, heat and chemicals.

Absorption. The relationship of the weight of the water absorbed by a material specimen subjected to prescribed immersion procedure, to the weight of the dry specimen, expressed in percent.

AC. Initials for alternating current. Alternating current flows alternately in one direction, then the opposite, completing 60 cycles every second.

Acceleration. Rate of change of velocity.

Accelerator. A substance which, when added to concrete, mortar, or grout, increases the rate of hydration of the hydraulic cement, shortens the time of setting, or increases the rate of hardening of strength development, or both.

Access Door. A door or panel creating a means of access for the inspection or repair of concealed equipment.

Access Stair. A stair system to provide specific access to roofs, mechanical equipment rooms or as a means of exit in an emergency.

Accessory. An object or device aiding or contributing in a secondary way.

Accessory, Concrete. An implement or device used in the formwork, pouring, spreading, finishing, or facilitate in the installation of masonry or concrete reinforcing.

Acid Etch. The use of acid to cut lines into metal or glass. The use of acid to remove the surface of concrete.

Acoustical Treatment. The act or process of applying acoustical materials to walls and ceilings.

Acrylic. A general class of resinous polymers derived from esters, amides or other acrylic aid derivatives. A transparent plastic material used in sheet form for window glass and skylights.

AD Plywood. A designation or gradation of plywood. The "A" and the "D" designate quality of surface layers.

Adapter. A device for connecting two different parts.

Additive. A term frequently used as a synonym for addition or admixture.

Adhesive. A substance that dries or cures and binds two surfaces together. A substance capable of holding materials together by surface attachment.

Adjustable Shelf. A shelf that can be adjusted to different heights.

Admixture. A material other than water, aggregates, and hydraulic cement used as an ingredient of concrete or mortar, and added immediately before or during its mixing. A chemical additive used to alter the normal properties of concrete.

Aggregate. Inert particles such as sand, gravel, crushed stone, or expanded materials, in a concrete or plaster mixture.

Air Compressor. A tool which takes air and forces it at a high pressure into a storage tank. The air is released through a regulator and a hose to power small tools.

Air Distribution. To force air to desired locations in a building or facility.

Air Eliminator. A mechanical device that expels excess air.

Air Entraining Agent. A substance added to concrete, mortar or cement that produces air bubbles during mixing, making it easier to work with and increasing its resistance to frost and freezing.

Air Handling. Single or variable-speed fans pushing air over hot or cold coils, through dampers and ducts to heat or cool a building or structure.

Air Powered Hoist. A hoist that is operated by compressed air.

Air Tool. Attachments using compressed air to saw, spray-paint, sand, drill or nail, etc.

Air Vent. An opening in a building or structure for the passage of air.

Amperes (Amp). A unit of measure of electrical flow in a conductor.

Anchor. Irons or metals of special form and shapes used to fasten together and secure timbers or masonry.

Anchor Bolt. A bolt embedded in concrete for the purpose of fastening a building frame to a concrete or masonry foundation.

Anchor, Rafter. A bolt or fastening device which attaches the rafters used to support a roof to the walls or rafter plate.

Anchor Slot. A groove in an object into which a fastener or connector is inserted to attach objects together.

Anchorage. An attachment for resistance to movement. The movement can be a result of overturning, sliding or uplift. The most common anchorage for these movements are tie-downs (hold-downs) for overturning and uplift, and anchor bolts for sliding.

Anchoring Cement. Grout used in sleeves to anchor tubing in place.

Angle. A structural section of steel which resembles an "L" in cross section.

Angle Valve. A valve in which the shut-off in the pipe openings are set at right angles to each other.

Apron. The lower trim member under the sill of the interior casing of a window. An upward or downward vertical extension of a sink or lavatory. A paved area immediately adjacent to a building, structure or facility.

Arch. A curved structure in which compression is the essential cause of internal stresses.

Arch Culvert. A curved shaped drain under a roadway, canal or embankment.

Architectural Equipment. The implements, apparatus, or equipment used in the construction and initial outfitting of a building.

Armor Plate. A kick-plate made of metal installed on the bottom of a door to protect it from denting and scratching.

Asbestos. A fibrous mineral that was used for its noncombustible, non-conducting or chemically resistant properties.

Ashlar Veneer. An ornamental or protective facing of masonry composed of squared stones.

Asphalt. A residue in petroleum or coal-tar refining that is used for pavements and as a waterproofing cement.

Astragal Weatherstripping. Fabric, rubber or plastic strips attached to the molding that is attached to one of a pair doors or casement windows to cover up the joint between the two stiles.

Attenuation Blanket. Material utilized to help in the reduction of the energy or intensity of sound.

Auger. An instrument or device used for boring or forcing through materials or soil.

Auxiliary Switch. A standby device for switching.

Awning Window. A window hinged at the top.

B

Back-Up Block. Block which is used for the inner load bearing, structural portion of a masonry wall.

Back-Up Brick. A load bearing or structural portion of a masonry wall constructed of brick against which a veneer is attached.

Backfill. Earth or earthen material used to fill the excavation around a foundation; the act of filling around a foundation.

Backflow. The unintentional flow of water or other substances into the distribution pipes of a water drainage system from a source other than the intended source.

Backhoe. An excavating machine with a bucket rigidly attached to a hinged stick on a boom that is drawn toward the machine in operation.

Backhoe/Loader. An excavation machine combining a bucket on a hinged stick on a boom on one end, and a bucket or scoop at the other.

Ball Valve. A valve in which a ball regulates the opening by its rise and fall due to fluid pressure, a spring, or its own weight.

Ballast, Roof. Crushed rock or gravel which is spread on a roof surface to form its final surface.

Baluster. Vertical members that extend from a stair, or floor, to a handrail. A small pillar or column used to support a rail.

Band. A low, flat molding.

Bank Run Gravel. Excavated material that is generally 1/4 inch minimum to 6 inches maximum.

Bar, Reinforcing. A manufactured usually deformed steel bar, used in concrete and masonry construction to provide additional strength.

Barbed Wire. Wire that is twisted with barbs or sharp points.

Barge. A floating platform or vessel from which construction activities may be performed. Often used in rivers to install bridge piers and also used extensively in waterfront construction.

Barge Board. The installation of ornamental boards to conceal roof timbers projecting over gables.

Barricade. An obstruction to prevent passage or to prevent access.

Barrier, Vapor. A type of plastic sheeting that both eliminates drafts and keeps moisture from damaging a building or structure.

Base Cabinet. Case, box, or piece of furniture which sets on floor with sets of drawers or shelves with doors, primarily used for storage.

Baseboard Heater. Heating strips that are installed at the juncture of the wall and floor and may be either recessed or surface-mounted; generally along the outside walls of rooms.

Baseplate. A steel plate inserted between a column and the foundation, used to level the column and to spread the load of the column to a larger area of the foundation.

Batt Insulation. Loosely matted plant or glass fibers with one or both sides faced with kraft paper or aluminum foil available in specifically sized sections.

Batten Siding. Vertical siding which has narrow strips of metal or wood covering the joints.

Batter Pile. Pile driven at an angle to brace a structure against lateral or horizontal thrust.

Bay. A rectangular area of a building defined by four adjacent columns; a portion of a building that projects from a facade.

Bead. A narrow line of weld metal or sealant; a strip of metal or wood used to hold a sheet of glass in place; a narrow, convex molding profile; a metal edge or corner accessory for plaster.

Beam. A straight structural member that acts primarily to resist transverse loads. A structural element which sustains transverse loading and develops internal forces of bending and shear in resisting the loads. An inclusive term for joists, girders, rafters, and purlins.

Beam, Grade. An end-supported horizontal load-bearing foundation member that supports an exterior wall.

Beam, Reinforcing. A horizontal member installed to strengthen and support the load of a structure.

Bearing Wall. A wall which supports any vertical loads in addition to its own weight.

Bedding. A layer of crushed stone, to secure a firm bearing.

Bell and Spigot. Cast iron pipe joints formed with sections that have a wide opening (bell) at one end and a narrow end (spigot) at the other. They are then fitted by caulking with oakum and lead.

Bend, Soil. A piece of short, curved pipe, like an elbow, used to connect two straight links of conduit in a sewage system.

Bevel. An end or edge cut at an angle other than a right angle.

Bibb, Hose. A water spigot or faucet with its nozzle threaded or coupling attached to accept a hose.

Bitumen. A tar based mixture, such as asphalt or coal tar.

Bituminous. Resembling, containing, or impregnated with various mixtures of hydrocarbons (like tar) together with their nonmetallic derivatives.

Bituminous Sidewalk. A walkway constructed with an impregnated mixture of hydrocarbons together with aggregate such as sand or stone. Commonly called "blacktop".

Blanket Insulation. Thermal insulating material made of fibrous glass or mineral wool, sometimes offered with paper or foil surfacing, formed in batts or rolls.

Block, Concrete. A hollow concrete masonry unit constructed a composite material consisting of sand, coarse aggregate, cement and water.

Block, Glass. A hollow masonry unit made of glass.

Block, Granite. A masonry unit consisting of a very hard natural igneous rock used for its firmness and endurance.

Block, Grout. Mortar mixes used in block walls to fill voids and joints.

Block, Splash. A small masonry block placed in the ground beneath a downspout to receive roof drainage and prevent standing water or soil erosion.

Blocking. Pieces of wood inserted tightly between joists, studs, or rafters in a building frame to stabilize the structure, inhibit the passage of fire, provide a nailing surface for finish materials, or retain insulation.

Bluestone. A sandstone of a dark-greenish to bluish-gray color that splits into thin slabs, commonly used to pave surfaces for pedestrian traffic.

Board. Lumber less than 2 inches thick.

Board Foot. A unit of lumber volume, a rectangular solid nominally 12" x 12" x 1". The equivalent of a board 1 foot square and 1 inch thick.

Board Siding. A type of lumber installed on the exterior walls of a building or structure to act as the finish sheathing.

Bollard. Short steel post (usually filled with concrete) set to prevent vehicular access to or to protect property from damage by vehicular encroachment. Steel or cast iron post to which ships are tied.

Bolster. A long wire type "chair" used to support reinforcing bars in a concrete slab while the concrete is being placed.

Bolt. A fastener consisting of a cylindrical metal body with a head at one end and a helical thread at the other, intended to be inserted through holes in adjoining pieces of material and closed with a threaded nut.

Bolted Steel. Steel members bolted with a metallic pin or rod having a head at one end and threads on the other for attaching the nut.

Bond Beam. A horizontal grouted element within masonry in which reinforcement is embedded.

Bond Breaker. A material used to prevent adhesion of newly placed concrete to other surfaces.

Bonding Agent. A substance applied to a suitable substrate to create a bond between it and a succeeding layer as between a subsurface and a terrazzo topping or a succeeding plaster application.

Borrow. Excavated material that has been taken from one area to be used as fill at another location.

Box Culvert. A concrete drainage structure rectangular shaped, reinforced and cast in place or made of precast sections.

Bracing. Diagonal members, either temporary or permanent, installed to stabilize a structure against lateral loads. Structural member used to prevent buckling or rotation of wood studs.

Bracket. A projecting support for a shelf or other structure.

Brass Fitting. Threaded pipe connector constructed of brass, used to join two pieces of pipe together.

Brick. A solid masonry unit having the shape of a rectangular prism. Usually made from clay, shale, fire clay, or a mixture of these.

Bridge Crane. A hoisting device spanning two overhead rails. The hoisting device moves laterally along the bridge with the bridge moving longitudinally along the rails.

Bridge Deck. The slab or other structure forming the travel surface of a bridge.

Bridging. Pieces fitted in pairs from the bottom of one floor joist to the top of adjacent joists, and crossed to distribute the floor load; sometimes pieces of width equal to the joists and fitted neatly between them. Diagonal or longitudinal members used to keep horizontal members properly spaced, in lateral position, vertically plumb, and to distribute load.

Bronze. An alloy of copper and tin and sometimes other elements.

Broom Finish. A finish applied to an uncured concrete surface, to provide skid or slip resistance, made by dragging a broom across the freshly placed concrete surface.

Brown Coat. The second of three coats of a plaster or stucco application.

BTU. British thermal unit; measurement of the heat energy required to raise one pound of water one degree Fahrenheit.

Building, Metal. A building or structure constructed of a structural steel frame covered by metal roof and wall panels. Commonly prefabricated in a factory and assembled at the site.

Building Paper. Water repellent paper used to assist in shedding incidental moisture what may penetrate exterior finishes of exterior wall construction. Thick paper, used to insulate a building before the siding or cladding is installed. Occasionally used in flooring between double floors.

Built-Up Roof. A roof covering made of continuous rolls or sheets of saturated or coated felt. They are then cemented together with bitumen and may have a final coating of gravel or slag.

Bulb Tee. Rolled steel in the form of a "T" with a formed bulb on the edge of the web.

Bulk Excavation. The digging out of large amounts of dirt and debris.

Bulkhead Formwork. The temporary formwork that blocks fresh concrete from a section of forms or closes the end of a form at a construction joint.

Bulldozer. A tractor driven machine with a horizontal blade for clearing land, road building, or similar work.

Burlap Rub. A finish obtained by rubbing burlap to remove surface irregularities from concrete.

Bush Hammer. A hammer used to dress concrete or stone with its serrated face and may pyramidal points. A hammer that has a rectangular head with serrated or jagged faces. Used for roughing concrete.

Bushed Nipple. A pipe threaded at both ends to connect two pipes of different dimensions.

Butt, Pile. The large end of a pile. The small end of the pile is called the tip. The pile butt is the end of the pile which the pile driver impacts or "hits" and after completion, the part which the structure is connected to.

Butterfly Valve. A valve constructed with a disc that rotates 90 degrees within the valve body.

Butyl Caulk. Caulking that is made from various synthetic rubbers derived from butanes.

Butyl Membrane. Pliable thin sheets or layers made from synthetic rubber.

C

Cage Ladder. A vertical device with rungs for climbing that has, for safety, a surrounding structure to prevent the climber from falling off.

Caisson. A cylindrical, sitecast concrete foundation that penetrates through unsatisfactory soil to rest upon an underlying stratum of rock or satisfactory soil. A type of drilled or augured piling.

Cant Strip. A strip of material, usually treated wood or fiber, with a sloping face used to ease the transition from a horizontal to a vertical surface at the edge of a flat roof. This prevents the roofing material from abruptly stopping at the parapet wall and also helps prevent leakage at that juncture.

Cap. A trim tile with a convex radius on one edge. Used for finishing the top of a wainscot or for turning an outside corner, a bullnose.

Cap, Pile. A structural member usually fastened to, and placed on the top of a slender timber, concrete or structural element. Used to transmit loads into the pile or group of piles and to connect them.

Carbon Steel. Low carbon or mild steel.

Carpentry, Rough. The preliminary framing, boxing and sheeting of a wood frame building.

Cast Iron. Iron with a high carbon content, which cannot, because of the percentage of carbon, be classified as steel.

Cast-In-Place. Mortar or concrete which is deposited in the place where it is required to harden as part of the structure, as opposed to precast concrete.

Catch Basin. Formed pan, usually constructed of masonry, which collects run-off and debris. The basin usually includes a drain connected to plumbing or stormwater system.

Caulking. Waterproof material used to stop up and fill seams to make watertight.

Cavity Wall. A masonry wall that includes a continuous airspace between its outermost wythe and the remainder of the wall.

CDX. A grading system mark for plywood which means: grade C and D, exterior glue.

Cement. Usually refers to portland cement which when mixed with sand, gravel, and water forms concrete. Generally, cement is an adhesive; specifically, it is that type of adhesive which sets by virtue of a chemical reaction.

Cement Grout. A cementitious mixture of portland cement, sand or other ingredients and water which produces a uniform paste used to fill joints and cavities between masonry units.

Cement, Mortar. A mixture of cement, lime, sand, or other aggregates, and water, used for plastering over masonry or to lay block, brick or tile.

Cement, Plaster. Plaster used on exterior surfaces or in damp areas. Plaster having Portland Cement as a binder.

Cementitious Topping. A compound that is capable of setting like concrete and is applied on a concrete base to form a floor surface.

Centering. Temporary formwork for an arch, dome, vault, or other overhead surface.

Centigrade. A scale of temperature which features 0 degrees and 100 degrees as the freezing and boiling points of water respectively. To convert centigrade to Fahrenheit multiply by 1.8 and add 32, e.g. (100 degrees x 1.8)+32 = 212 degrees F.

Centrifugal Pump. A pump which draws water into the center of a high speed impeller and forces the fluid outward with velocity and pressure.

CFM. Cubic feet per minute.

Chain Trencher. A self-propelled machine with blades attached to a continuous chain, used to excavate trenches.

Chair, Reinforcing. Metal supports made of fabricated wire, made to hold reinforcing steel in place until concrete is poured.

Chamfer. A beveled surface cut on the edge of a piece of wood. A strip of wood or other substance cut on an angle and placed in concrete formwork to create a beveled finished surface on the corners of the final concrete shape (beam, column, etc.).

Channel. A U-shaped steel or aluminum section shaped like a rectangular box with one side removed.

Chicken Wire. Thin, woven wire mounted on an exterior wall as a base for stucco plaster to cling to, often backed by paper.

Chord. One of the main members of a truss braced by web members of the truss. Perimeter member of a building or structure which resists lateral forces.

Clay Pipe. Pipe used for drainage systems and sanitary sewers made of earthenware and glazed to eliminate porosity.

Clay Tile. Earthenware tile that is fired and is used on roofs. Known as quarry tile when used for flooring.

Cleaning Masonry. The final removal of excess grout, excess concrete, etc. from an exterior masonry structure.

Cleft, Natural. A natural V-shaped channel, space, opening or fissure in a material.

Closer. The last masonry unit laid in a course; a partial masonry unit used at the corner of a course to adjust the joint spacing. A hydraulic device used to close doors.

CMU. Concrete Masonry Unit (concrete block).

Cofferdam. A watertight enclosure from which water is pumped to expose the bottom of a body of water and permit construction.

Cold Formed Steel. Process of shaping steel without using heat.

Collar. A compression ring around a small circular opening.

Column. A structural member used primarily to support axial compression loads and with a height of at least three times its least lateral dimension. An upright structural member acting primarily in compression. A square, rectangular, or cylindrical support for roofs, ceilings, and so forth, composed of base, shaft, and capital.

Column Capital. The uppermost member of a column crowning the shaft and taking the weight of the beam or girder.

Column Footing. Commonly known as individual footing, generally square or rectangular in shape.

Combined Footing. A concrete footing which supports two or more columns.

Common Bond. Brickwork laid with each five courses of alternating stretchers followed by one course of headers.

Compaction. The process whereby the volume of freshly placed material is reduced or flattened by vibration or tamping, or some combination of these.

Composite Beam. A beam that is composed of two different materials; for example, a wood and a steel beam, or a steel beam and concrete slab, in which the two act as one.

Compression. Force which tends to crush adjacent particles of a material together and cause overall shortening in the direction of its action. Stress which tends to shorten a member.

Compression Fitting. Bends, couplings, crosses, elbows, tees, unions, etc., which use a force when connecting that pushes together and squeezes a metal or rubber gasket.

Compressor. A machine that compresses gases or air.

Concealed Z Bar. A hidden z-shaped bar that is used as a wall tie.

Concrete. A composite material which consists essentially of a binding medium within which is embedded particles or fragments of aggregate; in portland cement concrete, the binder is a mixture of portland cement and water. A mixture of cement, water, fine aggregate (sand), and coarse aggregate (gravel). An artificial building material made by mixing cement and sand with gravel, broken stone, or other aggregate, and sufficient water to cause the cement to set and bind the entire mass.

Concrete Masonry Unit (CMU). A block of hardened concrete, with or without hollow cores, designed to be laid in the same manner as a brick; a concrete block.

Concrete Pipe. Pipe manufactured from concrete. The manufacturing is done in a plant under controlled conditions. Concrete pipe is usually used for drainage but may be used for sanitary sewers also.

Concrete Placement. The placing and finishing of concrete during a continuous operation. Also known as pouring.

Concrete Plank. A solid or hollow-core, flat-beam used for floor or roof decking. Usually precast and prestressed.

Concrete, Precast. Cast and cured concrete manufactured in a plant under controlled conditions. Examples are; precast concrete slabs, precast reinforced lintels, beams, columns, piles, parts of walls and floors, etc.

Concrete, Prestressed. Concrete in which internal stresses of such magnitude and distribution are introduced so that the tensile stresses resulting from the structure's loads are counteracted to a desired degree. The stresses are usually developed by inserting tendons (cables) through pre-formed tubes in the concrete member, at which time the tendon in stressed (tightened) and then grouted in place.

Concrete Pump. An apparatus which forces concrete to the placing position through a pipeline or hose.

Concrete Reinforcement. Steel rods that are embedded in wet concrete to give additional strength.

Concrete Testing. Testing to determine the plasticity or strength of concrete.

Concrete Topping. A rich mixture of fine aggregate concrete used to top concrete floor surfaces for durability, safety and appearance.

Conduit. A protective sleeve or pipe commonly used for individual electrical conductors.

Connector, Compression. A connecting device which when attaching uses a force that pushes together and squeezes.

Connector Set Screw. A screw on a connector fitting that when tightened connects two components together.

Continuous Footing. A concrete footing that supports a wall or two or more columns. The footing can vary in width and depth. Sometimes called a strip footing.

Continuous High Chair. A rigid wire device used to hold steel reinforcements off the bottom of the slab. As opposed to a single chair, the continuous chair is manufactured in strips.

Control Joint. An intentional linear discontinuity in a structure or component, designed to form a plane of weakness where cracking or movement can occur in response to various forces so as to minimize or eliminate cracking elsewhere in the structure.

Coping. The material or units used to form a cap or finish on top of a wall, pier, pilaster, or chimney. A protective cap at the top of a masonry wall.

Corbel. A projection from the face of a beam, girder, column, or wall used as a beam seat or a decoration. A spanning device in which masonry units in successive courses are cantilevered slightly over one another; a projecting bracket of masonry or concrete. A masonry unit such as brick or stone which projects beyond the unit below.

Core Drilling. The process of drilling which extracts a cylindrical sample of concrete, rock or soil. Sometimes used to install pipe or conduit in or through an existing concrete or masonry wall.

Corner Bead. A metal or plastic strip used to form a neat, durable edge at an outside corner of two walls of plaster or gypsum board.

Cost Estimate. Determined by one of the following methods: a. Area and volume method: The estimate is based on a cost per square foot of the building and/or a cost per cubic foot of the building. b. Unit use: The estimate is based on the cost of one unit multiplied by the number of units in the project. For example, in a hospital, the cost of one patient unit multiplied by the number of patient units in the project. c. In-place unit: The estimate is based on the cost in-place of a unit, such as doors, cubic yards of concrete etc..

Counterflashing. An inverted L-shaped metal strip built into a wall to overlap flashing and make a roof or wall watertight. A flashing turned down from above to overlap another flashing turned up from below so as to shed water.

Course. A horizontal layer of masonry units, one unit high; a horizontal line of shingles.

Coursed. Laid in courses with straight bed joints.

Cover, Manhole. A heavy, usually round, steel or iron cover used to gain access to underground work through a manhole.

Crane. A machine for raising, shifting and lowering heavy weights by means of a projecting swinging arm or with the hoisting apparatus supported on an overhead track.

Crane, Traveling. A crane which is connected to tires or tracks and can be readily moved.

Creosote. A brownish oily liquid obtained by distillation of coal tar, used as a wood preservative.

Crew Trailer. A trailer provided on a job site for use by employees.

Cross Brace. Bracing with two intersecting diagonals. Slender diagonal member within framed wall or partition, to support wall or partition and to withstand structural loads imposed by wind and suction loads, building loads, movement and deflection of structure.

Crusher Run. Gravel, rock, boulders, or blasted rock that has been reduced in size by a machine, but has not been sorted for size.

Culvert. A drain pipe or masonry crossing made of concrete, galvanized corrugated metal, aluminum or steel constructed under a road or embankment to provide a waterway.

Curb Inlet. The opening in a curb through which water flows and drains.

Curing. Maintenance of humidity and temperature of freshly placed concrete during some definite period following placing, casting, or finishing to assure satisfactory hydration of the cementitious materials and proper hardening of the concrete. The hardening of concrete or plaster.

Cut and Fill. Excavated material removed from one location and used as fill material in another location.

Cylinder Test. A test to determine the compressive strength of concrete.

D

Dampproofing. The treatment of concrete or mortar to help prevent the passage or absorption of water. Applied with a suitable coating to exposed surfaces or applied with a suitable admixture of treated cement.

Deck. The form on which concrete for a slab is placed, also the floor or roof slab itself.

Deformed Reinforcement. Deformed reinforcing bars, bar and rod mats, and deformed wire.

Demolish. To raze or tear down a building or structure.

Dewatering. To remove water from the ground or excavations with pumps, wellpoints or drainage systems.

Diaphragm Pump. A water pump used to continuously remove water from excavations containing mud and small stones.

Dimension Lumber. The most common type of framing lumber. For example, lumber which is 2 inches (nominal) thick and commonly is called "2 by 4", "2 by 6", etc.

Direct Burial Conduit. Electric conduit suitable for burial in exterior applications with an outer surface that resists moisture, fungus, and corrosion.

Domestic Marble. Marble which comes from the country or area where the structure in which it used is built.

Door, Roll-up. A door which raises and rolls into a coiled configuration and lowers on tracks on either side.

Douglas Fir. A species of framing lumber. A tree which supplies framing lumber because of its straight growth.

Dovetail Anchor Slot. A matching interlocking strip or slot which is used with a dovetail fastener.

Dowel. A steel bar, which extends into two adjoining portions of a concrete construction, as at a joint in a pavement slab, so as to connect the portions and transfer shear loads.

Downspout. The vertical part of a drainage system from a roof which forms a channel or pipe to remove water from the gutters.

Dozer. A term used in the trade for a bulldozer.

Dragline. A bucket attachment for a crane commonly used in a marsh or marine area, that digs soft materials that must be excavated at some distance from the crane, and draws the bucket towards itself using a cable.

Drain. A trench, ditch, or pipe designed to carry away waste water.

Drain Field. A system of trenches filled with sand, gravel or crushed stone and a series of pipes to distribute septic tank effluent into the surrounding soil.

Drop Panel. A thickening of a two-way concrete structure at the head of a column.

Dry Stone Wall. A wall of stone that has been constructed without the use of mortar or concrete in its joints.

Drywall. An interior facing panel consisting of a gypsum core sandwiched between paper faces, also called gypsum board or plasterboard. Different types are available for standard, fire-resistant, water-resistant, and other applications.

Ductile Iron Pipe. Iron pipe that is manufactured so as to render it able to be flexed. Ductile iron pipe has the non-corrosive qualities of cast iron but is not brittle and has the handling characteristics of steel.

Dump Truck. A truck used for transporting and dumping loose materials.

Dumpster. A large, heavy metal container used to hold and haul rubbish.

E

Earthwork. An embankment or other construction made of earth. Any work involving movement or use of soil and other earthen material.

Edge Form. A forming member used to limit the horizontal spread of fresh concrete on flat surfaces.

Elbow. A pipe fitting which joins two pipes at 90 degree angles.

Elevated Floor. A floor system not supported by a subgrade.

Elevated Slab. A roof slab or floor supported by structural members.

Elevated Slab Formwork. The system of support for a freshly poured or placed concrete elevated slab.

Elevated Stairs. A stair system not supported by the subgrade.

EMT Conduit. "Electrical Metallic Tubing" or "thin wall". For running of electrical wires, that is not threaded, easier to handle than "rigid", and installed more rapidly because of the type of non-threaded fittings used with it.

Encased Burial Conduit. Metal or plastic conduit EB for outdoor wiring with type TW wires encased, or type UF (underground feeder) cable.

End Bell. Pipe with a wide opening at one end, commonly cast iron.

Engineered Fill. Earth compacted into place in such a way that it has predictable physical properties, based on laboratory tests and specified, supervised, installation procedures.

Equipment Mobilization. The assembly and movement of equipment to a jobsite.

Equipment Pad. A thick slab-type stone or precast concrete block placed under mechanical devices to spread the weight and load of the machinery evenly and to prevent excessive vibration.

Estimate. A prediction of the cost of performing work. A value judgement based on experience. An approximation of construction costs.

Etch. The art of producing designs on metal or glass by the use of the corrosive action of an acid. The use of acid to cut lines into metal or remove the surface of concrete.

Excavation. A cavity formed by cutting, digging or scooping.

Expanded Metal Lath. Open mesh cut and drawn from solid sheet of ferrous or non-ferrous metal. Made in various patterns and metal thicknesses with uneven or flattened surface.

Expansion Joint. A joint through tile, mortar, concrete or masonry down to the substrate, intended to allow for gross movement due to thermal stress or material shrinkage.

Exposed Aggregate Finish. A concrete surface in which the coarse aggregate is revealed. A decorative finish for concrete work achieved by removing, generally before the concrete has fully hardened, the outer skin of mortar and exposing the coarse aggregate.

Extrusion. The process of squeezing a material through a shaped orifice to produce a linear element with the desired cross section; an element produced by this process.

F

Face Brick. A brick selected on the basis of appearance and durability for use in the exposed surface of a wall.

Fahrenheit. A scale for registering temperature where freezing is 32 degrees above zero and boiling is 212 degrees.

Fan, Exhaust. An electrical powered device used to withdraw fumes, dusts, or odors from an enclosure.

Fastener. Generic term for welds, bolts, rivets, screws, and other connecting devices.

Fence. A barrier used to prevent escape or intrusion or to mark a boundary, usually made of posts, boards or wire.

Fertilizing. The act or process of adding a substance (as manure or a chemical mixture) in order to make soil more fertile.

Fibrous Concrete. A light concrete made from a fibrous aggregate, like sawdust or asbestos, for increased tensile strength and making it easy to nail.

Field Engineer. An engineer who works primarily at the jobsite as opposed to the home office.

Field Welded Truss. A truss fabricated and welded at a jobsite.

Filler, Joint. Compressible material used to fill a joint to prevent the infiltration of debris and to provide support for sealants.

Fire Hydrant. A discharge fitting or apparatus with a valve and spout at which water may be drawn from the mains of waterworks. Used primarily in fire-fighting, also used to service water mains and systems.

Firewall. A wall constructed to prevent or slow down the spread of fire.

Flange. A rib or rim on an object for strength, for guiding, or for attachment to another object. A projecting hard ring, ridge, or collar placed on a pipe or shaft to strengthen, prevent sliding, or accommodate attachments.

Flashing. A thin, continuous sheet of metal, plastic, rubber or waterproof paper used to prevent the passage of water through a joint in a wall, roof or at a chimney. The material used and the process of making watertight the roof intersections and other exposed places on the outside of a structure.

Flat Seam. A sheet metal roofing seam that is formed flat against the surface of the roof.

Flat Slab. A reinforced concrete slab that is designed to span, without any beams or girders, in two directions to supporting columns.

Flat Washer. A washer which goes under a bolt head or a nut to spread the load, prevent loosening, and protect the surface.

Floor Joist. A support beam, commonly installed in parallel with other beams to create a structural floor system, after which floor sheathing is fastened.

Floor Pedestal. A member, such as a short pier, used as a base for a floor system.

Floor Slab. A reinforced concrete slab on grade or elevated, used as a floor.

Flooring. Any type of material used to create a final floor surface.

Flue Liner. Heat-resistant firebrick or other fire clay materials that make up the lining of a chimney.

Flush. Adjacent surfaces even, or in same plane (with reference to two structural or finish pieces).

Footing. The widened base of a foundation that spreads a load from the building across a broader area of soil. An enlargement at the lower end of a wall, pier, or column, to distribute the load.

Foreman. An specially trained workman/manager who works with and usually leads a crew or gang.

Form Deck. Thin, corrugated steel decking that serves as permanent formwork for a reinforced concrete deck.

Form Fabric. Welded-wire fabric used to reinforce concrete while it is setting and gaining sufficient strength to be self-supporting.

Form Oil. Oil applied to interior surface of formwork to promote easy release from the concrete when forms are removed.

Form Tie. A steel rod with fasteners on each end, used to hold together the formwork for a concrete wall.

Formwork. Temporary structures of wood, steel or plastic that serve to give shape to poured concrete.

Foundation. The portion of a building that has the sole purpose of transmitting structural loads from the building into the earth. That part of a building or wall which supports the superstructure.

Foundation Vent. Opening in foundation wall to provide natural ventilation to foundation crawl spaces.

Framing. The rough wooden structural skeleton of a building, including interior and exterior walls, floor, roof, and ceilings.

Framing Lumber. Wood members of framing systems which are manufactured by sawing, resawing, passing lengthwise through standard planing machine, crosscutting to length, and matching, but without further manufacturing.

Friction Connection. Two or more structural steel members clamped together by high-strength bolts with sufficient force that the loads on the members are transmitted between them by friction along their mating surfaces.

Front End Loader. A tractor or bulldozer with a bucket which operates from the front of the vehicle.

Furring. Wood or metal strips used to build out a surface such as a studded wall. Narrow strips fastened to the walls and ceilings to form a straight surface upon which to lay the lath or other finish.

Fused Reducer. A pipe coupling with a larger size at one end that the other and is attached to a length of pipe by welding.

G

Gabion. A metal or wire cage filled with ballast or stone, used in large scale retaining walls.

Galvanized. Zinc plated for corrosion protection achieved by hot dipping into molten zinc or by electrolysis.

Galvanized Mesh. Mesh screening that has been galvanized. Commonly used as wire lath, reinforcing, or fencing.

Gate, Swing. The operable member of a fence system that is hinged for opening and closing.

Gate Valve. A valve utilizing a wedge-shaped gate, which allows fluid flow when the gate is lifted from the seat.

Gauging Plaster. A gypsum plaster formulated for use in combination with finish lime in finish coat plaster.

General Conditions. A written document, supplementing the specifications, which indicates and defines areas of the project relating to other than specific building trades.

Geotextile. Synthetic fabrics used to separate backfill materials for proper drainage. Used in high retaining walls and landscape design.

Girder. A beam that supports other beams; a very large beam, especially one that is built up from smaller elements. A timber used to support wall beams or joists.

Glass Pipe. Glass and glass-lined pipe used in process piping and in laboratories.

Glass, Wire. Glass in which a wire mesh was embedded during manufacture.

Glass-Fiber-Reinforced Concrete (GFRC). Concrete with a strengthening admixture of short alkali-resistant glass fiber.

Glazed Block. Concrete blocks with a surface produced by fusing it with a glazing material.

Glazed Brick. Brick or tile with a surface produced by fusing it with a glazing material.

Globe Valve. A valve with a rounded disc that shuts off the flow when closed, and seats to prevent fluid flow.

Glue Laminated Timber. A timber made up of a large number of small strips of wood glued together.

Gradall. A hydraulic, wheel-mounted backhoe often used with a wide bucket for dressing earth slopes.

Grade Beam. A reinforced concrete beam that transmits the load from a bearing wall into spaced foundations such as pile caps or caissons.

Grading. The act or process of leveling earth.

Granite Block. A masonry unit consisting of a very hard natural igneous rock used for its firmness and endurance.

Granolithic Topping. A covering layer consisting of an artificial stone of crushed granite and cement.

Grating. Open grid of metal bars structurally formed.

Grating Frame. Frame, usually metal, to contain floor grating and provide means to anchor to floor construction.

Gravel. Small rock particles resulting from natural disintegration and weathering such as river gravel. Or a loose term for mechanically crushed stone.

Gravel Stop. A metal flange or strip with a vertical lip placed around a built-up roof to prevent loose gravel from falling off the roof.

Gravity Wall. A retaining wall which depends solely on its weight to resist lateral forces of retained earth.

Grease Trap. A device to trap and retain the grease content of wastewater and sewage.

Ground. A strip of wood assisting the plasterer in making a straight wall and in giving a place to which the finish trim of the room may be nailed. A strip attached to a wall or ceiling to establish the level to which plaster should be applied.

Ground Hydrant. A water hydrant for the use in fighting fires, installed in the ground.

Grounding. The act or process of making an electrical connection with the earth. A large conduction body (as the earth) used as a common return for an electric circuit.

Grout. A rich or strong cementitious or chemically setting mix used for filling masonry or tile joints and/or voids. A mixture of portland cement, aggregates, and water, which can be poured or pumped into cavities in concrete or masonry.

Grout Lift. An increment of height that grout is poured.

Grubbing. The act or process of clearing and digging up roots and stumps.

Guardrail. A safety device used as a barrier to prevent encroachment. In street or highway construction, a barrier to keep vehicles in their lanes. A device for protecting a machine part or the operator of a machine.

Gunite. A term used to describe a concrete material applied by pumping through a hose.

Gusset Plate. A flat steel plate to which the chords of a truss are connected at a joint; a stiffener plate.

Guy Cable. Cable anchored at one end and supporting or stabilizing an object at other end.

Guy Rod. A metal rod with a cable or rope attached, leading to an object to support and stabilize it.

H

H Beam. A steel beam which in cross section resembles the letter "H". Commonly used in earthwork as a retaining structure or piling.

Hand Excavation. The act or process of digging out earth using hand tools.

Head, Pop-Up. A watering device in an irrigation or sprinkling system that pops up when the system is charged with water.

Headwall. A wall, usually constructed of concrete or masonry, that is placed at the outlet side of a drain or culvert to protect fill from scouring, undermining or to divert flow.

Heavy Timber. Construction requiring noncombustible exterior walls with a minimal fire-resistance rating of two-hours, laminated or solid interior members, heavy plank or laminated wood floors and roofs.

Hollow Block. Concrete blocks that can be filled with insulation or reinforced.

Hose Bibb. An outdoor water faucet, usually at sill height, used as a hose connection.

Hub Union. A pipe fitting used to join two pipes without turning either pipe.

HVAC. Heating, ventilation, and air-conditioning system.

Hydrant. A connection with a valve to a water main, used for the dispensation and delivery of water.

I

I Beam. An obsolete term; an American Standard designation for a particular section of hot-rolled steel which in cross section is shaped like a capital "I".

IMC (Conduit). Intermediate Metal Conduit.

Industrial Equipment. Mechanical or non-mechanical devices used in an industrial setting.

Industrial Wood Floor. A heavy duty wood floor made of decking (2 inches thick) or of wooden blocks laid on end. Also called a "factory" floor. Very resistant to heavy loads and traffic.

Inlet. An opening for intake.

Insulated Block. Hollow masonry block filled with insulation.

Insulation. Any material used to reduce the effects of heat, cold or sound transmission and to reduce fire hazard. Any material used in the prevention of the transfer of electricity, heat, cold, moisture and sound.

Interceptor. A device that collects foreign matter and prevents it from reaching the sewer system.

Invert, Manhole. The lowest inside surface of a manhole. A channel in the manhole through which wastewater or stormwater flows.

Irregular Stone. Stone cut to or quarried in different shapes and/or sizes.

Irrigation, Lawn. To supply water to grassy areas by artificial means.

J

Jacking Pipe. Forcing pipe through the ground in a tunnel created by the pipe itself. The pipe is generally jacked horizontally in short lengths.

Jib Crane. A crane which has a projecting arm of its derrick boom.

Job Built Form. A temporary structure or mold constructed on a jobsite, for the support of concrete while it is setting and gaining sufficient strength to be self-supporting.

Job Requirements. A list of specific, necessary and essential tasks to bring to completion a building or structure. Sometimes called "General Requirements".

Joint, Contraction. Formed, sawed, or tooled groove in a structure to create a weakened plane and regulate the location or cracking resulting from the dimensional change of different parts of the structure.

Joint, Expansion. A joint through tile, mortar, concrete or masonry down to the substrate, intended to allow for gross movement due to thermal stress or material shrinkage. A discontinuity or break intended to allow movement.

Joist. One of a group of light, closely spaced beams used to support a floor deck or flat roof. Timbers supporting the floorboards.

Joist Girder. A light steel truss used to support open-web steel joists.

Joist, Wood. A piece of lumber used horizontally as a support for a floor, measuring two to four inches thick and six or more inches wide.

K

Keene's Cement. A cement composed of finely ground, anhydrous, calcined gypsum, the set of which is accelerated by the addition of other materials, used in areas subjected to moisture. A hard, strong finishing plaster that is made from gypsum and maintains a high polish.

Kerf. The cut made by a saw.

Key. A slot formed into a concrete surface for the purpose of interlocking with a subsequent pour of concrete; a slot at the edge of a precast member into which grout will be poured to lock it to an adjacent member.

Kiln Dried. Lumber that has been heated in a kiln to dry and control the amounts of moisture.

King Post. In a roof system, the member placed vertically between the center of the horizontal tie beam at the lower end of the rafters and the ridge, or apex of the inclined rafters. A term often used to refer to a certain type of fabricated truss.

L

Ladder. Frame consisting of two parallel side pieces connected by rungs at suitable distances to form steps on which persons may climb up or down.

Lag Rod. A large diameter rod with a square or hexagonal head.

Lawn Irrigation. A system where water is transported and distributed to water grass.

Lb. Symbol for pound or pounds.

Leaching Pit. An excavated hole (pit) that can hold solids but allows liquids to pass through and leach into the ground.

Let-In Bracing. Diagonal bracing nailed into notches cut in the face of the studs so as to avoid an increase in the thickness of the wall.

LH Joist. A type of long-span high strength bar joist.

Lift-Slab Construction. A method of building multi-story sitecast concrete buildings by casting all the slabs in a stack on the ground, then lifting them up the columns with jacks and welding them in place.

Lightweight Aggregate. Aggregate of low specific gravity, such as expanded or sintered clay, shale, slate, diatomaceous shale, perlite, vermiculite, or slag; natural pumice, scoria, volcanic cinders, tuff, and diatomite, sintered fly ash or industrial cinders; used to produce lightweight concrete. Aggregate with a dry, loose weight of 70 pounds per cubic foot or less.

Lightweight Block. A concrete unit constructed of lightweight materials and used to reduce the weight of walls.

Lightweight Concrete. Concrete that achieves a significant reduction in weight by the substitution of lighter materials for the concrete aggregate.

Limestone Panel. A limestone slab, relatively thin with respect to other dimensions, and rectangular in shape.

Liner, Flue. Heat-resistant firebrick or other fireclay materials that make up the lining of a chimney.

Liner Panel. A panel used for interior finish.

Lintel. The horizontal beam placed over an opening.

Load Bearing. Supporting a superimposed weight or force.

Loader. An excavating machine with a movable bucket or scoop, used to transport earth, crushed stone, or other construction materials.

Lockwasher. A flat, split ring of metal or steel that when tightened with a nut is used prevent loosening.

Low-Lift Grouting. A method of constructing a reinforced masonry wall in which the wall is grouted in increments not higher than 4 feet.

Lumber. Sawed parts of a log such as boards, planks, and timber. Wood members which are manufactured by sawing, resawing, passing lengthwise through standard planing machine, crosscutting to length, and matching, but without further manufacturing.

M

Malleable Iron. Iron that can be hammered or bent without breaking.

Manhole. A hole through which a person may go to gain access to an underground or enclosed structure.

Marble. Limestone that is more or less crystallized by metamorphism, that ranges from granular to compact in texture, that is capable of taking a high polish, and that is used in architecture and sculpture.

Masonry. Brickwork, blockwork, and stonework.

Masterformat. The copyrighted title of a uniform indexing system for construction specifications, as created by the Construction Specifications Institute and Construction Specifications Canada, commonly (but inaccurately) called the CSI format or numbering system.

Mat, Concrete. A grid of metal reinforcement for concrete foundations, slabs, or mats.

Mat Foundation. A concrete slab used as a building or equipment foundation.

Material Handling. The act or process of transporting materials on or to a jobsite.

Mechanical. Of or relating to machinery or tools. Relating to, governed by, or in accordance the principles of mechanics. Loosely used as a term for anything in the plumbing, heating, air-conditioning, or fire sprinkler trades.

Membrane. A sheet material that is impervious to water or water vapor.

Membrane Waterproofing. A membrane, usually made of built-up roofing or sheet material, to provide a positive waterproof floor over the substrate, which is to receive a tile installation using a wire reinforced mortar bed.

Mesh, Slab. Welded-wire fabric in sheets or rolls used to reinforce concrete slabs.

Mesh Wire. A series of longitudinal and transverse wires arranged at right angles to each other in sheets or rolls, used to reinforce mortar and concrete. Welded-wire fabric.

Metal Fabrication. The building, construction or manufacture of metal structures or metal devices.

Metal Framing. The construction of a building or structure by using steel. Loosely used term to denote the construction of frame houses and partitions by using light gauge metal studs and members.

Metal Joist. Horizontal cold formed metal framing member of floor, ceiling or flat roof to transmit loads to bearing points. Often refers to a bar joist.

Metal Lath. A steel mesh used primarily as a base for the application of plaster.

Mild Steel. Steel containing less than three-tenths of one percent carbon, not used as structural steel because of its low strength.

Millwork. Interior components such as trim work, cabinets, doors and windows, etc. but not including floors, siding and ceilings. Often made of wood or plastic laminates and produced in a shop or factory.

Mixer. A machine used for blending the constituents of concrete, grout, mortar, cement paste, or other mixtures.

Mobilization. The act of putting into movement or circulation. The assembly and movement of equipment to a jobsite.

Moisture Barrier. A membrane used to prevent the migration of liquid water through a floor or wall.

Moisture Protection. The act or process of retarding the seepage of moisture.

Monolithic Concrete. Concrete cast with no joints other than construction joints or as one piece, generally, the term is used on larger structures.

Mop Sink. A deep well plumbing fixture with a faucet and a drain used for collecting and dispensing water for mopping/janitorial purposes.

Mortar. A mixture of cement paste and fine aggregate; in fresh concrete, the material occupying the interstices among particles of coarse aggregate; in masonry construction, mortar may contain masonry cement, or may contain hydraulic cement with lime (and possibly other admixtures) to afford greater plasticity and workability than are attainable with standard hydraulic cement mortar. A substance used to join masonry units consisting of cementitious materials, fine aggregate, and water.

Moss, Peat. Moss containing partially carbonized vegetable tissue formed by the partial decomposition of water in that moss.

Mud. A term used in the trade for mortar.

Mulch. A mixture, as of leaves and compost, that covers or is mixed with the earth, often to help enrich the soil. Bark, crushed stone or other material used to cover planting beds, retain moisture, reduce weeds, and improve appearance.

N

Nailable Concrete. Concrete, usually made with suitable lightweight aggregate, with or without the addition of sawdust, into which nails can be driven.

Natural Cleft Slate. Slate which has been split into thinner pieces along its natural cleft or seam. It is rougher in appearance than machined slate.

Neat Cement. Hydraulic cement in the unhydrated state.

Net, Safety. A meshed fabric that is spread below activity, to protect materials or people that may fall from dangerous heights.

Nipple, Bushed. A pipe threaded at both ends to connect two pipes of different dimensions.

Nipple, Offset. A fitting, threaded at both ends, that is a combination of elbows or bends which brings one section of pipe out of line with, but into a line parallel with, another section.

No-Hub Pipe. Pipe usually manufactured of cast iron, which is fabricated without hubs and coupled together by a stainless steel and rubber fastener.

Nominal Size (Lumber). The commercial size designation of width and depth, in standard sawn lumber and glued-laminated lumber grades; larger than the standard actual net size of the finished, dressed lumber.

Nonbearing. Not carrying a load.

Non-shrink Grout. Cementitious or epoxy based mix used to fill gap created between bearing components or base plates and foundation or other supporting element.

Nut, Hex. A six-sided, short metal nut with a threaded hole for receiving a rod or threaded bolt.

O

Oak. A strong, hard, heavy wood.

Office Trailer. A highway vehicle parked on a job site, designed to serve wherever needed, as an office and a place to carry out business.

Offset Connector. A fitting that is a combination of elbows or bends which brings one section of pipe out of line with, but into a line parallel with, another section.

Open-Web Steel Joist. A prefabricated, welded steel truss used at closely spaced intervals to support floor or roof decking.

Oriented Strand Board (OSB). A building panel composed of long shreds of wood fiber oriented in specific directions and bonded together under pressure. Commonly called "strand board".

OS&Y Valve. A type of valve, with external exposed threads supported by a yoke, indicating the open or closed position of the valve.

Overhead Door. A door, commonly used in garages and warehouse, that opens upward from the ground.

P

Pad, Bearing. A thick slab-type stone or precast concrete block placed under a structural member of a building to spread the load or weight evenly.

Pad, Equipment. Typically, a cast-in-place or precast concrete block placed under mechanical devices to spread the weight and load of the machinery evenly and to prevent excessive vibration.

Pad Eyes. Metal rings mounted vertically on a plate for tying small vessels.

Pan. A form used to produce a cavity between joists in a one-way concrete joist system.

Panel Cladding. Metal sheathing panels used to provide durability, weathering and corrosion, or impact resistance.

Parapet. The region of an exterior wall that projects above the level of the roof.

Parging. Portland cement plaster applied over masonry.

Parking Barrier. A structure either temporary or permanent, that is placed to prevent the encroachment of vehicles.

Particleboard. A building panel composed of small particles of wood and resins bonded together under pressure. Flat sheet material producing durable and dimensionally stable product which is often used in dry conditions in place of plywood.

Partition. A permanent interior wall which serves to divide a building into rooms.

Partition Toilet. Pre-finished panels used in a toilet enclosure.

Patio Block. Lightweight concrete paving slabs installed in lightly used foot traffic areas.

Pavement Cutting. The process of scoring or cutting through pavement surfaces with a power saw with a specific blade for that purpose.

Pavement Marking. The act or process of applying painted lines or necessary instructional signage on pavement surfaces for pedestrians or vehicle drivers.

Paver, Masonry. Shaped or molded units, composed of stone, ceramic brick or tile, concrete, or cast-in place concrete used for driveways and patios.

Paver, Stone. Blocks of rock processed by shaping, cutting or sizing, used for driveways and patios.

Paving, Concrete. The use of a mixture of portland cement, fine aggregate (sand) coarse aggregate (gravel or crushed stone) and water, to make a hard surface in areas such as walks, roadways, ramps, parking areas, etc..

Pea Gravel. Screened gravel particles most of which would pass through a 3/8 inch (9.5mm) sieve.

Peat Moss. Moss containing a partially carbonized vegetable tissue formed by the partial decomposition in that moss.

Pedestal. A short compression member of reinforced concrete that is placed between a column and the footing to distribute the load to the footing.

Perforated Pipe. Pipe with one or more rows of uniform holes along the length. Buried in the ground alongside building foundations or structures, to aid in drainage of groundwater and moisture.

Performance Bond. A bond, secured by the general contractor, which guarantees that the contract will be performed. An undertaking by a surety that a contractor will perform a contract.

Perlite. Expanded volcanic rock, used as a lightweight aggregate in concrete and plaster, and as an insulating fill.

Personnel Lift. An elevator for use by persons at a job site, a building, or structure.

Pest Control. The act or process of the placement of devices or spraying of chemicals or powders to control the spread of insects and pests.

Photographs. Used as a loose term for the pictures or photographs taken before a job commences to provide an accurate representation of what the site was like before construction.

Pickup Truck. A light truck having an open body with low sides and tailboard.

Pier. Timber, concrete, or masonry supports for girders, posts, or arches. Intermediate supports for a bridge span. Structure extending outward from shore into water used as a dock for ships.

Pilaster. A vertical, integral stiffening rib in a masonry or concrete wall. A portion of a square column, usually set within or against a wall.

Pile. A long, slender, piece of material driven into the ground to act as a foundation. A member embedded into the ground that supports vertical loads, can be made of wood, steel or concrete.

Pipe. A long tube or hollow body for conducting a liquid, gas, or finely divided solid. May be used for structural elements as well.

Pipe Cleanout. A pipe fitting placed in a drain waste (DWV) system for access to remove blockages.

Pipe Flange. Projecting ring, ridge or collar placed on pipe to strengthen, prevent sliding, or accommodate attachments.

Pipe Insulation. Insulation that covers pipes to help in the reduction of heat loss or gain.

Pipe Jacking. Forcing pipe through the ground in a tunnel created by the pipe itself. The pipe is generally jacked horizontally in short lengths.

Pipe, No-Hub. Pipe manufactured in cast iron, which is fabricated without hubs for coupling.

Pipe Sleeve. Cylindrical insert cast into concrete wall or floor to provide for later passage or anchorage of pipe.

Pipe, Structural. Pipe used in a structure to transfer imposed loads to the ground.

Pitch Pocket. An opening between growth rings of a tree which usually contains resin, bark, or both. In roof construction, a metal container placed around a roof penetration at roof level to receive hot bitumen or caulking and provide a roof seal. Commonly found at columns or plumbing stacks.

Plank. A wide piece of sawed timber, usually 1-1/2 to 4-1/2 inches thick and 6 inches or more wide.

Plaster. A cementitious material, usually based on gypsum or portland cement, which is applied to lath or masonry in paste form, to harden into a finished surface. A mixture of lime, hair, and sand, or of lime, cement, and sand, used to cover exterior or interior wall surfaces.

Plastic Coated Conduit. A type of conduit for electrical wiring that is used around moist areas and highly corrosive fumes.

Plastic Pipe. Pipe manufactured from hard plastic to resist corrosion and rust.

Plate. The top horizontal piece of the wall of a frame building upon which the roof or other structural elements rests.

Plate, Toe. A metal bar fastened to the outer edge of a grating; rear of a tread; the bottom rail of a door.

Platform. Horizontal landing in stair either at the end of a flight or between flights, either at floor level or between floors.

Ply. A term used to denote a layer or thickness of building or roofing paper as two-ply, three-ply, etc.

Plywood. A wood panel composed of a number of layers of wood veneer bonded and glued together under pressure. Flat sheet material built up of sheets of veneer called plies, united under pressure by bonding agent to create a panel with adhesive bond between plies as strong as or stronger than the wood alone.

Plywood, Finish. The finest grade of plywood.

Plywood Sheathing. A flat panel made up of a number of thin sheets, or veneers, of wood used to close up side walls, or roofs preparatory to installation of finish materials on the surface.

Pneumatic Concrete. Concrete that is delivered by equipment powered by compressed air.

Pneumatic Tool. A tool powered by compressed air.

Pointing. The process of applying and compacting mortar to the surface of a mortar joint after the masonry has been laid, either as a means of finishing the joint or to repair a defective joint.

Pole, Utility. A vertical pole usually with cross arms, where utilities install their service or supply lines.

Polyethylene Pipe. Pipe manufactured from a thermoplastic, high-molecular-weight organic compound.

Polyethylene Vapor Barrier. A sheet form thermoplastic membrane, high molecular weight, organic compound, used as a protective cover to prevent the passage of air and/or moisture.

Polypropylene Pipe. A tough plastic pipe with resistance to chemicals and heat.

Pop-Up Head. The part of a sink drain assembly that is operated by a linkage to open or close the drain. In lawn irrigation systems, a watering head that retracts when not in use, flush with the ground.

Portland Cement. The element used as the binder in concrete, mortar, and stucco.

Post. A timber set on end to support a wall, girder, or other member of the structure.

Post-Tensioning. A method by which concrete is compressed after it has been cast by stressing the steel reinforcing. The compressing of the concrete in a structural member by means of tensioning high-strength steel tendons against it after the concrete has cured.

Pour. To place concrete; a continuous increment of concrete casting carried out without interruption.

Power Tool. An apparatus or device used in construction, powered by electric current.

Precast. A concrete component or member cast and cured in other than its final position.

Precast Specialty. Special shapes or ornamental objects of precast masonry or concrete. A loose term which may refer to any manufacturing of precast members.

Preservative Treated. Applied or pressurized chemical treatment of wood or plywood to make it resistant to deterioration from moisture and insects.

Pressure Reducing Valve. A valve which maintains fluid pressure uniformly on its outlet side as long as pressure on the inlet side is at or above a design pressure.

Pressure-Treated Lumber. Lumber that has been impregnated with chemicals under pressure, for the purpose of retarding either decay or fire.

Prestressed Concrete. Concrete that has the reinforcing pretensioned prior to placement.

Prestressing. Applying compressive stress to a concrete structural member to increase its strength.

Pretensioning. A method by which the design tensile force is applied to the steel reinforcing before the concrete is set.

Professional Fee. The amount of money charged by a person hired to perform a service.

Profit. The excess of returns over expenditures in transactions or series of transactions.

Protective Board. A board or sheet of material that is installed next to a waterproofing membrane and then backfilled against thus protecting the membrane from puncture or damage.

PSF. Pounds per square foot.

PSI. Pounds per square inch.

Pump. A device that raises, transfers, or compresses fluids or attenuates gases by suction or pressure or both.

Pump, Sump. A small capacity pump that empties pits receiving water, sewage, or liquid waste.

Pumped Concrete. Concrete that is pumped through a hose or pipe.

Purger, Air. A mechanical device that removes unwanted air.

Purlin. A timber supporting several rafters at one or more points. Beams or struts that span across a roof to support the roof framing system.

PVC. Polyvinyl Chloride.

PVC Waterstop. A nonmetallic synthetic resin prepared by polymerization of vinyl chloride inserted across a joint to obstruct the seepage of water through the joint.

Q

Quadruplex Cable. Four wire cable.

Quarry Tile. A large clay floor tile, usually unglazed.

Quarter Round. Wood molding in the shape of a quarter circle.

R

Raft. A footing or foundation, usually a large thick concrete mat.

Rafter. A framing member that runs up and down the slope of a pitched roof. The beams that slope from the ridge of a roof to the eaves and make up the main body of the roof's framework.

Rafter Anchor. A bolt or fastening device which attaches the parallel beams used to support a roof covering to the rafter plate.

Rafter, Hip. The diagonal outside intersecting rafter at the meeting planes in a hip roof. A structural member of a roof forming the junction of an external roof angle or, where the planes of a hip roof meet.

Rafter, Jack. A rafter that fills the space between the hip rafter and the top of the wall plate.

Rafter, Valley. The rafter extending from an inside angle of the plates toward the ridge or center line of the house.

Rafters, Common. Those which run square with the plate and extend to the ridge.

Rail. The horizontal members of the balustrade or panel work.

Railing. An open fence or guard for safety, made of rails and posts. A banding in cabinetwork. On plywood, the solid wood band around one or more edges.

Railing, Pipe. A metal railing made of pipe.

Rake. The sloping edge of a pitched roof. The trim of a building extending in an oblique line, as rake dado or molding.

Reducer. A pipe fitting which connects pipes of different sizes. A tile trim unit used to reduce the radius of a bullnose or a cove to another radius or to a square.

Reflectorized Sign. A sign made with highly reflective material.

Refractories. Heat-resistant non-metallic ceramic materials.

Refrigeration. The cooling of a material or space.

Reinforced Concrete. Concrete containing adequate reinforcement (pre-stressed or not prestressed) and designed on the assumption that concrete and steel act together in resisting forces. Concrete work into which steel bars have been embedded to impart tensile strength to the member.

Reinforced Grouted Masonry. Wall construction consisting of brick or block that is grouted solid throughout its entire height and has both vertical and horizontal reinforcing.

Reinforcement. The action or state of strengthening by additional assistance, material, or support. The action or state of strengthening or increasing by fresh additions. A term used for reinforcing steel.

Reinforcement, Mesh. A series of longitudinal and transverse wires arranged at right angles to each other in sheets or rolls, used to reinforce mortar and concrete. Welded-wire fabric.

Reinforcing Accessory. The items used to install reinforcing in concrete. These include but are not limited to; chairs, couplings, tie wire etc.

Release Agent. Material used to prevent bonding of concrete to a form surface.

Reline Pipe. To install new linings in pipes. Commonly includes the cleaning of built-up scale or debris from the existing pipe and relining with a compatible material.

Remove. The act or process of demolishing, dismantling, and carrying away an old fitting or component.

Restoration. The act or process of bringing a structure back to its former position or condition. The act of restoring to an original condition.

Retaining Wall. A wall that is designed to resist the lateral pressures of retained soil. A wall that holds back a hillside or is backfilled to create a level surface.

Retardant, Fire. A material or treatment which effects a reduction in flammability and in spread of fire.

Return. The ending of a small splash wall or a wainscot at right angle to the major wall. The continuation of a molding or finish of any kind in a different direction. In HVAC, a term for the return-air duct of a forced air heating/cooling system.

Revolving Door. Typically a four panel door attached at 90 degrees to each other that turns on a center axis.

Ribband. The support for the second-floor joists of a balloon-frame house. The horizontal member of a wood frame wall which supports floor joists or roof rafters.

Ridge. The top edge or corner formed by the intersection of two roof surfaces. The board against which the tips of rafters are fastened.

Ridge Cut. Any cut made in a vertical plane; the vertical cut at the top end of a rafter.

Ridge Vent. A construction element mounted along the ridge of a roof to aid in ventilating an attic space.

Rigid Insulation. Thermal insulating material made of polystyrene, polyurethane, polyisocyanurate, cellular glass, or glass fiber, sometimes offered with a skin surfacing, formed to flat board shape of constant thickness.

Riprap. A foundation or sustaining wall of stones placed together without order, as in deep water or on an embankment, to prevent erosion.

Rise. The vertical distance through which anything rises, as the rise of a roof or stair.

Riser. A single vertical increment of a stair; the vertical face between two treads in a stair; a vertical run of plumbing, wiring or ductwork.

Rivet. A structural fastener on which a second head is formed after the fastener is in place.

Rock Anchor. A post-tensioned rod or cable inserted into a rock formation for the purpose of anchoring a structure to it.

Rock Drilling. The act or process of boring holes into rock.

Rodding. A method of using a straightedge to align mortar with the float strips or screeds. This technique also is called dragging, pulling, floating, or rodding off.

Roll Roof. A roof that has been covered with an asphaltic material that comes in rolls.

Roll-Up Door. A door which raises and rolls into a coiled configuration and lowers on tracks on either side.

Roof. The top cover of a building or structure.

Roof Ballast. Crushed rock or gravel which is spread on a roof surface to form its final surface.

Roof, Built-Up. A roof covering made of continuous rolls or sheets of saturated or coated felt embedded in a bituminous coating. The roofing may have a final ballast coating of gravel or slag.

Roof, Curb. A roof with a slope that is divided into two pitches of each side. Also known as a gambrel or mansard roof.

Roof, Fluid-Applied. A roof coated with an asphalt-based liquid.

Roof Insulation. Materials used in between rafters or roof supports for the protection from heat and cold. Solid sheets of insulating material installed on a flat roof.

Roof Scupper. A fitting cut through a roof system, acting as a waterway so water can be discharged off the roof.

Roof Sheathing. The first layer of covering on a roof, fastened to the rafter boards, used to support the roofing material.

Roof Shingle. Roof covering pieces of asphalt, asbestos, wood, tile, slate, etc. Pieces of wedge shaped wood or other material used in overlapping courses to cover a roof.

Roof Slab. The flat section of a reinforced concrete roof, supported by beams, columns, or other framework.

Roof Specialties. A fitting or piece of trim used in the installation of a roof, such as gravel stop, flashing, vent strips, etc..

Roof Walkway. A permanent aisle for safe access across a roof. Also serves as a protection for the roofing material when maintenance is being done.

Roofing. The material put on a roof to make it weatherproof.

Rosewood Veneer. A thin layer of cabinet wood with a dark red or purplish color streaked and variegated with black, applied to give a superior and decorative surface.

Rough Opening. The clear dimensions of the framed opening that must be provided in a wall to accept a given door or window unit.

Roughing In. The act of preparing a surface by applying tar paper and metal lath (or wire mesh). Sometimes called "wiring". In plumbing and electrical, the pre-wiring or pre-piping of a building before the finished walls, fixtures, and devices are installed.

Rowlock. A brick laid on its long edge, with its end exposed in the face of the wall.

Rubble Masonry. Uncut stone, used for rough work, foundations, backfilling, and the like.

Run. Horizontal dimension in a stair. The length of the horizontal projection of a piece such as a rafter when in position.

Runner Channel. A steel member from which furring channels and lath are supported in a suspended plaster ceiling.

Running Bond. Brickwork consisting entirely of stretchers.

Rusticated Concrete. Beveled edges of concrete making the joints conspicuous.

S

Safety Chain. Chain installed horizontally in railing assembly to provide for ease in providing temporary opening in railing.

Safety Glass. Specific type of glass having the ability to withstand breaking into large jagged pieces.

Safety Net. A woven meshed fabric that is spread below activity to protect materials and people that may fall from dangerous heights.

Safety Nosing. Stair nosing with abrasive non-slip strip surface flush with tread surface.

Salamander. A portable source of heat, customarily kerosene or oil-burning, used to temporarily heat an enclosure. Commonly used around newly placed concrete to prevent freezing.

Sample Soil. A representative specimen of soil from a site.

Samples. Material samples requested by the architect of the general contractor or materials specified for the project.

Sampling. The method of obtaining small amounts of material for testing from an agreed-upon lot.

Sandblast. A system of cutting or abrading a surface such as concrete or metal by a stream of sand ejected from a nozzle at high speed.

Sandstone. A sedimentary rock formed from sand.

Sanitary Piping. Drain, waste, and vent plumbing systems.

Sash. A frame that holds glass. The framework which holds the glass in a window.

Saw Cut, Concrete. A cut in hardened concrete utilizing diamond or silicone-carbide blades or discs.

Scab. A short piece of lumber used to splice, or to prevent movement of two other pieces.

Scaffold or Staging. A temporary structure or platform enabling workmen to reach high places.

Scale. A short measurement used as a proportionate part of a larger, actual dimension. The scale of a drawing is expressed as 1/4 inch = 1 foot.

Scarifier. A piece of thin metal with teeth or serrations cut in the edge. It is used to roughen fresh mortar surfaces to achieve a good bond for the tile. A scarifier also can be used to roughen the surface of concrete.

Scratch Coat. The first coat in a three-coat application of plaster.

Screed. A strip of wood, metal, or plaster that establishes the level to which concrete or plaster will be placed. To strike or plane off wet mortar or concrete which is above the desired plane or shape.

Scupper. An opening through which water can drain from the edge of a flat roof. An opening cut through a roof system, acting as a waterway so water can be discharged off the roof.

Sealant. An elastomeric material that is used to fill and seal cracks and joints. At expansion joints, this material prevents the passage of moisture and allows horizontal and lateral movement.

Seated Connection. A connection in which a steel beam rests on top of a steel angle fastened to a column or girder.

Security Grille. Steel grille to prevent intrusion or entry through openings.

Self Tapping. Creates its own screw threads on the inside of a hole.

Septic Tank. A private sewage system holding tank, installed together with a leaching field, that collects sewage and allows the solid waste to settle to the bottom of an area while the liquid particles of the sewage drain into the leaching field area.

Service Sink. A deep basin with a faucet used in janitorial applications.

Service Wye. A pipe fitting which joins three pipes at a 45 degree angle. A drainage fitting in a plumbing system.

Set Screw. A headless screw used to secure two separate parts in a relative position to one another, preventing the independent motion of either part. A screw to adjust the tension of a spring.

Shear Plate. In heavy timber construction, a round steel plate used for connecting wood to non-wood materials.

Sheathing. The rough covering applied to the outside of the roof, wall, or floor framing of a structure. In clay tile or wood shingle/shake roofs, roofing boards generally installed as narrow boards laid with a space between them, according to the length of a shingle or tile exposed to weather.

Shed. A building or dormer with a single pitched roof.

Sheet Metal Roof. A roof covering of aluminum, copper, stainless steel, or galvanized metal sheets. In high-finish applications, loosely referred to as an "architectural roof".

Sheet Piling. Planking or sheeting made of concrete, timber, or steel that is driven in, interlocked or tongue and grooved together to provide a tight wall to resist the lateral pressure of water, adjacent earth or other materials.

Sheetrock. Plasterboard sheets, an interior facing panel consisting of a gypsum core sandwiched between paper faces, also called gypsum board or plasterboard. Different types are available for standard, fire-resistant, water-resistant, and other applications.

Shield, Expansion. Device inserted in predrilled holes (usually in concrete or masonry) which expands as screw or bolt is tightened within it, used to fasten items to concrete or masonry.

Shingle. A small thin piece of building material often with one end thicker than the other for laying in overlapping rows as a covering for the roof or sides of a building or structure.

Shiplap. A board with edges rabbeted so as to overlap flush from one board to the next.

Shop Painted. Coating(s) of paint applied in shop, usually a primer coat to protect metal from corrosion, which may or may not receive additional coats in the field.

Shotcrete. A low-slump concrete mixture deposited by being blown from a nozzle at high speed with a stream of compressed air.

Shower Receptor. The floor and side walls of the shower up to and including the curb of the shower.

Sidewalk. A walk for pedestrians at the side of a street.

Siding. The outside finish of an exterior wall.

Signage. Any of a group of posted commands, warnings or directions.

Sills. The horizontal timbers of a house which either rest upon the masonry foundations or, in the absence of such, form the foundation.

Sink. A basin with a drainage system and water supply, used for washing and drainage. A pit or pool used for the deposit of waste.

Sitecast Concrete. Concrete that is cast-in-place.

Sizing. Working material to the desired size; a coating of glue, shellac, or other substance applied to a surface to prepare it for painting or other method of finish.

Skim Coat. A thin coat of plaster over any base system. May be the final or finish coat on plaster base or a certain type of drywall.

Slab Bolster. Continuous, individual support used to hold reinforcing bars in the proper position.

Slab Form. The formwork used for the pouring or placing of a concrete slab. A type of manufactured metal decking which is made expressly to receive a final layer of poured concrete.

Slate. A form of geologically hardened clay, easily split into thin sheets.

Sleeper. A timber laid on the ground to support a floor joist. A framing system, usually for a wood floor system, which is fastened directly to a concrete floor thus facilitating the installation of the finished floor.

Slot. An opening in a member to receive a connection with another part.

Slot, Anchor. A groove in an object into which a fastener or connector is inserted to attach objects together.

Slump Test. The procedure for measuring slump with a slump cone.

Slurry. A mixture of water and any finely divided insoluble material, such as portland cement, slag, or clay in suspension. A watery mixture of insoluble materials.

Soffit. The underside part of a member of a structure, such as a beam, stairway, roof, or arch. The undersurface of a horizontal element of a building, especially the underside of a stair or a roof overhang.

Soil. A generic term for unconsolidated natural surface material above bedrock.

Soil and Waste Pipe. Plastic, copper, cast iron or DWV drainage, water and vent.

Soldier Course. Oblong tile or brick laid with the long side vertical and all joints in alignment.

Sole Plate. The horizontal piece of dimension lumber to which the bottom of the studs are attached in a wall of a light frame building.

Solid Block. A block with small or no internal cavities.

Sound Absorption. The process of dissipating or removing sound energy. The property possessed by materials, objects, and structures such as rooms, of absorbing sound energy.

Sound Attenuation. A process in which sound is reduced as its energy is converted to motion or heat.

Spall. A fragment, usually in the shape of a flake, detached from a larger mass by a blow, by the action of weather, by pressure, or by expansion within the larger mass.

Span. The distance between supports for a beam, girder, truss, vault, arch or other horizontal structural device; to carry a load between supports. The distance between the bearings of a timber or arch.

Spandrel. The wall area between the head of a window on one story and the sill of a window on the floor above; the area of a wall between adjacent arches.

Spandrel Beam. A beam that runs along the outside edge of a floor or roof.

Special Door. A door that has a unique use, such as a bank vault door.

Specialties. A designation of construction materials or components which commonly typifies items that furnish and finish off the structure. A loosely used term to denote any item which aids in the installation of a structure's parts (i.e. roof specialties).

Specifications. The written or printed directions regarding the details of a building or other construction.

Spiral Reinforcing. Continuously wound reinforcing in the form of a cylindrical helix.

Splash Block. A small precast block of concrete or plastic used to divert water at the bottom of a downspout.

Splice. Joining of two similar members in a straight line.

Stability. The ability to remain unchanged. Ability to restore to original condition after being disturbed by some force.

Staging. Temporary scaffolding with platforms placed in and around a building to create elevated work areas.

Stair, Access. A stair system to provide specific access to roofs, mechanical equipment rooms etc.

Stair, Concrete. A stair system constructed solely from concrete.

Stairs, Box. Stairs built between walls, and usually with no support except the wall.

Standing Seam Roof. A sheet metal roof system that has seams that project at right angles to the plane of the roof.

Static. State of being at rest, having no motion.

Steel. Iron compounded with other metals to increase strength and wearing or rust resistance.

Steel Angle. An L-shaped member constructed of steel, often used as a lintel or carrying shelf for masonry.

Steel Joist. Open web, parallel chord, load-carrying members suitable for direct support of floors and roof decks, utilizing hot rolled or cold formed steel.

Steel Plate. Sheet steel of a heavier thickness.

Steel, Structural. Steel that is rolled in a variety of shapes and manufactured for use as structural load-bearing members.

Steel Stud. Steel pin or rod having head at one end for driving into material used for holding members or parts of members together. In light gage construction, a vertical framing member such as a "2 by 4".

Stile. A vertical framing member in a panel door.

Stirrup. A vertical loop of steel bar used to reinforce a concrete beam against diagonal tension forces.

Stone. Earthy or mineral matter of indeterminate size or shape such as rock, etc.

Stone Paver. Blocks of rock processed by shaping, cutting or sizing, used for driveways, patios and walkways.

Stop, Gravel. A metal flange or strip with a vertical lip placed around the edge of a built-up roof to prevent loose gravel from falling off the roof.

Stop Notice. A charge against construction funds in the hands of a property owner or a construction lender for the value of work or materials incorporated into a construction project.

Storefront. The facade which is constructed on the street side of a building or structure into which persons can enter and transact business. A loosely used term to denote a steel or aluminum tube frame and glass wall.

Strap Tie. A metal plate that fastens two parts together as a post, rod or beam.

Stressed-Skin Panel. A panel consisting of two face sheets of wood or metal bonded to perpendicular spacer strips.

Stretcher. A masonry unit laid with its length horizontal and parallel with the face of a wall or other masonry member.

Striking Joints. A process of removing excess grout from the joints by wiping with a sponge or cloth or scraping, compacting or rubbing with a curved instrument.

Structural Lumber. Wood members of a structural system which are manufactured by sawing, resawing, passing lengthwise through standard planing machine, crosscutting to length, but without further manufacturing.

Structural Pipe. Pipe used in a structure to transfer imposed loads to the ground.

Structural Plywood. The highest grade of plywood. Plywood of exterior structural grade, secured to top side of floor joists used to create rigidity in building superstructure and also to provide smooth and even surface to receive finish floor covering.

Structural Steel. Steel hot-rolled into variety of shapes for use as load-bearing structural members.

Structural Tube. A hollow metal product used to carry imposed loads in a building or structure.

Strut. A compression member, a column, usually implying that it can be placed at any angle, not just vertically.

Stucco. A cement plaster used for coating exterior walls and other exterior surfaces of buildings. Portland cement plaster used as an exterior cladding or siding material. A plaster used for interior decoration and finish work; also for rough outside wall coverings.

Stucco Lath. Wood or metal lath strips which form the base for the application of a cement plaster on an exterior wall surface.

Stud. One of a set of small vertical elements, usually wood or steel, used to produce a framed wall. An upright beam in the framework of a building. Vertical member of appropriate size (2x4 to 4x10 in.) (50x100 to 100x250 mm) and spacing (16 to 30in.) (400 to 750 mm) to support sheathing of concrete forms; also a headed steel device used to anchor steel plates or shapes to concrete members.

Sub-Flooring. Certain material, like plywood, that is installed on the floor joists of a building or structure, onto which the walls and finished flooring is attached.

Subfloor. Typically a wood floor which is laid over the floor joists and on which the finished floor is laid.

Subpurlin. A small roof framing member that spans between joists and purlins.

Sump Pump. A small capacity pump that empties pits receiving groundwater, sewage, or liquid waste.

Superintendent. An individual who is at the top level of a construction team in the field. A person with executive oversight, often of a board or building complex.

Surcharge. An increase in the lateral earth pressure of a retaining wall, caused by a vertical load behind the wall. A load placed over an area to compact it or change its characteristics.

Swale Excavation. The digging up of low-lying land such as a small meadow, swamp, or marshy depression.

Swing Gate. The operable member of a fence system that is hinged for opening and closing.

T

T-Beam. A reinforced concrete beam that contains of a portion of the slab above and which the two act together.

Tack Coat. Application of material made from asphalt on old surface to insure its bond to new construction.

Tapping Valve. A device to open or close a duct, pipe, or other passage, or to regulate flow. It is inserted into an existing pipeline by piercing the wall of the pipe and thus tapping into the flow.

Tarpaulin. A waterproofed canvas or other material used for protecting construction projects, athletic fields, goods or other exposed objects or areas.

Tee. A metal or precast concrete member with a cross section resembling the letter "T". A pipe fitting which joins three pipes at 90 degree angles.

Tee, Bulb. Rolled steel in the form of a "T" with a formed bulb at the end of the web.

Tee Weld. Weld in joint between two members located approximately at right angles to each other in form of "T".

Temporary Facility. A structure erected for temporary use.

Tendon. A steel strand used for prestressing a concrete member.

Tension Ring. A structural element, forming a closed curve in plan, which is in tension because of the action of the rest of the structure. A concrete or masonry dome commonly has a tension ring.

Terne. An alloy of lead and tin, used to coat sheets of carbon steel, stainless steel, or copper for use as metal roofing sheet.

Terra Cotta. Hard baked clayware, or tile, of variable color, averaging reddish red-yellow in hue and of high saturation.

Terrazzo. A finish floor material consisting of concrete with an aggregate of marble chips selected for size and color, which is ground and polished smooth after curing.

Test Pile. A pile which is driven before the final design is done to see what bearing is developed and what length of pile is actually needed. A pile which is tested by placing a predetermined load on it, commonly done by erecting a crib on the pile and then filling it with ballast.

Thin Coat. A loosely used term for a one coat plaster system over gypsum board.

Thin Set. A term used to describe the bonding of tile with suitable materials, applied approximately 1/8" thick.

Threshold. The wood or metal beveled floor piece at door openings which commonly separates non-continuous floor types.

Threshold, Door. A beveled piece of floor trim over which a door swings.

Tie. A device for holding components together, a structural device that acts in tension.

Tie Rod. A steel rod that acts in tension and commonly holds together wall forms while concrete is being poured.

Tilt-Up Construction. A method of constructing concrete walls in which panels are cast and cured flat on the floor slab, then tilted up into their final positions.

Timber. Lumber with cross-section over 4 by 6 inches, such as posts, sills, and girders.

Timekeeper. A clerk who keeps records of the time worked by employees.

To the Weather. A term applied to any part of the structure which faces the elements. A shingled roof is "to the weather" the framing system is not.

Toggle Bolt. A bolt and nut assembly to fasten objects to hollow construction assembly from only one side. Nut has pivoted wings that close against spring when nut end of assembly is pushed through hole and spring open on other side in void of construction assembly.

Toothed Ring. A timber connector used in the manufacturing of large member wood trusses.

Top Bars. Steel reinforcing bars near the top of reinforced concrete.

Top Plate. The horizontal member at the top of a stud wall, usually supporting rafters.

Topping, Concrete. Concrete layer placed to form a floor surface on a concrete base.

Topsoil. Surface soil at and including the average plow depth, soil which is used as a planting or growing medium.

Torsion. The rotation of a diaphragm caused by lateral forces and whose center of mass does not coincide with the center of rigidity.

Tort. A negligent or intentional wrongful act that damages the person or property of another, the wrongful nature of which is independent of any contractual relationship.

Total Cost. A method of computing damages sustained by a contractor because of breaches of contract causing the contractor to operate in an inefficient or unproductive manner.

Tower. A tall structure, constructed of frames, braces, and accessories rising to a greater height than the surrounding area.

Trailer. A vehicle designed to be hauled or to serve parked as dwelling or place of business.

Translucent Panel. A building panel that permits the passage of light but not vision.

Transom. A transverse piece in a structure, lintel. A horizontal crossbar in a window, over a door, or between a door and a window or a fanlight above it. A window above a door or other window built on and commonly hinged to a transom.

Trap. Located at a plumbing fixture, designed to hold a quantity of water that prevents gasses in the sewer system from entering a room.

Trash Chute. A device either constructed on the inside of a building or structure or hung outside for the removal of waste materials from upper floors.

Traveling Crane. A tower crane mounted on tires, rails or crawlers.

Tread. One of the horizontal planes of a stair. The horizontal part of a step.

Treated Lumber. Lumber infused or coated with stain or chemicals to retard fire, decay, insect damage or deterioration due to weather.

Tremie. A large funnel with a tube attached, used to deliver concrete into deep forms or beneath water or slurry. A tremie slows down the concrete and resists segregation of the aggregates.

Trench. A long, thin excavation.

Trencher. A mechanical device used to dig narrow channels in the ground.

Trenching. The act or process of digging narrow channels in the ground.

Trespass. Unauthorized entry upon the real property of another.

Trim. A term sometimes applied to outside or interior finish woodwork and the finish around openings.

Trowel Finish. The final finish of concrete, plaster, stucco, etc., by the use of a hand trowel.

Truck. A wheeled vehicle for moving heavy articles. A strong automotive vehicle for hauling. A small wheelbarrow consisting of a rectangular frame having at one end a pair of handles and at the other end a pair of small, heavy wheels and a projecting edge to slide under a load.

Truck Crane. A mechanical device for hoisting or lifting materials mounted on the bed of a truck.

Truss. A triangular arrangement of structural members that reduces nonaxial forces on the truss to a set of axial forces in the members. Structural framework of triangular units for supporting loads over long spans.

Tube, Structural. A hollow metal or plastic product used to carry imposed loads in a building or structure.

Tunneling. The act or process of digging a horizontal passageway through or under an obstruction.

Turnout. A widened space in a highway for vehicles to pass or park. A railroad siding.

Two-Way Flat Slab. A reinforced concrete framing system in which columns with mushroom capitals and/or drop panels directly support a two-way slab that is planar on both its surfaces.

Type X Gypsum Board. A gypsum board used where increased fire resistance is required.

U

U Bolt. A U-shaped, bent iron bar that has bolts and threads at both ends.

U Stirrup. An open-top, U-shaped loop of steel bar used as reinforcing against diagonal tension in a beam.

Underdrain. Perforated pipe drain installed in crushed stone under a slab to intercept ground water and drain it away from the structure.

Underpinning. The process of placing new foundations beneath an existing structure.

Underslab Drainage. The process of continuous interception and removal of ground water from under a concrete slab with the installation of perforated pipe.

Union. A type of pipe fitting used to join two pipes in line without turning either pipe.

Union T. A pipe tee with a fitting on one end that joins two pipes without turning either pipe.

Unit Masonry. Manufactured or natural building units of concrete, burned clay, glass, stone, gypsum, etc..

Unit Substation. A freestanding assembly including a transformer, switchgear and meter(s).

Unlined Stack. Chimney or vent fabricated of single pipe.

Unrated Door. A door which has not been rated for any ability to withstand the spread of fire.

Unreinforced. Constructed without steel reinforcing bars or welded wire fabric.

Urethane Board. Rigid form of plastic foam of polyurethane.

Urinal. A plumbing fixture used to collect urine and which can be flushed.

Urinal Screen. A privacy panel that separates urinals from each other. The screens are either floor, wall or ceiling mounted.

Utility Excavation. The act or process of either digging up existing cable buried in the ground, or trenching to lay new cable.

Utility Pole. A vertical pole with cross arms, where utility company lines are carried.

V

V Beam Roof. A roof with corrugated sheeting with flat, V-angled surfaces.

Vacuum Breaker. An electrical breaker with a space that contains reduced air pressure.

Valley. A trough or internal angle formed by the intersection of two roof slopes. The internal angle formed by the two slopes of a roof.

Valley Flashing. Thin sheet metal used to line the valley of a roof.

Valley Rafter. A diagonal rafter that supports a roof valley.

Valve. Numerous mechanical devices by which the flow of liquid, gas, or loose material in bulk may be started, stopped, or regulated by a movable part that opens, shuts, or partially obstructs one or more passageways.

Vapor Barrier. Waterproof membrane placed under concrete floor slabs that are on grade.

Varnish. A colorless, clear, resinous product dissolved in oil, alcohol, or a number of volatile liquids, applied to wood to provide a hard, glossy, protective film.

Vault. An arched surface. An arch translated along an axis normal to the plane of its centerline curve. A room to store valuable items.

Veneer. A thin layer, sheet or facing.

Veneer, Ashlar. An ornamental or protective facing of masonry composed of squared stones.

Veneer Plaster. A wall finish system in which a thin layer of plaster is applied over a special gypsum board base.

Vent. A vertical pipe connected to a waste or soil distribution system that prevents a vacuum that might suck the water out of a trap. Vertical pipe to provide passageway for expulsion of vent gases from gas-burning equipment to outside air.

Vent, Foundation. Opening in foundation wall to provide natural ventilation to foundation crawl spaces.

Vent Stack. A plumbing vent (pipe) in a multistory building, a separate pipe used for venting, that either connects with a stack vent above the highest fixture, or extends through the roof.

Vermiculite. Expanded mica, used as an insulating fill or a lightweight aggregate.

Vermont Slate. A fine grained thin-layered rock used for roofing, paving, etc..

Vertical Bar. An upright reinforcing bar in a reinforced concrete shape.

Vertical Siding. Exterior wall covering attached vertically to the wood frame of a building or structure.

Vestibule. An entrance to a house; usually enclosed.

Vibration. A periodic motion which repeats itself after a definite interval of time.

Vinyl Sheet. The rolled form of vinyl.

Vinyl Sheetrock. Plasterboard with a thin layer of vinyl as the finished surface.

Vision Panel. Glass placed in an opening of a door.

Vitrified Clay Pipe. Pipes used especially for underground drainage, that are made of clay baked hard.

Voltage. Electricity is caused by creating a higher electric charge at one point in a conductor than at another. This potential difference is called voltage.

W

Waferboard. A building panel made by bonding together large, flat flakes of wood.

Waffle Slab. A two-way concrete joist system. Two-way slab or flat slab made up of a double system of narrow ribs or joists, usually at right angles to each other, forming a pattern of waffle-like coffers.

Wainscoting. A wall facing, usually of wood, cut stone, or ceramic tile, that is carried only part way up a wall. Matched boarding or panel work covering the lower portion of a wall.

Wale. A horizontal beam.

Walk. A path specially arranged or paved for walking.

Wall. A member, usually vertical, used to enclose or separate spaces.

Wall Blocking. Framing lumber cut in short lengths and installed horizontally between wall studs as filler pieces to stabilize the framing or to provide a backing for fastening a finish item.

Wall Flange. A ridge on a wall that prevents movement. A supporting rim on a wall for attachments.

Wall Footing. A continuous spread footing that supports a uniform load from a wall.

Wall Hydrant. A connection to a water main cut through and mounted on a wall.

Wall, Retaining. A wall that is designed to resist the lateral pressures of retained soil.

Wall Sheathing. The first layer of covering on an exterior wall, fastened to the wall studs.

Wall Tie. A mechanical metal fastener which connects wythes of masonry to each other or to other materials.

Wash. The slant upon a sill, capping, etc., to allow the water to run off easily.

Washer. A flat thin ring or a perforated plate used in joints or assemblies to ensure tightness or relieve friction.

Washer, Flat. A washer which goes under a bolt head or a nut to spread the load and protect the surface.

Watchman. A guard who keeps watch over a certain area.

Water Closet. A plumbing fixture for the disposal of human wastes, a toilet.

Water Meter. A device for measuring the flow of water.

Water Reducing Admixture. Material added to cement or a concrete mix to cut down on its water content.

Water Slide. A sloping trough down which water is carried by gravity.

Water Softener. A device attached to a water system to remove unwanted minerals and substances.

Water Valve. A device to regulate the flow of water in a pipe or other passage.

Waterproof Membrane. A membrane, which can be made of built-up roofing or an elastomeric sheet, to provide positive waterproofing for a floor wall.

Waterproofing. The act or process of making something waterproof. A coating capable of stopping penetration of water or moisture.

Waterstop. A synthetic rubber strip used to seal joints in concrete foundation walls.

Watertight Manhole. A cover for a vertical access shaft that prevents the elements from coming in.

Wearing Course. A topping or surface treatment to increase the resistance of a pavement or slab to abrasion.

Weathering Steel. A steel alloy that forms a tenacious, self-protecting rust layer when exposed to the atmosphere.

Weed Control. The act or process of spraying chemicals or placing powders to control the spread of weeds.

Weep Hole. A small opening, the purpose of which is to permit drainage of water that accumulates inside a building component.

Weld. A joint between two pieces of metal formed by fusing the pieces together, usually with the aid of additional metal melted from a rod or electrode. Join two pieces of metal together by heating until fusion of material either with or without filler metal.

Welded Pipe. Piping where connections and fittings are welded.

Welded Railing. Railing sections with the components fastened with welds.

Welded Truss. Trusses with components fastened together with welds.

Welded Wire Fabric. A series of longitudinal and transverse wires arranged substantially at right angles to each other in sheets or rolls, used to reinforce mortar and concrete.

Well. A pit or hole sunk into the earth to reach a supply of water. An open space extending vertically through floors of a structure.

Wellpoint. A perforated pipe surrounded by sand to permit the pumping of ground water.

Wellpoint System. A series of vertical pipes in the ground connected to a header and pump to drain marshy areas or to control ground seepage.

Wet Sprinkler System. A sprinkler system that is filled with water at design pressure for immediate use upon activation.

Wheel Barrow. A small vehicle with handles and one or more wheels used for carrying small loads.

Wide Stile Door. Wider than normal vertical members forming the outside framework of a door.

Wide Flange Section. Any of a wide range of steel sections rolled in the shape of a letter "T" or "H".

Window Guard, Steel Bar Grille. Guard fabricated from a steel bar grille.

Window Guard, Woven Wire. Guard fabricated of woven wire..

Window Header. A horizontal construction member placed across the top of a window opening to support the load above.

Window Sub-Sill. Component anchored to wall construction and located just below window sill to receive window sill.

Windowwall. The opening in a wall surface which contains a window assembly or wall of assemblies. Often referred to loosely as tube-framing or storefront.

Wire, Aluminum. Electrical conductors and cable manufactured from aluminum.

Wire, Chicken. Thin, woven wire mounted on an exterior wall as a base for stucco plaster.

Wire Glass. Glass in which a wire mesh was embedded during manufacture.

Wire Hanger. A wire that supports or connects material.

Wire Mesh. A series of longitudinal and transverse wires arranged substantially at right angles to each other sheets or rolls, used to reinforce mortar and concrete.

Wire, Stranded. Fine wires twisted together in a group to create a larger stronger cable or wire.

Wood Anchor. A bolt or fastening device which attaches wood to wood or wood to another material.

Wood and Plastics. A category of the CSI Masterformat which is represented in Division 6 of the format. Commonly called just "Wood" or "Carpentry".

Wood Batten. Wood strips covering vertical joints on boards used as exterior siding.

Wood Beam. Horizontal wood structural member that supports uniform and/or concentrated loads.

Wood Buck. Wood frame typically built into concrete or masonry wall to accommodate finish door frame.

Wood Bumper. Wood component used to absorb impact and prevent damage to other surfaces.

Wood Cap. Wood member used on top of an assembly to provide termination and/or finish.

Wood Carriage. Sloping beam installed between stringers to support steps of wood stair.

Wood Casing. Wood exposed millwork or trim molding around doors, windows, beams, etc.

Wood Chip Mulch. Wood chips, spread on the ground to prevent erosion, control weeds, minimize evaporation and improve the soil.

Wood Decking. Plywood, lumber, or glued laminated member placed over roof or floor structural members for structural rigidity of building frame and to provide a surface for traffic or substrate for roofing or flooring system.

Wood Fiber Insulation. Wood particles and fiber used to reduce heat transfer.

Wood Fiber Panel. Form sheathing manufactured from glued and bonded wood particles.

Wood Finish Concrete. The act or process of using a wood float to smooth irregularities left in curing concrete, work the surface or compact the concrete.

Wood Frame. Floors, roofs, exterior and bearing walls of a building or structure constructed with wood.

Wood Nailer. Strip of wood attached to steel or concrete to facilitate making nailed connections.

Wood Roof Curb. Wood member elevated above plane of roof surface used for mounting of equipment or other elements.

Wood Roof Nailer. Wood strip (usually in plane of roof insulation) secured to structural roof deck used for securing roofing membrane.

Wood Saddle. Short horizontal wood member set on top of wood column to serve as seat for a girder.

Wood Screw. A screw for fastening objects in wood.

Wood Sleeper. Wood member laid on concrete floor to support and receive fastening of wood subfloor or finish flooring.

Wood Treatment. The act or process of applying a variety of stains or chemicals to retard fire, decay, insect damage or deterioration, due to the elements.

Wood Truss. A structural component formed of wood members in a triangular arrangement, often used to support a roof.

Woodwork, Architectural. A higher than average feature of finish work using wood for ornamental design.

Wye. A pipe fitting which joins three pipes at a 45 degree angle.

Wythe. A vertical layer of masonry one masonry unit thick.

X,Y,Z

X Bracing. That form of bracing wherein a pair of diagonal braces cross near mid-length of the bracing members.

X-Ray. Electromagnetic radiations of a short wavelength that can penetrate various thicknesses of all solids.

X-Ray, Weld. To examine, treat, or photograph the connection of surfaces that have been welded together.

Y Strainer. A device in the shape of a "Y" for withholding foreign matter from a flowing liquid or gas.

Z Tie, Wall. A Z-shaped reinforcing strip used as a support bracket from the structural wall to the masonry veneer.

Zero Slump Concrete. A concrete mixed with so little water that it has a slump of zero when tested.

A

B

C

D